The purpose of this book is to introduce Fourier descriptors as a method for measuring the shape of whole, or parts, of organisms. Fourier descriptors refer to the utilization of Fourier analysis, primarily the Fourier series as a curve-fitting technique, that can numerically describe the outline (shape) of irregular structures such as are commonly found in living organisms. The quantitative characterization of irregular forms is often a first step toward elucidation of the underlying biological processes, whether they be genetic, evolutionary, or functional.

The first five chapters discuss the theory behind the use of Fourier descriptors and the remaining chapters show case studies of how they can be used in various fields of biology such as anatomy, cell biology, medicine, and dentistry. These chapters cover a broad spectrum of biological materials analyzed with the use of Fourier descriptors and demonstrates the utility and diversity of Fourier descriptor applications in a biological context.

This is the first book solely devoted to this subject and will be of interest to all those involved in biological morphometrics.

FOURIER DESCRIPTORS AND THEIR APPLICATIONS IN BIOLOGY

FOURIER DESCRIPTORS AND
THEIR APPLICATIONS IN BIOLOGY

Edited by
PETE E. LESTREL
University of California, Los Angeles

CAMBRIDGE
UNIVERSITY PRESS

CAMBRIDGE UNIVERSITY PRESS
Cambridge, New York, Melbourne, Madrid, Cape Town, Singapore, São Paulo

Cambridge University Press
The Edinburgh Building, Cambridge CB2 8RU, UK

Published in the United States of America by Cambridge University Press, New York

www.cambridge.org
Information on this title: www.cambridge.org/9780521452014

© Cambridge University Press 1997

First published 1997
This digitally printed version 2008

A catalogue record for this publication is available from the British Library

Library of Congress Cataloguing in Publication data

Fourier descriptors and their applications in biology / edited by Pete
E. Lestrel.
        p.   cm.
    Includes bibliographical references.
    ISBN 0-521-45201-5 (hc)
      1. Morphology—Mathematical models.   2. Fourier analysis.
    3. Biology—Mathematical models.   I. Lestrel, Pete E.
    QH351.F685   1997
    574′.01′51—dc20   96-28577
                                                              CIP

ISBN 978-0-521-45201-4 hardback
ISBN 978-0-521-05573-4 paperback

Contents

List of contributors   *page* vii

Preface   ix

Part one: Theoretical considerations   1

1 Introduction   *Pete E. Lestrel*   3

2 Introduction and overview of Fourier descriptors   *Pete E. Lestrel*   22

3 Growth and form revisited   *Dwight W. Read*   45

4 Methodological issues in the description of forms   *Paul O'Higgins*   74

5 Phase angles, harmonic distance and the analysis of form
    Roger L. Kaesler   106

Part two: Applications of Fourier descriptors   127

6 Closed-form Fourier analysis: A procedure for extraction of
    ecological information about foraminiferal test morphology
    Nancy Healy-Williams, Robert Ehrlich and William Full   129

7 Fourier descriptors and shape differences: Studies on the upper
    vertebral column of the mouse   *David Johnson*   157

8 Application of the Fourier method on genetic studies of
    dentofacial morphology   *Lindsay Richards, Grant Townsend*
    and Kazutaka Kasai   189

9 Fourier analysis of size and shape changes in the Japanese skull
    Fumio Ohtsuki, Pete E. Lestrel, Teruo Uetake, Kazutaka Adachi
    and Kazuro Hanihara   210

10 Craniofacial variability in the hominoidea   *Burkhard Jacobshagen*   227

11 Heuristic adequacy of Fourier descriptors: Methodologic aspects
    and applications in morphology   *Vittorio Pesce Delfino,*
    Teresa Lettini and Eligio Vacca   250

12 Analyzing human gait with Fourier descriptors   *Teruo Uetake*   294

13 Elliptical Fourier descriptors of cell and nuclear shapes
    Giacomo Diaz, Corrado Cappai, Maria Dolores Setzu,
    Silvia Sirigu and Andrea Diana   307

14 Cranial base changes in shunt-treated hydrocephalics: Fourier
    descriptors   *Pete E. Lestrel and Jan A. Huggare*   322

15 A numerical·and visual approach for measuring the effects of
    functional appliance therapy: Fourier descriptors   *Won Moon*   340
16 Size and shape of the rabbit orbit: 3-D Fourier descriptors
    Pete E. Lestrel, Dwight W. Read and Charles Wolfe   359
17 From optical to computational Fourier transforms: The natural
    history of an investigation of the cancellous structure of bone
    Charles E. Oxnard   379
18 Epilogue: Fourier methods and shape analysis   *Neal Garrett*   409
    Appendix   415
    Glossary   435
    Index   460

Contributors

Kazutaka Adachi
*Laboratory of Applied Anatomy, Institute of Health and Sports Sciences,
University of Tsukuba, 1-1-1 Tennoudai, Tsukuba-shi, Ibaraki-ken 305, Japan*
Corrado Cappai
Department of Cytomorphology, Cagliari University, I-09124 Cagliari, Italy
Vittorio P. Delfino
*Consorzio di Ricerca Digamma, C. so Alcide de Gasperi, 449/A, 70125 Bari,
Italy*
Andrea Diana
Department of Cytomorphology, Cagliari University, I-09124 Cagliari, Italy
Giacoma Diaz
Department of Cytomorphology, Cagliari University, I-09124 Cagliari, Italy
Robert Ehrlich
*Department of Geological Sciences, The University of South Carolina
Columbus, South Carolina, USA*
William Full
Department of Geology, Wichita State University, Wichita, Kansas, USA
Neal Garrett
*Oral Biology Research, West Los Angeles Medical Center, Los Angeles,
California, USA*
Kazuro Hanihara
*International Research Center for Japanese Studies, 3-2 Goryo-Ocyama-cho
Nishikyo-ku, Kyoto 510-11, Japan*
Nancy Healy-Williams
*Department of Geological Sciences, The University of South Carolina,
Columbus, South Carolina, USA*
Jan Å. Huggare
*Department of Orthodontics, School of Dentistry, Karolinska Institute,
S-141 04 Huddinge, Sweden*
Burkhard Jacobshagen
*Anthropologisches Institutim FB Biologie, der Justus-Liebig-Universität
Giessen, D-35392 Giessen, Germany*

David R. Johnson
Centre for Human Biology, University of Leeds, Leeds, United Kingdom
Roger L. Kaesler
Paleontological Institute, University of Kansas, Lawrence, Kansas, USA
Kazutaka Kasai
Department of Orthodontics, Nihon University School of Dentistry at Matsudo, 2-870-1 Sakaecho-nishi, Matsudo-shi, Chiba-ken 271, Japan
Teresa Lettini
Consorzio di Ricerca DIGAMMA, Corso Alcide de Gasperi, 449/A 70125 Bari, Italy
Won Moon
Private Dental Practice, 1031 Rosecrans Blvd. Suite #105, Fullerton, California, USA
Paul O'Higgins
Department of Anatomy and Developmental Biology, University College London, London, United Kingdom
Fumio Ohtsuki
Department of Exercise and Sport Science, Tokyo Metropolitan University, 1-1 Minamiohsawa, Hachioji-shi, Tokyo 192-03, Japan
Charles E. Oxnard
Department of Anatomy and Human Biology, The University of Western Australia, Nedlands, Perth, Western Australia 6009, Australia
Dwight W. Read
Department of Anthropology, University of California, Los Angeles, Los Angeles, California, USA
Lindsay C. Richards
Department of Dentistry, The University of Adelaide, South Australia 5005, Australia
Maria Dolores Setzu
Department of Cytomorphology, Cagliari University, I-09124 Cagliari, Italy
Silvia Sirigu
Department of Cytomorphology, Cagliari University, I-09124 Cagliari, Italy
Grant C. Townsend
Department of Dentistry, The University of Adelaide, South Australia 5005, Australia
Teruo Uetake
Laboratory of Health and Amenity Science, Department of Regional Ecosystems, Tokyo University of Agriculture and Technology, Fuchu-shi, Tokyo 183, Japan
Eligio Vacca
Consorzio di Ricerca DIGAMMA, Corso Alcide de Gasperi, 449/A 70125 Bari, Italy
Charles A. Wolfe
Wolfe Software Associates, 13376 Dronfield Avenue, Sylmar, California, USA

Preface

The essays in this volume broadly represent contributions to one of the fundamental issues facing the biological sciences—the ability to quantify, in a thorough and complete manner, the visual information inherent in all biological organisms, information readily apparent yet remaining inexorably difficult to adequately characterize. One such technique is Fourier descriptors. This volume does not claim to be an exhaustive survey of Fourier descriptors, nor does it represent the last word. However, it is a first attempt in the direction of a more complete representation (or model) of the biological form.

This volume brings together a number of investigators who have used Fourier descriptors in very diverse contexts for the numerical description of the biological form. The term *Fourier descriptors* in the title refers to the utilization of Fourier analysis, primarily the Fourier series as a curve-fitting technique. *Applications* focuses more on the utilization of these analytical methods with actual data sets.

Although the majority of this book deals with data analysis as distinct from theory, there is a certain amount of chapter overlap with respect to a number of theoretical issues on the one hand, and techniques on the other. This situation seems to be inevitable with an edited volume of this type and can be desirable from the reader's point of view.

Part one, *Theoretical Considerations*, consisting of five chapters, is intended to (1) introduce Fourier descriptors and (2) highlight some of the theoretical issues. The first two chapters consist of an introductory chapter, and a relatively straightforward treatment of the mathematics that underlie Fourier descriptors. Chapter 3 (Read), Chapter 4 (O'Higgins), and Chapter 5 (Kaesler) deal with some of the more formal issues related to the analysis of growth and form and their numerical characterization.

Part two, *Applications of Fourier Descriptors*, is composed of twelve chapters followed by an epilogue which are largely devoted to applications. The initial seven chapters deal with conventional Fourier descriptors. Chapter 6 (Healy-Williams, *et al.*) represents a study of ecological pressures on the shape of

Foraminifera (Protozoa), while Chapter 7 (Johnson) and Chapter 8 (Richards, *et al.*) deal with questions of genetic interest. Anthropological data forms the content of Chapter 9 (Ohtsuki, *et al.*), Chapter 10 (Jacobshagen), and Chapter 11 (Delfino, *et al.*). Chapter 12 (Uetake) is a specialized study of human gait.

The subsequent four chapters are based on a comparatively new Fourier descriptor termed Elliptical Fourier Functions which is applied to anatomical structures. Chapter 13 (Diaz, *et al.*) deals with cell shape and Chapter 14 (Lestrel and Huggare) examines cranial base shape changes present in a craniofacial abnormality. Chapter 15 (Moon) assesses mandibular changes due to dental treatment, and Chapter 16 (Lestrel, *et al.*) focuses on the rabbit orbit. The last application in Chapter 17 (Oxnard) represents a departure in that it is concerned with Fourier Transforms.

These essays cover a broad spectrum of biological materials analyzed with the use of Fourier descriptors and demonstrate the utility and diversity of Fourier descriptor applications in a biological context.

Pete E. Lestrel

Acknowledgments

This volume represents the culmination of some twenty-odd years of work on Fourier descriptors since I completed my 1975 Ph.D dissertation at the University of California at Los Angeles. Collecting chapters by authors scattered over Europe, Asia, and the United States represented an initially intimidating challenge. The cooperation of all participants involved and the use of the now ubiquitous FAX machine considerably alleviated this rather large task.

Thus, first and foremost, I owe an immeasurable debt of gratitude to the contributors to this volume. Without their collaboration this compendium would not have been possible.

Special thanks go to Dr. Neal Garrett and Charles Wolfe who unselfishly devoted many hours assisting me with the editing process. Thanks also go to Dr. Robin Smith, and his associates at Cambridge University Press for their assistance and careful attention to detail in the publication process of this venture.

Finally, and most important of all, gratitude goes to my family, especially to my wife Dagmar, who patiently tolerated the numerous late nights and early morning vigils for the last three years which were required to bring this 'labor of love' to fruition.

Part one

Theoretical Considerations

1

Introduction

PETE E. LESTREL

UCLA School of Dentistry

Entia non sunt multiplicanda praeter necessitatem.
Entities are not to be multiplied beyond necessity (Ockham's Razor).

William of Ockham (1300–1349)

The study of form may be descriptive merely, or it may become analytical. We begin by describing the shape of an object in simple words of common speech: we end by defining it in the precise language of mathematics; and the one method tends to follow the other in strict scientific order and historical continuity.

D'Arcy Thompson (1915)

1.1 Introduction

Advances in the biological sciences often proceed on a number of fronts, which include (1) development of theoretical frameworks (formal model building); (2) development of appropriate tools; and (3) applications based on various techniques (using data) to solve problems. All three of these approaches are conceptually linked and need to proceed simultaneously if progress is to be made. Generally, specific data-oriented problems tend to spearhead the need for new techniques leading to new algorithms. These in turn may lead to a reevaluation of accepted theory or assist in the development of new formal models. Often, applications and algorithms tend to outrun the development of formal models.

There also seems to be a distinction between practitioners who might be called "theoreticians" for the lack of a better word, and those who are "practitioners". The former tend to be individuals, often with a mathematical bent, who are primarily concerned with the building of formal models and devising the requisite mathematical solutions toward that end. The latter often tend to be more focused on questions of smaller scope that need a timely response and are, therefore, more data-driven. Although this distinction into 'theoreticians' and 'researchers' is obviously artificial, and individuals often possess both 'theoretical' and 'practical' qualities to a greater or lesser degree, it is nevertheless suggested that this dichotomy has a certain amount of validity.

For the last century or more, there has been little questioning of the validity of the measurement systems used to describe the biological form. In fact, for early biologists the very act of taking measurements was viewed as an unattractive endeavor. To quote D'Arcy Thompson: "The introduction of mathematical concepts into natural science has seemed to many men no mere stumbling-block, but a very parting of the ways" (Thompson, 1942:11). Clearly, times have changed and the use of quantitative methods is now widespread. Moreover, with the rapidly increasing utilization of powerful computers (even at the PC level), the ability to analyze large data sets has become routine and extensive developments since the 1960s have occurred in an attempt to simplify these large data sets. These developments have lead to an increasing application of multivariate statistics consisting of such methods as discriminant functions, cluster analysis, principal components, etc., to biological data sets. The success of these methods is now undoubted. However, even those who routinely apply these multivariate procedures to data rarely question the appropriateness of the measurements that compose the data sets being utilized. Thus, a number of basic questions can be raised about the utility of the measurement system being used to derive the data sets that are routinely analyzed with various statistical procedures.

1.2 Some basic aspects of form

The idea of form is one of the most fundamental concepts underlying all of the biological sciences. All biological forms consist of a large number of shared aspects that include size, shape, color, patterning, etc. In the most basic sense, the human ability to readily discriminate forms by noting these differences is so well integrated that the required behavior responses are often unconscious. The eye detects subtle differences by detecting contrasts, the greater the contrast between comparisons, the more readily can such differences be distinguished. However, the exact physiochemical processes involved in the reception, recognition, and especially, integration of visual stimuli still remain incompletely understood (Spoehr and Lehmkuhle, 1982).

In contrast to the human capability of rapidly identifying and classifying the visual information present in the biological form, the mathematical description of the content of these visual images has been slow in forthcoming. Two issues emerge: (1) the need for analytic procedures that can precisely measure forms and the differences between them; and (2) the nature of the relationship of these measurements to the underlying biological processes that are ultimately responsible for the form. The latter issue, in part, is taken up in Chapter 3. It is argued here that without an adequate model for the representation of form, it may be premature to claim a precise understanding of those underlying biological processes. The difficulties inherent in attempting to formulate models that can account, even in a limited way, for the diverse morphological structures involved, from conception to the adult organism, are immense. Rene Thom in discussing his catastrophic model of morphogenesis, indicated that: "In biology, if we make exceptions of the theory of population and formal genetics, the use of mathematics is confined to *modeling* a few local situations (transmission of nerve impulses, blood flow in the arteries, etc.) of slight theoretical interest and of limited practical value" (Thom, 1983:114; italics added). A situation that has not materially changed much in spite of the fact that ". . . the problem of the integration of local mechanisms into a global structure is the central problem of biology. . ." (Thom, 1983:154). This "global structure" can be viewed as a model of biologic process. Related, but applicable in a wider context, to the model of morphogenesis (whether viewed in deterministic, stochastic, or catastrophic terms) is the need for an adequate numerical characterization of form which becomes, in one way or another, the raw data for model building (see Section 1.6). Thus, with respect to the study of form, the subject matter of this volume, there is no generally accepted theoretical foundation. That is, *a unified science of form*, (morphometrics) exists only in a very rudimentary sense.

1.3 The development of morphometric methods

The quantitative description of form is now called morphometrics (Blackith and Reyment, 1971; Reyment et al., 1984; Reyment, 1991; Lestrel, in press). Morphometrics consists of procedures which facilitate the mapping of the visual information of form into a mathematical (symbolic) representation (Read, 1990). Morphometrics as viewed here encompasses four separate and distinct approaches dealing with the numerical description or representation of form:

> *multivariate morphometrics*, typically applied to data sets composed of distances and angles,
> *coordinate morphometrics* with the focus on deformations, including biorthogonal grids, finite elements, and thin-plate splines,
> *boundary morphometrics* such as Fourier descriptors, eigenshape analysis, medial axis analysis, and elliptical Fourier functions, and
> *Textural morphometrics* consisting of techniques such as Fourier transforms and coherent optical processing (see Chapter 17 for an example; also Davis, 1971; Lugt, 1974).

These approaches tend to be treated independently from each other because of a lack of a unifying underlying model. An extended discussion of the merits and limitations of these morphometric approaches is beyond the scope of this chapter. A discussion of the various morphometric techniques can be found in Lestrel (in press).

Of the four approaches above, multivariate morphometrics is the oldest and most familiar from a biological context. The data of multivariate morphometrics are most often based on the universally used linear distances, angles, and ratios. I have defined this measurement system as the *conventional metrical approach* or CMA (Lestrel, 1974; Lestrel, 1982; Lestrel et al., 1991). CMA is discussed in some detail in the next section.

To handle the insufficiencies of CMA, two alternative representations of form have been developed over the last three decades. Although both have constraints and as formal models are incomplete, they exemplify improvements over CMA. These are (1) homologous point approaches, and (2) boundary outline approaches. The former share two elements in common with CMA: a reliance on landmarks and an exclusion of the curvature between points (Bookstein, 1978; Bookstein, 1991; Richtsmeier et al., 1992). The boundary representations focus on the outline of forms (Lestrel, 1974; Lohmann and Schweitzer, 1990; Blum and Nagel, 1978; Lestrel et al., 1993; Lestrel, in press). Clearly, both methods are different ways of looking at the same morphology. Approaches that combine both representations, landmark and outline, have been occasionally suggested, although little work has appeared (Read and Lestrel, 1986; Ray, 1990; and see Chapter 14 for such an approach). Given the almost universal appeal of the use of landmarks for distances and angles (CMA), further comments are warranted.

1.4 The conventional metrical approach

The success of CMA is predicated on the fact that extensive phenomena, especially in the physical sciences, seem to be adequately described with CMA. Examples abound such as the structure of crystals, snowflakes, etc. Biological structures may also exhibit regularity and for those, CMA can be profitably applied.

With respect to shape,[1] if the object under consideration is *regular* in geometric form,[2] then an accurate quantitative representation, subject to measurement error, can be readily attained with the use of CMA.[3] Examples of regular forms in-

---

[1] Shape is defined here as the boundary of a form (see Section 1.6).

[2] Simple examples are circles, rectangles, triangles, etc., but they can be quite complex when combined to form other objects; consider a geodesic dome.

[3] For purposes of discussion here, morphometric representations are presumed to not only include universally used measures such as distances, angles, and ratios, but also simple curve-fitting procedures such as lines, parabolas, etc.

clude, but are not limited to, man-made objects. When such man-made forms are composed of curves, they tend to be simple to accommodate manufacturing. The numerical characterization of such forms is based on a measurement system that has its roots in antiquity, but was greatly advanced by Descartes. In effect, this measurement system has been universally successful in the numerical character- ization of *regular* forms, hence its popularity and widespread use.

It is only when irregularities in form become predominant, a common situation in biology, that this leads to difficulties if *shape* is of primary concern. One of the primary drawbacks with CMA is that it is an incomplete mapping of the mea- surement domain with the form domain. That is, much of the visual information of potential biological significance is not being measured and is thereby lost. Thus, *natural* morphologies, in contrast to man-made objects, and excluding those classes that display regularities (for example crystal growth), are difficult, if not impossible, to adequately characterize with CMA. Although some information can always be extracted with CMA, it remains necessarily incomplete.

Moreover, other insufficiencies also begin to surface. CMA was often com- posed of a limited set of selected distances and angles connecting points located on a form. This approach initially led to the use of a large number of such mea- sures, in the hope that "more was better", a view favored in the last century and the early part of this century (McDonell, 1906; Martin, 1928). Moreover this over- measurement approach, although an unintentional consequence, was further le- gitimized with the rise of numerical taxonomy (Sneath and Sokal, 1973).

Although the points used to produce the measurement set of distances and an- gles are undoubtedly necessary for analysis, exclusive reliance on them, to the exclusion of other aspects of form, is unfortunate for a number of reasons. First, the choice of CMA generally contains a large subjective element which is often not recognized. Second, the use of CMA precludes the ability to subsequently vi- sually reproduce the form. Third, the use of homologous points in a CMA dataset represent a very small percentage of the information present in the biological form. This raises questions about the validity of the analyses as they pertain to the growth process, the effect of environmental factors, functional correlates, or issues of classification (taxonomy) to name just a few. If, as it is suggested here, the over- whelming majority of the biological content does not reside at the points (or land- marks), where does it reside? That question is not difficult to answer. It resides with a host of other aspects such as the surface texture, the boundary outline, its color, etc. For example, it is entirely possible that in a specific case the variables most sensitive to discrimination may be color or surface texture, suggesting that the routine metrical measurements entered into the analysis may be superfluous or at the very least insensitive for discrimination.

I hasten to add that I do not suggest that all analyses conducted in the past are incorrect; far from it. In fact, the use of more complete and appropriate measurement sets in some cases may not, *a priori*, materially improve previously accepted results. However, there is the lingering doubt that in other cases, a systematic revision may be called for. The point to be stressed here is that we are not yet in a position to know.

In sum, it would seem that we have "placed the cart before the horse". That is, too little emphasis was initially placed on the measurement system *per se*, in contrast to other legitimate and more pressing concerns; whether taxonomic in nature, issues of process such as growth, or functional considerations. Unfortunately, an imperfect measurement system may lead to a flawed end product.

This leads to the question of what should such a measurement set consist of, if it is to adequately address some of these concerns? A partial answer is given in Section 1.6, where an attempt is made to provide a formal, if heuristic, model of form.

1.5 Representation of forms

Even if the analysis is limited to the shape of the form (other aspects being momentarily excluded), and given the insufficiencies of CMA for shape analysis, focus needs be placed on the representation being used to numerically describe the form (Lestrel, 1980). The following list is not necessarily inclusive, but intends to reflect the challenges associated with development of a formal model of form (see Section 1.6). The numerical representation should:

> be a computationally unique description of the size, shape, and structure of the form, and allow for the visual recreation of the form with high precision,
>
> be efficient in terms of computation; that is, allow for a reduction of the observed dataset to a smaller sub-set of variables, with a minimum loss of accuracy (Ockham's Razor),
>
> reflect global features (measuring large changes in the form) as distinctly recognizable from finer levels of detail in the form. These differences should be accurately mirrored in the values of the computed parameters,
>
> provide for the extraction of a significant percent of the variability present,
>
> allow for the uncorrelated assessment of the components of size and shape (property of orthogonality),
>
> display numerical differences which can be related to actual biological changes in the form and,
>
> be invariant with respect to translation, rotation, reflection, and uniform scaling.

The coordinate transformations identified as *uniform scaling, translation, rotation*, and *reflection*, must each be invariant (Mokhtarian and Mackworth, 1986;

Xu and Yang, 1990). That is, these properties must not result in shape changes (distortions in a visual sense) under each of these transformations. Of the four properties, only three, translation, rotation, and reflection, are "identical" in the sense that both size and shape remain unchanged. With the fourth property, uniform scaling, only shape remains invariant. Ultimately, it is to be emphasized that the technique used to numerically describe the form must be *information preserving*; that is, the size, shape, and structural information must not be lost (*i. e.*, is recoverable from the technique at a later date) (van Otterloo, 1991; Lestrel, in press).

With the exception of Chapter 17, this volume focuses on size and shape considerations. Other structural aspects are excluded from consideration.

1.6 A heuristic model of form

The rather abstract notion of form has engendered considerable confusion leading to the question of what exactly constitutes a "form"? According to *Webster's Unabridged Dictionary* (1983), form is defined as "the shape or outline of anything; figure; image; structure, excluding color, texture and density." Upon looking up "shape" we find that things get hopelessly confusing with shape defined as "outline or external surface" or "the form characteristic of a particular person or thing." According to these definitions, "shape" and "form" are interchangeable, to be viewed as identical. Clearly, this unsatisfactory state of affairs is in need of re-definition.[4]

1.6.1 Earlier models

An early attempt formally defined form as: Form = Size + Shape (Needham, 1950; Penrose, 1954). This linear formulation was subsequently extended to Form = Size + Shape + Structure (Lestrel, 1980, Lestrel, in press). Each of the aspects—size, shape, and structure—were initially defined as follows:

> Size—a quantity that depends on dimensional space. In a one- dimensional world, difference in size can be viewed as a difference in vector length. In two dimensions, linear measurements in combinations (such as ratios) have proven to be inadequate, and area becomes one definition of size (perimeter, although influenced by boundary considerations, has also been utilized). In three dimensions, volume would be the appropriate quantity.
>
> Shape—a quantity that is difficult to adequately define but has been characterized as "residual" or what is left after size has been controlled. It is defined here as the boundary of the form (but see below).

---

[4] It is of some interest that D'Arcy Thompson also viewed form and shape interchangeably.

Structure—a quantity that describes the "within boundary" aspects. It can refer to both internal and external considerations. For example, the external "roughness" on the outside of an object, or the internal patterning or "orientation" of the cancellous bony spicules (trabeculae) that make up the femur head. Techniques for the analysis of structure include coherent optical processing and Fourier transforms.

Although this linear formulation, Form = Size + Shape + Structure, was initially considered as a step toward a formal model of form, it only represents a fraction of the attributes that define all forms, especially those found in nature. Clearly, the lagging developments leading to such a theoretical model of form are due to its formidable difficulties. These include: (1) problems of definition, that is, a vocabulary that precisely conveys information about form; (2) problems of description, in quantitative terms, of the various attributes; and, (3) the global, that is, the inclusion of the set of attributes into a useful multivariate model. It seems apparent that at the moment there is no single numerical value which can adequately describe a form in all its aspects.

1.6.2 A new model of form

Clearly a redefinition is called for, and form has been redefined and broadened using an order-independent notation which makes no assumptions about linearity: A form, **F,** can be defined using the notation:[5]

Form = F = [State, Size, Shape, Orientation, Surface, Interior, Substance].

Each of these seven attributes (each of which may be further composed of multiple properties) is defined as follows:[6]

State—this attribute defines whether it is a solid, liquid, or gas. A three-item vector may be applicable to indicate percentages of each quantity. For the purposes of this volume, it is equal to a solid. [1.0, 0.0, 0.0].
Size—is a measure of spatial dimension(s) and can be viewed as a displacement in space such as: (1) volume in 3-space; (2) area in 2-space; and (3) length in 1-space.
Shape—is a boundary phenomena. It refers to the boundary outline of a form in 2-D or 3-D. It is a geometric property, as defined here, and not related to the no-

---

[5] This notation is formally known as an *n*-tuple, here a 7-tuple. Although perhaps unfamiliar to the reader, the concept is an extension of such ideas as a triplet (a 3-tuple). The notation is widely used in such fields as computer science and computational linguistics.

[6] I am indebted to discussions with C. Wolfe, N. Garrett, and D. Read for this section; however, any errors of omission or otherwise, are solely my own.

tion of surface as such (see Surface). Its focus is on curvature. In 1-space, all point forms and all line forms have the same shapes.

Orientation—refers to location in space, generally with respect to a reference system. Orientation is considered static at any point in time for a specific form.
Surface—is characterized with zero thickness and by two primary properties: texture and color. Texture refers to the presence of vascularities, smoothness, pits, bumps, roughness, etc.
Interior—is also characterized by textural and color properties but has thickness in contrast to surface. It can be usefully viewed as normal ($\perp$) everywhere to the surface of the form.
Substance—based on physical properties such as thermal expansion, mass or weight, density, conduction, modulus of elasticity, hardness, etc.

For a discussion of the "size" attribute the reader is referred to Koenderink (1990) with his use of the terms "blob" and "volume". In a fundamental sense, everything emanates from the notion of the displacement of space by a volume. If no space is displaced, there is no object, hence no form. These notions, as mathematical constructs, are dimensionally independent. Modeling, to describe real objects, requires simplification, leading to the use of points and lines with infinitesimal diameters and thickness is one such example; while the definition (above) of a surface having no thickness, is another.

Defining S as the number of surfaces, a form can have a single surface, S, for example, a Mobius strip or a Klein bottle, although multiple surfaces are the rule in nature. For example, a soap film can be considered as composed of two surfaces with minimal thickness between them (see Chapter 16 for an application). Most forms have multiple surfaces each possessing specific properties of texture and color (thus S may be $= 1$ or some $n > 1$). For example, each face of a 4-sided pyramid could be considered as a "sub-" surface and further, the face of each "building block" comprising a face could itself be another level of surface. Each building block might have to be broken down into subregions composed of unique color and/or texture attributes.

The texture attribute is also intended to be represented by a set of numerical values, such as those derived from Fourier transforms, or these may require the use of some sort of topological complexity measure, or a chaos/fractal formulation which could give an objective quantitative characterization. The color attribute also has complexity since few forms are of single or uniform color over an entire surface.

Considerably more challenging is the embedding of the above attribute data into a proper representation (or model) of the form. A digression into modeling and properties of models may be instructive.

1.6.3 Models of biological form

A biological system can be envisaged as composed of sub-structures that are dependent on each other (for example, consider the lower jaw and its interconnections, consisting of muscles, etc., with the cranial base). One is interested in not only how such a system works but also how it adapts to changes imposed on it (often over time). Ultimately, a good (effective) model will allow for prediction, in a generalized way, of those system changes. Whereas mathematical models have been remarkably effective in the physical sciences, they have not been all that successful in the biological sciences, undoubtedly due to the complexity of living organisms. A model has been defined as consisting of two elements: (1) the choice of variables to be included; and (2) the relationship between these variables (Saaty and Alexander, 1982). Fig. 1.1 depicts the modeling process as a closed system (Giordano and Weir, 1985). In brief, we start with a biological system from which we gather sufficient data to formulate a model. Simplification is generally indicated, otherwise it may become impossible to build the model.[7] Next the model is analyzed and conclusions are drawn. These conclusions form the basis for interpretations about the biological process under consideration. Finally, these conclusions about the biological process are tested or verified against new observations and new data. This leads to changes, if necessary, refinement, and improvement of the model.

Returning to the basic issue of this volume; namely, the numerical representation of the biological form, a number of questions need to be resolved before one can legitimately build effective models that deal with biological processes. Two of these are: (1) which variables constitute the "correct" set for the complete representation of form; and (2) how is the representation or model to be built. Ideally, one aims for a minimalist solution which contains the essential variables as implied in the principle of parsimony of Ockham's Razor (i.e., all non-relevant variables are removed in some stepwise fashion — as with the use of discriminant functions).

1.6.4 Modeling biological processes

To eventually apply the model to biological processes, a form as a quantitative construct composed of the above attributes needs to be defined at a given instant in time. Presumably, some of its attributes may or may not change significantly as a function of time (i.e., viewed in a longitudinal context). At each time inter-

---

[7] Nevertheless, although simplification is desired, the opposite may also be problematic; that is, building a model with an insufficient number of variables, for example, the use of CMA to characterize form, as alluded to in Section 1.4.

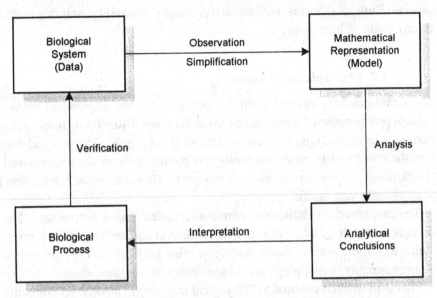

Fig. 1.1 The modeling of a biological process as a closed system (see text).

val the form would be characterized as a unique entity. A combination of linked-list and a tree structure might be a fruitful approach in this regard.

Differences between such entities are the data from which theories of biological processes are developed. That is, the precise quantification of a form with all its attributes must be considered as the initial step *prior* to the subsequent steps leading toward the identification of those biological processes involved in the alterations of forms over time. Clearly, model building of those aspects of the form that determine biological processes remains an uncharted area, involving experimentation, development of hypotheses, and trial and error. Thus, little in the way of answers can be expected or offered at this moment.

Different metrics have been developed to attempt to numerically determine these form alterations with time, such as the ones based on deformations. The problem with most of these approaches to delineate process is that they are never complete. That is, they are based on only one or two of the above attributes. This constraint also applies to the studies in this volume. The implementation of a majority of the attributes in the above model of form will, in all probability, require multidisciplinary cooperation and can be considered as one of the major challenges for the next century.

Finally, even if one limits the comparison between forms to the boundary, and excludes all other attributes, this still requires some measure of correspondence between the representations serving as a model of form. Ideally, what is required for comparison is a 1:1 correspondence, or mapping, between related structural aspects. Biological forms, unfortunately, contain variability, and this leads to the thorny issue of homology.

1.7 The problem of homology

The term homology as used in the current literature has engendered considerable controversy because of the tendency to confuse two distinctly different definitions currently in vogue. One of these is derived from purely mathematical concepts, and the other one is based on biological premises. Homology as derived from Greek roughly refers to *the study of sameness*. However, such a definition is not operationally very useful.

The mathematical definition, commonly called "point homology", has been brought into biology as a means to make certain characteristics of the form, namely landmarks, more precise. Such landmarks must stand in a one-to-one relationship (mathematically, a mapping) across specimens of the same class.

This is in marked contrast to "biological homology" as derived originally from the work of Owen (1848), who defined it in rather vague, non-evolutionary terms as "the same organ in different animals." He also defined a parallel term, analogue, which referred to "an organ that has the same function in different animals," thereby attempting to relate biological structures to their function. These definitions are not very satisfactory since it is not at all apparent what he meant by "same". Nevertheless, a discussion of the problems attendant to definitions of homology is beyond scope here. Suffice it to say that after Darwin, homology was redefined as "resemblance due to inheritance from a common ancestry" (Simpson, 1961:78). Nevertheless, even this approach has been criticized on the grounds that it is based on circular reasoning (Sneath and Sokal, 1973). They propose instead a definition termed *operational homology* (see Chapter 4 for an extended discussion of this concept). In any case, homology, in either the Owenian or Darwinian sense, tends to be global in nature and very distant from the notion of point homology.

What has led to confusion is that very few of the data that comprise "biological homology" are composed of points that satisfy the criterion of a one-to-one mapping (point homology). Conversely, very few of the data of "point homology" (landmarks) have much to do with the curvature/surface between such points which represent the basis of biological structure (biological homology). These differences between the two datasets are not trivial and have led to and are continuing to elicit considerable debate and polemics.

The use of such homologous points to the exclusion of the boundary or surface (shape) has the effect of neglecting essential biological information. The application of deformation models (Bookstein, 1978; 1991), although undoubtedly useful in other contexts, can shed only limited light on shape changes arising from evolutionary/functional considerations, since these deformation models also tend to exclude boundary data from consideration and are often based on few landmarks.

The continuing focus on easily obtained homologous points such as the landmarks *nasion*, *basion*, etc., as used in cephalometrics, to the exclusion of other information, is regrettable. This arises because it is generally not the landmark *per se* that is of interest from a growth or functional point of view, but rather the information that is *not* being utilized or measured. Namely, it is the local curvature in the vicinity of the landmark which may be undergoing shape changes due to growth or functional considerations and 'carrying' the landmark with it. Additional discussions dealing with the issue of homology are contained in Chapters 3, 4, and 5. A particularly detailed treatment can be found in Chapter 4.

1.8 The boundary problem

It is apparent that the human visual system treats the boundary outline as one of the primary features for discrimination, hence the need to precisely elicit boundary outline information as an aspect of the form. However, the boundary outline (defined here as the shape of a form), has represented formidable challenges which are difficult to handle with simple metrics.

One alternative technique available to surmount some of the difficulties is Fourier analysis. The use of a convergent series, specifically Fourier's series, in various guises, represents a curve-fitting approach particularly suited for the characterization of boundary outline data of complex irregular morphologies commonly encountered in the physical, biological, and clinical sciences. Chapter 2 briefly develops the mathematics underlying Fourier descriptors and readers intending to utilize these analytical tools for the first time are strongly encouraged to peruse it. Some of the software available is discussed in the Appendix.

1.9 Applications of Fourier descriptors

Fourier descriptors have been employed as an effective tool in scientific inquiry and applications can be expected to increase. They allow for a far more complete assessment of the information that is always present in morphological forms than has been heretofore possible. Since Fourier descriptors are curve-fitting approaches, they are particularly useful for the purpose of describing *global* characteristics of the form, characteristics which are not dependent on homologous points. Nevertheless, a procedure has been recently developed which allows for the incorporation of the homologous point set into the boundary outline. Once the homologous point set has been embedded into the boundary, conventional measures such as distances from a neutral center (centroid or center of gravity) to specific aspects on the boundary can be computed, providing a mechanism for the analysis and comparison of *local* aspects of the form (see Chapters 14 and 16 for examples).

The results obtained with the use of Fourier descriptors may, in certain cases, also provide insights into functional interpretations that underlie observed data, results not easily obtainable with other methods. Chapter 15 is particularly instructive in this regard. Numerous dental researchers working in a clinical setting have suggested that functional appliance therapy was somehow responsible for the accelerated "growth" as seen in the increased length of the mandible. The use of Fourier descriptors clearly displayed a posterior bending in the condyle which very nicely accounts for this increased length, a result solely due to treatment and not growth! Fourier descriptors facilitated the visualization of the mandible as a single complete morphological structure rather than as composed of separate "pieces" described by isolated distances and angles. This study is illustrative of some of the pitfalls attendant with CMA, in this case reliance on single measures such as mandibular length.

Fourier descriptors are applicable to data in diverse disciplines indicating the generalized applicability of the approach. The many fields of study are grouped below into five major areas to illustrate the varied nature of the numerous actual studies and the potential for future applications.

1. *Biological sciences* (biology, physical anthropology, primatology, zoology, botany, entomology, paleontology, etc.) are based on morphological forms such as plants, insects, fossils, and cells, which are of particular interest for capturing differences in size and shape and for classification purposes.
2. *Dental and medical sciences* (orthodontics, anatomy, growth and development, etc.) are concerned with the differences that arise with the implementation of different treatment modalities.
3. *Earth sciences* (geology, soil science, petrology, remote sensing, etc.) also deal with organisms, cell packing of fossil bryozoans being an example.
4. *Engineering sciences* (pattern recognition, electrical engineering, cybernetics, etc.) represent the original development of Fourier descriptors. Typical examples are identification of handwritten characters, samples of speech, pictures and scenes, waveforms and radar signals.
5. *Social sciences* (anthropology, archeology, etc.) for example, the study of archeological features such as projectile points.

1.10 Literature on Fourier descriptors

Whereas the literature on Fourier descriptors is diverse, much of it is confined to the traditional fields of electrical engineering and pattern recognition. Comparatively little has filtered into biology, although the fields of geology, paleontology, and geography have had early adherents.

One of the deficiencies in the current state of development of Fourier descriptors is the lack of an integrated approach to the utilization of the method with actual data. The literature, although voluminous, is difficult to easily isolate. Specific papers that focus on Fourier descriptors are scattered across very diverse disciplines. Significant papers can be found in fields ranging from engineering to anthropology.

Terminology also presents problems. Many of the concepts such as harmonics, phase angle, amplitude, and power have their roots in engineering and may not translate easily into a biological or medical context. The concept of "power" and its relationship to variability is a good example of this. A glossary is provided at the end of the book to clarify the terms that arise with so many varied disciplines involved in the application of Fourier descriptors.

If one looks for comprehensive treatments of Fourier descriptors, there are none. In a geological context, only Davis (1986) has material devoted to Fourier analysis. On the other hand, a large number of treatises are available which deal with Fourier analysis from a theoretical point of view (Carslaw, 1950; Franklin, 1949; Tolstov, 1962). However, their presentation is neither introductory nor elementary; in fact, a considerable mathematical sophistication is required, at a level beyond most practitioners in the biological or medical sciences. In the area of morphometrics, the only comprehensive work currently available is the book *Multivariate Morphometrics* by Blackith and Reyment (1971) where there is no mention of Fourier methods, and the slimmer second edition by Reyment et al., (1984) in which there is only a hint of such methods.

The following journals, from time to time, have contained articles related to Fourier descriptors. Note the diversity of these sources:

1. Biometrics
2. Transactions on Computers
3. Transactions on Medical Imaging
4. Transactions on Pattern Analysis and Machine Intelligence
5. Pattern Recognition
6. Proceedings of the Society of Photo-Optical Instrumentation Engineers (SPIE)
7. Applied and Computational Harmonic Analysis
8. Mathematical Geology
9. Journal of Sedimentary Petrology
10. Journal of Paleontology
11. Marine Geology
12. Marine Micropaleontology
13. Geographical Analysis
14. Journal of Embryology and Experimental Morphology
15. Experientia

16. Systematic Zoology
17. Journal of Zoology (Lond.)
18. Folia Primatologia
19. Human Biology
20. American Journal of Human Biology
21. American Journal of Physical Anthropology
22. Computer Programs in Biomedicine
23. Journal of Craniofacial Genetics and Developmental Biology
24. Journal of Dental Research
25. Forma (Japanese)

A brief list of papers dealing with both conventional Fourier descriptors and elliptical Fourier functions spanning almost 30 years of work (some of which can now be considered as "classic") are set forth below. These are intended to serve as an introduction to Fourier descriptors and are by no means exhaustive.

Papers dealing with conventional Fourier descriptors include Lu (1965), Schwarcz and Shane (1969), Ehrlich and Weinberg (1970), Medalia (1970), Anstey and Delmet (1972), Kaesler and Waters (1972), Kaye and Naylor (1972), Christopher and Waters (1974), Lestrel (1974), Gervitz (1976), Younker and Ehrlich (1977), Waters (1977), Gero and Mazzullo (1984), Johnson et al., (1985), Mok and Boyer (1986), O'Higgins and Williams (1987), Kieler et al., (1989), and Inoue (1990) to name only a few.

Papers dealing with the comparatively newer Fourier descriptor, elliptical Fourier functions, include the now classic paper by Kuhl and Giardina (1982), who developed elliptical Fourier functions, Rohlf and Archie (1984), Ferson, et al., (1985), Lin and Hwang (1987), White et al., (1988), Lestrel (1989a), Lestrel (1989b), Lestrel et al., (1990) Lestrel et al., (1993), and Nafe et al., (1992).

References

Anstey, R. L. & Delmet, D. A. (1972). Genetic meaning of zooecial chamber shapes in fossil bryozoans: Fourier analysis. *Science*, **177**, 1000–2.

Blackith, R. E. & Reyment, R. A. (1971). *Multivariate Morphometrics*. New York: Academic Press.

Blum, H. & Nagel, R. N. (1978). Shape description using weighted symmetric axis features. *Pattern Recog.* **10**, 167–80.

Bookstein, F. L. (1978). *The Measurement of Biological Shape and Shape Change. Lecture Notes in Biomathematics*. Vol. 24. New York: Springer-Verlag.

Bookstein, F. L. (1991). *Morphometric Tools for Landmark Data*. Cambridge: Cambridge University Press.

Carslaw, H. S. (1950). *An Introduction to the Theory of Fourier's Series and Integrals* (3rd Ed). New York: Dover.

Christopher, R. A. & Waters, J. A. (1974). Fourier series as a quantitative descriptor of miosphore shape. *J. Paleont.* **48**, 697–709.

Davis, J. C. (1971). Optical processing of microporous fabrics. In *Data processing in Biology and Geology*, ed. J. L. Cutbill. New York: Academic Press.

Davis, J. C. (1986). *Statistics and Data Analysis in Geology*. (2nd ed.) New York: John Wiley.

Ehrlich, R. & Weinberg, B. (1970). An exact method for the characterization of grain shape. *J. Sed. Petrol.* **40**, 205–12.

Ferson, S., Rohlf, F. J. & Koehn, R. K. (1985). Measuring shape variation of two-dimensional outlines. *Syst. Zool.* **43**, 59–68.

Franklin, P. (1949). *An Introduction to Fourier Methods and the Laplace Transformation.* New York: Dover.

Gero, J. & Mazzullo, J. (1984). Analysis of artifact shape using Fourier series in closed form. *J. Field Archeol.* **11**, 315–22.

Gervitz, J. L. (1976). Fourier analysis of bivalve outlines: Implications on evolution and autoecology. *Math Geol.* **8**, 151–63.

Giordano, F. R. & Weir, M. D. (1985). *A First Course in Mathematical Modeling.* Monterey, California: Brooks/Cole.

Inoue, M. (1990). Fourier analysis of the forehead shape of skull and sex determination by use of the computer. *Foren. Sci. Int.* **47**, 101–12.

Johnson, D. R., O'Higgins, P., McAndrew, T. J., Adams, L. M. & Flinn, R. M. (1985). Measurement of biological shape: A general method applied to mouse vertebrae. *J. Embryol. Exp. Morph.* **90**, 363–7.

Kaesler, R. H. & Waters, J. A. (1972). Fourier analysis of the ostracode margin. *Geol. Soc. Am. Bull.* **83**, 1169–78.

Kaye, B. H. & Naylor, A. G. (1972). An optical information procedure for characterizing the shape of fine particle images. *Pattern Recog.* **4**, 195–9.

Kieler, J., Skubis, K., Grzesik, W., Strojny. P., Wisniewski, J. & Dziedzic-Goclawska, A. (1989). Spreading of cells on various substrates evaluated by Fourier analysis of shape. *Histochem.* **92**, 141–8.

Koenderink, J. J. (1990). *Solid Shape.* Cambridge, Mass: MIT Press.

Kuhl, F. P. & Giardina, C. R. (1982). Elliptic Fourier features of a closed contour. *Comp. Graph Imag. Proc.* **18**, 236–58.

Lestrel, P. E. (1974). Some problems in the assessment of morphological size and shape differences. *Yearbk. Phys. Anthropol.* **18**, 140–62.

Lestrel, P. E. (1980). A quantitative approach to skeletal morphology: Fourier analysis. *Soc. Phot. Inst. Engrs.* (SPIE) **166**, 80–93.

Lestrel, P. E. (1982). A Fourier analytic procedure to describe complex morphological shapes. In *Factors and Mechanisms Influencing Bone Growth* ed. A. D. Dixon & B. G. Sarnat. New York: Alan R. Liss, Inc.

Lestrel, P. E. (1989a). Method for analyzing complex two-dimensional Forms: Elliptical Fourier functions. *Am. J. Hum. Biol.* 1:149–64.

Lestrel, P. E. (1989b). Some approaches toward the mathematical modeling of the craniofacial complex. *J. Craniofac. Genet. Dev. Biol.* **9**, 77–91.

Lestrel, P. E. (in press). Morphometrics of craniofacial form. In *Fundamentals of Craniofacial Growth*, ed. A. D. Dixon, D. A. N. Hoyte & O. Rönning. Boca Raton, Florida: CRC Press.

Lestrel, P. E., Engstrom, C., Chaconas, S. J. & Bodt, A. (1991). A longitudinal study of the human nasal bone in *Norma lateralis*: Size and shape considerations. In *Fundamentals of Bone Growth: Methodology and Applications* ed. A. D. Dixon, B. G. Sarnat & D. A. N. Hoyle. Boca Raton, Florida: CRC Press.

Lestrel, P. E., Bodt, A. & Swindler, D. R. (1993). Longitudinal study of cranial base changes in *Macaca nemestrina*. *Am. J. Phys. Anthrop.* **91**, 117–29.

Lin, C. S. & Hwang, C. L. (1987). New forms of shape invariants from elliptic Fourier descriptors. *Pattern Recog.* **20,** 535–45.

Lohmann, G. P. & Schweitzer, P. N. (1990). On eigenshape analysis. In *Proceedings of the Michigan Morphometric Workshop* ed. F. J. Rohlf & F. L. Bookstein. *Univ. Mich. Mus. Zool.* Special Pub. No. 2, 147–66.

Lu, K. H. (1965). Harmonic analysis of the human face. *Biometrics* **21,** 491–505.

Lugt, A. V. (1974). Coherent optical processing. *Proc IEEE* **62,** 1300–18.

Martin, R. (1928). *Lehrbuch der Anthropologie.* Stuttgart: G. Fischer.

McDonell, W. R. (1906). A second study of the English skull, with special reference to the Moorfields Crania. *Biometrika* **5,** 86–104.

Medalia, A. L. (1970). Dynamic shape factors of particles. *Powder Tech.* **4,** 117–38.

Mok, E. C. & Boyer, A. L. (1986). Encoding patient contours using Fourier descriptors for computer treatment planning. *Med. Phys.* **13,** 413–15.

Mokhtarian, F. & Mackworth, A. (1986). Scale-based description and recognition of planar curves and two-dimensional shapes. *IEEE Trans. Pattern Anal. Mach. Int.* PAMI-**8,** 34–43.

Nafe, R., Kaloutsi, V., Choritz, H. & Georgii, A. (1992). Elliptic Fourier analysis of megakaryocyte nuclei in chronic myeloproliferative disorders. *Anal. Quant. Cytol. Histol.* **14,** 391–7.

Needham, A. E. (1950). The form-transformation of the abdomen of the female peacrab, *Pinnotheres pisum. Proc. Roy. Soc. (Lond.) B,* **137,** 115–36.

O'Higgins, P. & Williams, N. W. (1987). An investigation into the use of Fourier coefficients in characterizing cranial shape in primates. *J. Zool. Lond.* **211,** 409–30.

Owen, R. (1848). Report on the archetype and homologies of the vertebrae skeleton. *Rep. 16th meeting Brit. Assoc. Adv. Sci.* 169–340.

Penrose, L. S. (1954). Distance, size and shape. *Ann. Eugen.* **18,** 337–43.

Ray, T. S. (1990). Application of eigenshape analysis to second-order leaf shape ontogeny in *Syngonium podophyllum* (Araceae). In *Proceedings of the Michigan Morphometric Workshop* ed. F. J. Rohlf & F. L. Bookstein. *Univ. Mich. Mus. Zool.* Special Pub. No. 2, 201–13.

Read, D. W. (1990). From multivariate to qualitative measurement: Representation of shape. *Hum. Evol.* **5,** 417–29.

Read, D. W. & Lestrel, P. E. (1986). A comment upon the uses of homologous-point measures in systematics: A reply to Bookstein et al. *Syst Zool.* **33,** 241–53.

Reyment, R. A. (1991). *Multidimensional Paleobiology.* Oxford: Pergamon Press.

Reyment, R. A., Blackith, R. E. & Campbell, N. A. (1984). *Multivariate morphometrics* (2nd ed.). New York: Academic Press.

Richtsmeier, J. T., Cheverud, J. M. & Lele, S. (1992). Advances in anthropological morphometrics. *Ann. Rev. Anthropol.* **21,** 283–305.

Rohlf, F. J. & Archie, J. W. (1984). A comparison of Fourier methods for the description of wing shape in mosquitos (*Diptera: culicidae*). *Syst. Zool.* **33,** 302–17.

Saaty, T. L. & Alexander, J. M. (1982). *Thinking with Models..* New York: Pergamon Press.

Schwarcz, H. P. & Shane, K. C. (1969). Measurement of particle shape by Fourier analysis. *Sedimentology* **13,** 213–31.

Simpson, G. G. (1961). *Principles of Animal Taxonomy.* New York: Columbia University Press.

Sneath, P. H. A. & Sokal, R. R. (1973). *Numerical Taxonomy* San Francisco: W. H. Freeman and Co.

Spoehr, K. T. & Lehmkuhle, S. W. (1982). *Visual Information Processing.* San Francisco: W. H. Freeman.

Thom, R. (1983). *Mathematical Models of Morphogenesis*. Trans. W. M. Brookes & D. Rand. New York: John Wiley.

Thompson, D. W. (1915). Morphology and mathematics. *Tran. Roy. Soc. Edinburgh* **50,** 857–95.

Thompson, D. W. (1942). *On Growth and Form*. (2nd ed.) Cambridge: Cambridge University Press.

Tolstov, G. P. (1962). *Fourier Series*. Englewood Cliffs, NJ: Prentice-Hall.

van Otterloo, P. J. (1991). *A Contour-Oriented Approach to Shape Analysis*. New York: Prenitce-Hall.

Waters, J. A. (1977). Quantification of shape by the use of Fourier analysis: the Mississippian blastoid genus *Pentremites. Paleobiol.* **3,** 288–99.

Webster, N. (1983). *New Universal Unabridged Dictionary*. (2nd ed.) Cleveland: Simon and Schuster.

White, R. J., Prentice, H. C. & Verwijst, T. (1988). Automated image acquisition and morphometric description. *Can. J. Bot.* **66,** 450–9.

Xu, J. & Yang, Y-H. (1990). Generalized multidimensional orthogonal polynomials with applications to shape analysis. *IEEE Trans. Pattern. Anal. Mach. Int.* **12,** 906–13.

Younker, J. L. & Ehrlich, R. (1977). Fourier biometrics: Harmonic amplitudes as multivariate shape descriptors. *Sys. Zool.* **26,** 336–42.

2

Introduction and Overview of Fourier Descriptors

PETE E. LESTREL
UCLA School of Dentistry

Nessuna humana investigazione si pio dimandara vera scienzia s'essa non pass per le matimatiche dimonstrazione.
No human investigation can be called real science if it cannot be demonstrated mathematically.

Leonardo da Vinci (1452–1519)

2.1 Introduction

Few mathematicians of the nineteenth century have had such a marked influence on both theoretical mathematics and the applied scientific realm as Jean Batiste Joseph Fourier (1768–1830). With Fourier, a distinct separation between pure and applied mathematics emerged. In terms of pure mathematics, his work led to revised formulations of the concepts of function, series, and the integral.[1] From an applied point of view, Fourier analysis became an essential part of mathematical physics, astronomy, acoustics, optics, electrodynamics, thermodynamics, hydrodynamics, geophysics, climatology, etc. More recent applications of Fourier analytic methods include the fields of pattern recognition, biology, and medicine.

Given the continuing importance of Fourier analysis, also known as harmonic analysis or spectral analysis, it is perhaps useful to briefly define the mathematical concepts that underlie a majority of the applications in this volume.

2.2 Boundary outline models

One of the most essential and visually recognizable aspects of the biological form is its outline. Methods used to model contours are curve-fitting procedures in-

---

[1] For the historical background of Fourier's work, I recommend Herivel (1975), Grattan-Guiness (1972) , and Ravetz and Grattan-Guiness (1981).

cluding the conventional Fourier series (see Chapters 5–12) and the comparatively recent development, elliptical Fourier functions (Chapters 13 through 16). The latter procedure is a parametric approach (described in detail below) which allows for the analysis of contours that cannot be simply represented as single-valued functions (a constraint with conventional Fourier analysis). A particularly attractive property of both of these approaches is that they are *information-preserving contour representations* (van Otterloo, 1991) in the sense that they not only allow for the precise reconstruction of the biological outline, but also allow for its re-creation, at any time, in the absence of the original specimen. Furthermore, both global as well as localized aspects of the contour are amenable to analysis. The major approach used to numerically describe boundary outline data is currently based on these Fourier descriptors.[2,3] This chapter is intended to serve as a brief introduction to the mathematics that underly these curve-fitting functions. The effective utilization of Fourier descriptors also requires some familiarity with what may be unfamiliar concepts, such as amplitude, harmonic, power, etc.; terms originally derived from electrical engineering (see Glossary).

Section 2.3 explores the conventional Fourier series in some detail. Section 2.6 presents Elliptical Fourier functions. Computer programs for both Fourier procedures are available (see Appendix).

The methods to be described in subsequent sections focus primarily on the boundary of outlines, rather than on landmarks, although they do not entirely ignore them (Chapters 14 and 16 provide an approach for embedding landmarks into the boundary outline). The importance of these boundary methods is predicated on the assumption that much of the biological information of interest (especially with respect to shape) is located on the outline and not necessarily confined to landmarks (homologous or otherwise). These approaches become increasingly useful when there are only a few, sparsely located, homologous landmarks and are essential in the absence of such points. They are indispensable if the boundary outline, as a totality, is of primary interest. However, these methods tend to be limited to planar forms (for a 3-D approach see Chapter 16).

Boundary outlines can be defined as either closed or open. Although open curves also lend themselves to analysis, the majority of investigations tend to deal with closed forms.

---

[2] Fourier descriptors, broadly defined as 'output' from the Fourier curve fit, consist of amplitudes, phase angles, etc. The Fourier spectrum and power computations have also been used in this sense. These parameters are obtained from the curve-fit of a Fourier series to data, usually consisting of a tabulated function (see Section 2.4.1).

[3] Although it is recognized that other methods are available for describing size and shape attributes of forms, such as Legendre and Chebyshev expansions and orthogonal polynomials, these have seldom been applied in a biological context.

2.3 Fourier's series

A brief development of Fourier's series is provided below with no attempt at rigor, and no proofs will be provided (see Carslaw, 1950; Franklin, 1958; Davis, 1989; and especially Tolstov, 1962, for detailed expositions of Fourier methods).

A trigonometric series of the form

$$a_0 + (a_1 \cos x + b_1 \sin x) + (a_2 \cos 2x + b_2 \sin 2x) + \ldots \tag{2.1}$$

is said to be a *Fourier series* if the constants a_0, a_1, b_1 . . . satisfy the following relations:

$$a_0 = \frac{1}{2\pi} \int_{-\pi}^{\pi} f(x) \, dx, \tag{2.2}$$

$$a_n = \frac{1}{\pi} \int_{-\pi}^{\pi} f(x) \cos nx \, dx \quad (n = 0, 1, 2, \ldots, \infty), \tag{2.3}$$

and

$$b_n = \frac{1}{\pi} \int_{-\pi}^{\pi} f(x) \sin nx \, dx \quad (n = 1, 2, \ldots, \infty), \tag{2.4}$$

The Fourier series is said to *correspond* to the function $f(x)$. This correspondence, in discrete form, is often shown as

$$f(x) \sim a_0 + \sum_{n}^{\infty} = 1 \, (a_n \cos nx + b_n \sin nx). \tag{2.5}$$

It is important to note that the correspondence (Eq. 2.5) does not imply convergence. Fourier recognized that any arbitrary function $f(x)$ over the interval $[-\pi, \pi]$ can be shown to correspond to the trigonometric series bearing his name.

The sign $\sim$ can be replaced by an equal sign only if the series can be shown to converge[4] and that its sum equals $f(x)$. This leads to the following theorem:

> **Theorem 1.** *If a function, $f(x)$, of period 2π can be expanded in a trigono-*
> **metric series which converges uniformly over the interval** $[-\pi, \pi]$*, then*
> *this series is the Fourier series of $f(x)$ (adapted from Tolstov, 1962).*

Although this theorem begins to define a Fourier series in a formal way, it is incomplete because it does not account for jump discontinuities. It presumes that the function $f(x)$ is continuous over the interval. This leads to the so-called

---

[4] Convergence can be defined in heuristic terms as follows: if, for sufficiently large n terms in the series, it is impossible to distinguish between the graph of the function, $f(x)$, and the series expansion used to approximate it, then the series is convergent.

Dirichlet conditions. Suppose $f(x)$ is defined over the interval $[-\pi, \pi]$; then the Dirichlet conditions on $f(x)$ are such that the function $f(x)$ is: (1) single-valued except at a finite number of points over the interval; (2) periodic over the interval; (3) piecewise-continuous with a finite number of discontinuities; (4) piecewise-monotone with a finite number of maxima and minima; and (5) absolutely integrable over the period (Spiegel, 1974; Hsu, 1967). These conditions imposed on $f(x)$ are considered mathematically as sufficient but not necessary. The incorporation of the Dirichlet conditions leads to the more general result:

> **Theorem 2.** *If an absolutely integrable function, f(x), of period 2π can be expanded in a trigonometric series which converges to f(x) everywhere, except at a finite number of points, over the interval $[-\pi, \pi]$, then this series is the Fourier series of f(x) (Tolstov, 1962).*

2.3.1 Even and odd functions

A number of useful properties arise from Fourier's series which are of considerable practical importance. Fourier's trigonometric expansion (Eq. 2.5), in common with other kinds of functions (that is, not limited to trigonometric relations), is composed of what are termed even and odd functions. The function $f(x)$ is defined as even if

$$f(x) = f(-x) \tag{2.6}$$

for every value of x. The graph of an even function is symmetric with respect to, in this case, the y-axis. A function $f(x)$ is odd if

$$f(-x) = -f(x) \tag{2.7}$$

for every value of x. The graph of an odd function is asymmetrical with respect to the y-axis.

It can be shown that the integral of an odd periodic function which is *symmetrical* about the origin is zero. Likewise, the integral of an even periodic function which is *asymmetrical* about the origin will also be zero. Consequently, if a function is even or symmetric about the origin, the bn coefficients (Eq. 2.4) will be zero and all sine terms will vanish, reducing Equation 2.5 to a cosine series:

$$f(x) \sim a_0 + \sum_{n=1}^{\infty} a_n \cos nx, \tag{2.8}$$

where the coefficients a_n calculated from Equation 2.3. The presence of symmetry allows for a reduction in the series in that all of the sine terms vanish. Similarly, if the function is odd, then the a_n terms will be zero for all n, resulting in a series with only sine terms.

2.3.2 Property of orthogonality

Another useful property of the trigonometric relations introduced in the previous section is that they are orthogonal. That is, the integral of the product of any two trigonometric functions over the interval $[-\pi,\pi]$ will vanish (Hamming, 1973). Consider a system such as

$$1, \cos x, \sin x, \cos 2x, \sin 2x, \ldots \cos nx, \sin nx. \tag{2.9}$$

The trigonometric relations (Eq. 2.9) are said to be pairwise orthogonal on the interval $[-\pi,\pi]$. Because they are also periodic, the interval need not be bounded by $[-\pi,\pi]$, but may be any multiple of 2π.

Using modern numerical methods, we can now view the Fourier approximation of a periodic function $f(x)$ over the period of $[-\pi,\pi]$ as

$$f(x) = \frac{a_0}{2} + \sum_{n=1}^{\infty} (a_n \cos nx + b_n \sin nx), \tag{2.10}$$

and since the trigonometric relations are orthogonal,[5] may be expressed as:

$$\int_0^{2\pi} \sin mx \cos nx\, dx = 0, \tag{2.11}$$

$$\int_0^{2\pi} \cos mx \cos nx\, dx = \begin{cases} 0 & m \neq n \\ \pi & m = n \neq 0 \\ 2\pi & m = n = 0 \end{cases} \tag{2.12}$$

and

$$\int_0^{2\pi} \sin mx \sin nx\, dx = \begin{cases} 0 & m \neq n \\ \pi & m = n \neq 0. \end{cases} \tag{2.13}$$

The coefficients a_n and b_n are given by

$$a_n = \int_0^{2\pi} f(x) \cos nx\, dx \qquad (n \geq 0) \tag{2.14}$$

and

$$b_n = \frac{1}{\pi} \int_0^{2\pi} f(x) \sin nx\, dx \qquad (n \geq 1) \tag{2.15}$$

---

[5] Whereas the definition of orthogonality commonly connotes perpendicularity in a graphical sense, the equations here illustrate another approach to orthogonality. In any case, orthogonality implies that the coefficients in the Fourier expansion are independent from each other, or uncorrelated. In a more technical sense this means that the covariance is equal to zero. Refer to Hamming (1973) and Tolstov (1962) for details.

2.3.3 Nyquist frequency requirements

It may be presumed that one merely needs to add terms to the series expansion until a perfect fit is obtained between the function $f(x)$ and the Fourier series representation. Unfortunately, things are not quite so simple, as the maximum number of terms taken in the series is subject to Nyquist frequency requirements (named after Harry Nyquist (1899–1976), a pioneer in communication theory), meaning that error will arise. This error, called aliasing, is the incorporation of irresolvable high frequencies into the lower frequencies. These high frequencies (harmonics), whose wavelengths are less than twice the spacing between sample points, cannot be detected (Davis, 1986).

Thus, the Nyquist frequency (or sampling interval) defines how closely together the points at which a signal is sampled (measured) must be taken in order for the representation to be unique. Viewed in a biological context this means that given k observations (or points sampled) on the outline of a form, the maximum number of harmonics attainable, N, is equal to either $(k - 1)/2$ if k is odd or $k/2$ if k is even.

2.3.4 The Gibbs phenomenon

The behavior of Fourier's series representation near a jump discontinuity,[6] as it converges onto the function $f(x)$, leads to an easily visualized 'overshoot' which becomes smaller as convergence is approached. This effect arises because the convergence is not uniform. This overshooting of the function is called the Gibbs phenomenon after the physicist and chemist Josiah Willard Gibbs (1839–1903). In 1899 Gibbs, in a letter to *Nature*, pointed out that Fourier's series representation behaved quite differently at points of discontinuity, although without proof (Carlsaw, 1950).

2.3.5 Goodness-of-fit considerations

At this point one may ask how many terms are required for a precise representation of the contour. The improvement in fit, as the Fourier series is expanded with the addition of terms (the maximum dictated by the Nyquist frequency requirements), can be assessed from the computation of the residual. This residual is the difference between the observed data points and the expected values derived from the series expansion. The mean residual over all data points can be computed as:

$$re = \frac{1}{k} \sum (r_i - r_i')^2, \tag{2.16}$$

---

[6] Jump discontinuities can be visualized as occurring when electric current is systematically turned on and off, creating the familiar square wave. Such discontinuities can also arise in biology with the presence of sharp "corners".

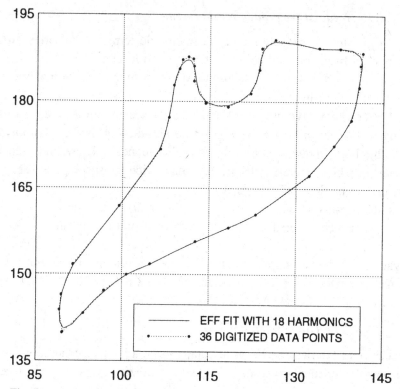

Fig. 2.1 An elliptical Fourier function fitted to the human cranial base in *Norma lateralis*. The cranial base outline is described with 36 points (dotted line) and fitted with an EFF curve composed of 18 harmonics (solid line). The mean residual was 0.23 mm over the whole outline. A slight loss of fit can be observed in regions of sharp curvature.

where r_i are the observed data points, r_i' are the expected values, and k is the total number of data points. Fig. 2.1 illustrates a human cranial base outline with 36 points, fitted with an elliptical Fourier function (to be described subsequently) with 18 harmonics. The mean residual was 0.23 mm which can be considered as a satisfactory fit except in regions of sharp curvature, an expected outcome based on the Gibbs phenomenon.

2.4 Harmonic analysis

The decomposition of a time series into separate components, or harmonics, is termed harmonic analysis or spectral analysis. The converse process of re-creation is called harmonic synthesis. The approach of spectral or harmonic analysis consists of transforming the data from one domain to another. In electronic engineering this transformation is commonly from the *time* domain into the *frequency* domain. In other fields that use Fourier descriptors (FDs), such as pattern recognition and the geological and biological sciences, the transformation is from the *spatial* domain, into the frequency domain. The spatial domain here, for example, can refer to the data points that determine the outline of the object of interest, whereas the frequency domain is defined by the harmonics that are computed with the FD. The frequency domain is usually defined in terms of amplitude and phase relationships. These, in turn, are obtained from the approximation of an arbitrary piecewise continuous function, $f(x)$, with the use of a finite sum composed of sine and cosine terms; hence the use of the term harmonic analysis (Eq. 2.10). A number of components associated with harmonic analysis are universally used and these will be briefly reviewed here. Five aspects of special interest are: (1) period; (2) frequency; (3) amplitude; (4) power; and (5) phase. Of these, the three properties *period*, *amplitude*, and *phase* provide a flexible system that can be used jointly to describe the biological form.

Consider a simple sinusoidal function such as a sine wave which repeats over an interval (along the x-axis). The period refers to one complete cycle from 0 to 2π, also called the wavelength. The frequency, on the other hand, is the reciprocal of this period or wavelength.

The sum of the lowest terms, $\sin x$ and $\cos x$, is called the fundamental frequency and is equal to the period of the arbitrary piecewise-smooth function, $f(x)$, that is being approximated. The fundamental frequency is given by the first harmonic which is from 0 to 2π. The next higher frequency is equal to the second harmonic, and is one-half the wavelength of the fundamental frequency, and so on. The subsequent terms in the series (higher harmonics) are integral multiples of the lowest term, the fundamental frequency. In Cartesian coordinates the amplitude refers to the maximum height of the waveform from the x-axis within a period. Phase refers to the placement of the starting point of the waveform from the origin along the x-axis. For example, the "shape" of the sinusoidal curve, $y = \cos x$, is identical to $y = \sin x$ except that it is shifted in phase by 90 degrees or $\pi/2$ radians.

The simplest periodic function of considerable practical importance is one which describes the motion of a body subject to restoring forces. Because this motion is repeated, it is periodic. Examples include a vibrating string, a swinging pendulum, etc. This motion can be precisely determined from

$$f(t) = A \sin(\omega t + \gamma), \tag{2.17}$$

where A, ω and γ are constants. A or $|A|$ is the amplitude or maximum displacement from a central or initial point, ω is the angular frequency (velocity in radians per unit time), and γ is the initial phase, or phase angle. The period of such a harmonic function is $T = 2\pi/\omega$. The amplitude, $|A|$, is defined as the maximum deviation of the path of motion, a pendulum for example, from rest. The quantity $1/T$ is the number of oscillations over a particular time interval, hence the term frequency. The constant γ, or initial phase, describes the initial position (or angle) of the pendulum when $t = 0$. Equation 2.17 above, can be rewritten in an alternative form using the trigonometric identity:

$$A \sin(\omega t + \gamma) = A(\cos \omega t \sin \gamma + \sin \omega t \cos \gamma). \tag{2.18}$$

Let

$$a = A \sin \gamma \tag{2.19}$$

and

$$b = A \cos \gamma. \tag{2.20}$$

Substituting, we can write

$$f(x) = a \cos \omega t + b \sin \omega t. \tag{2.21}$$

Let $T = 2L$ where $2L$ is the length of the period; then

$$\omega = \frac{2\pi}{T} = \frac{\pi}{L}. \tag{2.22}$$

Substituting π/L in Equation 2.21 and setting it up in terms of x instead of t, we obtain

$$f(x) = a_0 + \sum_{n=1}^{\infty} a_n \cos \frac{n\pi x}{L} + \sum_{n=1}^{\infty} b_n \sin \frac{n\pi x}{L}, \tag{2.23}$$

which is the representation of the infinite Fourier series in Cartesian coordinates where

> $f(x)$ = the dependent variable (y) or wave height for each x value,
> a_0 = coefficient of zeroth degree term or constant,
> a_n = coefficients of cosine terms where $n = 1, 2, 3, \ldots$,
> b_n = coefficients of sine terms where $n = 1, 2, 3, \ldots$,
> x = sampling points along the x-axis from $-L$ to $+L$,
> L = half the fundamental sampling length or period.

Finally, if the period is defined over a 2π interval and the limits changed, Equation 2.23 can be simplified to the familiar finite form

$$f(x) = a_0 + \sum_{n=1}^{N} a_n \cos nx + \sum_{n=1}^{N} b_n \sin nx \tag{2.24}$$

where N is the maximum harmonic number.

2.4.1 Tabulated functions

If the function $y = f(x)$ is known, then the coefficients can be obtained from Equations 2.3 and 2.4 or Equations 2.14 and 2.15 for all N. If $y = f(x)$ is unknown, a common occurrence, this precludes such an analytical solution and recourse must be made to numerical integration methods such as the trapezoidal rule. In the latter case the data (x_i, y_i) represent a tabulated list obtained by observation or experiment, instead of an analytical function.

Thus, one can represent any arbitrary (nonperiodic) tabulated function with a FD. This requires that the spatial form under consideration be made "periodic" in the sense that it is repeated over a set interval (the period). That is, the last data point on the outline is followed by the first point on the outline and the process repeated (Lestrel, 1989a). Much of the data discussed in this volume are tabulated lists.

If k observations along the boundary of a form are chosen with equal divisions over the interval $[-\pi, \pi]$, then the general solutions for the coefficients are derived from

$$a_0 = \sum_{i=0}^{k-1} f(x)/k, \tag{2.25}$$

$$a_n = \frac{2}{k} \sum_{i=0}^{k-1} f(x) \cos ix \qquad \left[n = 0, 1, 2, \ldots, \frac{k-1}{2} \right], \tag{2.26}$$

and

$$b_n = \frac{2}{k} \sum_{i=0}^{k-1} f(x) \sin ix \qquad \left[n = 1, 2, \ldots, \frac{k-1}{2} \right]. \tag{2.27}$$

The limits $i = 0$ and $i = k - 1$ are a consequence, in this case, of the trapezoidal rule (Harbaugh and Preston, 1968; Harbaugh and Merriam, 1968).

The trapezoidal rule (other methods are also available) satisfies the least squares criterion of minimizing the sum of the squared deviations. With tabulated functions, harmonic analysis consists of fitting, by a least squares estimation procedure, the regression equation (Eq. 2.24) to a set of tabulated data, where the Fourier coefficients a_0, a_n, and b_n, for all k points and n harmonics, are regression coefficients.

Thus, in the next section and for the remainder of this chapter, we will start with the assumption that we are dealing with an unknown function. That is, the data consist of points (x_i, y_i) on a closed boundary (open boundaries are also analyzable, for example, by using the property of reflection, but they will be ignored here). We will expand a Fourier series as a representation of this unknown and arbitrary function.

2.4.2 *Two-dimensional polar coordinates*

As many of the early studies using harmonic analysis are based on polar coordinates, a brief review is in order. If the observed two-dimensional (planar) data can be described with: (1) points on the boundary outline (equally spaced to avoid a weighted analysis); (2) distances (ranges) to those points from a predetermined center within the closed form; and (3) none of the distances being double-valued (i.e., where a single distance can lead to more than one point on the boundary—see Figure 2.2), then the structure can be represented as a tabulated function in polar coordinates (r_i, θ_i).

The familiar definite form of the Fourier series is now represented by

$$f(\theta) = A_0 + \sum_{n=1}^{N} a_n \cos n\theta + \sum_{n=1}^{N} b_n \sin n\theta, \qquad (2.28)$$

where the period is defined over a 2π interval and θ is in radians. The maximum degree or harmonic number is denoted by N. The a_n and b_n are the Fourier coefficients for the n harmonics. This is a convergent series that will provide a fit to any known piecewise-smooth single-valued function. The $a_n \cos n\theta$ terms (even) describe symmetric patterns whereas the $b_n \sin n\theta$ terms (odd) describe asymmetry. If total symmetry is present or can be imposed[7] with respect to the coordinate system, it can be shown that the sine terms vanish

$$f(\theta) = A_0 + \sum_{n=1}^{N} a_n \cos n\theta. \qquad (2.29)$$

The expansion of this series is

$$f(\theta) = A_0 + a_1 \cos \theta + a_2 \cos 2\theta \\ + a_3 \cos 3\theta + \ldots + a_n \cos n\theta, \qquad (2.30)$$

and the fit is improved as more terms are added.

One can readily plot the separate contributions that each Fourier component makes to the form. The first harmonic, $a_1 \cos \theta$ in Equation 2.29, describes the contribution of an offset circle. The second harmonic, $a_2 \cos 2\theta$, describes a figure eight or lemniscate (a two-leafed rose). The third, $a_3 \cos 3\theta$, describes a trefoil or three-leafed rose, and so on. In other words, any irregular single-valued planar form can be decomposed into a set of simpler components (harmonics), which when summed, will re-create the outline. Moreover, the larger the magnitude of the amplitude associated with the harmonic number, the greater the con-

---

[7] The imposition of symmetry by reflection has the attractive property that it may be possible to attach biological meaning to the first few harmonics. See Lestrel (1974) for an example using the human cranium.

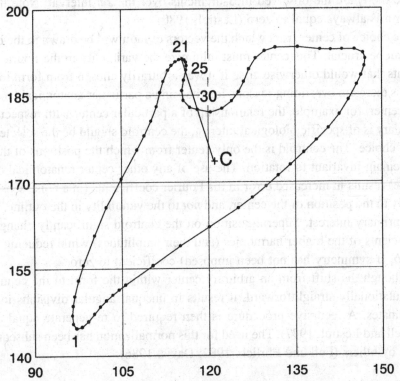

Fig. 2.2 Some of the difficulties encountered with morphologies which are not single-valued. This particular example of a human cranial base outline shows a vector (or radius) drawn from a center, C, which intersects the outline at three different locations (points 21, 25, and 30). Such morphologies are inadmissible with conventional FDs, but can be handled with EFFs (see text).

tribution of that particular harmonic to the total form. The constant or A_0 term is defined as the mean of all the observations, which is a circle in polar coordinates

$$A_0 = k^{-1} \sum_{i=0}^{k-1} r_i \tag{2.31}$$

where ri are observations or radii (distances from a predetermined origin) and k is the total number of measurements. The a_n and b_n coefficients are found by least squares estimation procedures, again subject to the Nyquist frequency:

$$a_n = \frac{2}{k} \sum_{i=0}^{k-1} r_i \cos n\theta \qquad \left[n = 0, 1, 2, \ldots, \frac{k-1}{2} \right], \tag{2.32}$$

and

$$b_n = \frac{2}{k} \sum_{i=0}^{k-1} r_i \sin n\theta \qquad \left[n = 1, 2, \ldots, \frac{k-1}{2} \right], \tag{2.33}$$

where the r_i are the observed measurements over the 2π interval. Note that the B_0 term is always equal to zero (Lestrel, 1980).

The choice of center from which the vectors or radii will be drawn to the boundary can be crucial. This center must minimize the variability in the Fourier coefficients that would otherwise arise if it was arbitrarily chosen from form to form. Unless there are overriding reasons for choosing a particular anatomical landmark as a center, for example, the relationship of a particular center with respect to the boundary is of specific biological interest, the centroid should be the selected center of choice. The centroid is the only center from which the positions of the vectors remain invariant to rotation. The use of any other center (anatomical or otherwise) results in increased error in the Fourier coefficients (in a noise sense) due strictly to the position of the center, and *not* to the variability in the outline, which is of primary interest. Superimposition on the centroid significantly changes the coefficients of the higher harmonics (and their amplitudes) while reducing the a_1 (and b_1 if symmetry has not been imposed) coefficient to zero.

Although the shift from an arbitrary center within the form to the centroid is computationally straightforward, it results in unequal angular divisions in polar coordinates. A recursive procedure is then required to re-generate equal angles (Parnell and Lestrel, 1977). The need for this normalization has been subsequently noted by others (Full and Ehrlich, 1982; Davis, 1986).

2.4.3 Amplitude, power, and phase

Once the Fourier coefficients have been calculated, a common approach is to compute the amplitudes associated with each harmonic number. These amplitude values can be considered as a measure of the influence (or weight) that each harmonic has on the form. This allows the creation of an amplitude versus harmonic number plot. The amplitude for the nth harmonic is

$$A_n = \sqrt{a_n^2 + b_n^2},\tag{2.34}$$

where a_n are the coefficients preceding the cosine terms and b_n are the coefficients preceding the sine terms. If symmetry about an axis is present or can be imposed, then the sine terms vanish and Equation 2.34 reduces to:

$$A_n = \sqrt{a_n^2}\tag{2.35}$$

for each harmonic (see Tolstov, 1962).

The raw power or variance of the nth harmonic is

$$S_n^2 = \frac{a_n^2 + b_n^2}{2} = \frac{amplitude_n{}^2}{2}.\tag{2.36}$$

If symmetry is involved, Equation 2.36 reduces to:

$$S_n^2 = \frac{a_n^2}{2}.$$ (2.37)

From either Eq. 2.36 or Eq. 2.37 we can plot the discrete spectrum or *power spectrum* as a graph of power versus harmonic number, n. This graph is also called a periodogram (Davis, 1986). The percentage of the variation explained by the nth harmonic is defined as

$$\% \; Explained = \frac{S_n^2}{S_T^2},$$ (2.38)

where S_n^2 is the variance accounted for by the nth harmonic and S_T^2 is the summation (or total variance) of all terms.

The remaining property, phase, deserves special attention because it is often misunderstood. Many studies seemingly ignore phase and emphasize only differences in amplitudes. Others have suggested that the phase angle information cannot be ignored (Rohlf, 1986; see also Chapter 5). With respect to the phase of the first harmonic, this can be especially critical if the outlines to be compared differ in orientation; that is, each with different starting points (r_1, θ_1) on the boundary. In that case, comparisons between outlines would be biased. However, problems with significantly differing *initial* phase angles can be minimized with a carefully chosen common orientation of the outlines *prior* to Fourier analysis. Ehrlich and coworkers have also demonstrated how the phase angle "locks" onto homologous features suggesting invariance of the phase angle within samples of similar specimens (Ehrlich et al., 1983). This, then, can be considered as evidence that phase and amplitude are largely independent of each other. Nevertheless, the phase angle, especially when associated with the higher harmonics, can contribute significant information and should not be arbitrarily ignored. The phase angle for the nth harmonic is

$$\Phi_n = \tan^{-1} \left[\frac{a_n}{b_n} \right],$$ (2.39)

where a_n and b_n are the coefficients for the nth harmonic.

2.4.4 Positional-orientation and size-standardization

Two other properties or normalizations need to be defined. These are termed *positional-orientation* and *size-standardization* (Parnell and Lestrel, 1977; Lestrel, 1980). Positional-orientation consists of the orientation of the outline in space (which has an effect on the initial phase angle). The first normalization, orientation of the boundary outline, can be controlled with judicious positioning of the outline prior to digitization. This entails precise positioning of the initial starting point (see Section 2.6 for a newer alternative procedure).

The second normalization, size-standardization, is predicated on the grounds that shape (defined as the boundary outline of form), in contrast to size, contains considerable informational content of biological importance. Thus, if size differences are at all appreciable, subtle shape differences may be confounded and possibly overwhelmed by the effect of size. Three scaling factors have been suggested as normalization approaches based on: (1) the A_0 term, or constant; (2) arc length or perimeter of the outline; and (3) the bounded area. The first normalization refers to the A_0 term which is defined as the mean of the vectors to the outline (Schwarcz and Shane, 1969 were probably the first to suggest this approach). Although this value is readily computed, it has the impediment that if the outline of interest departs significantly from a circle, the scaling factor will become increasingly inaccurate (Lestrel, 1980).

The second normalization, perimeter, has a serious drawback if used directly, as significant differences can arise. For example, consider two similar forms with the exception that one of the outlines is relatively smooth and the other has extensive invaginations. The latter case will produce an unduly large perimeter in contrast to the former (van Otterloo, 1991). The third approach, advocated here, is based on the area within the bounded outline. An algorithm for the calculation of the area within the boundary outline, in polar form, which is based on integration by parts, can be found in Lestrel (1980).

2.5 Conventional Fourier descriptors

Two Fourier descriptor approaches have been widely used. Both convert the data to polar coordinates (r_i, θ_i) prior to Fourier analysis. One is based on measurements from a center within the form, preferably the centroid. The other, the Zahn and Roskies algorithm (discussed subsequently), uses an angular function based on the points located on the outline.

2.5.1 A center-based algorithm

Many studies employing conventional FDs, usually in polar form, are now available and form part of the contributions to this volume. An early application of harmonic analysis to data was carried out by Lu (1965). He attempted to numerically describe the form of the human face. That paper fitted a Fourier series, in polar coordinates, to the outline of the face as seen from the frontal aspect. Unfortunately, this author was not able to replicate Lu's results, prompting an independent analysis (Lestrel, 1974). The work on periodic regressions by Bliss (1970) is another early contribution. The papers now available that are based on conventional FDs number in the hundreds; a sampling is included in Chapter 1 to show the breadth of disciplines involved.

2.5.2 An angular function algorithm

The Zahn and Roskies algorithm is based on an angular function $\phi^*(t)$ between segments composed of points on a closed planar outline (the outline is treated as a polygon) along its perimeter t. They then expand $\phi^*(t)$ as a Fourier series over the interval 0 to 2π

$$\phi^*(t) = A_0 + \sum_{n=1}^{\infty} (a_n \cos nt + b_n \sin nt). \tag{2.40}$$

In polar form this expansion yields

$$\phi^*(t) = A_0 + \sum_{n=1}^{\infty} A_n \cos(nt - \alpha_n) \tag{2.41}$$

which incorporates phase. The set $\{A_n, \alpha_n; n = 1, \ldots, \infty\}$ are the FDs for the curve. Note that for a circle $\phi^*(t) = 0$. The values, A_n and α_n are the amplitude and phase angle respectively for the nth harmonic (Zahn and Roskies, 1972). The function $\phi^*(t)$ is invariant under translation, rotation and changes of the perimeter, that is, length or scale (Persoon and Fu, 1977). Although the Zahn and Roskies Fourier descriptor contains all the information inherent in the curve outline, allowing for the recreation of the form, practical considerations require the truncation of the series. Applications of the Zahn and Roskies algorithm include Waters (1977); Ostrowski, et al., (1986); and Kieler et al., (1989). Also, according to Zahn and Roskies the coefficients of their function are independent of starting point. One problem with this descriptor is that because of truncation, the reconstructed curve may fail to close. Strackee and Nagelkerke (1983) provided a numerical approach to ensure closure. A second issue is that the coefficients are more sensitive to "noise" in the form under consideration (Rohlf and Archie, 1984).

2.5.3 Some constraints with conventional FDs

Conventional FDs can be thought of as a model which is a representation of the boundary of the morphological form under consideration. Its success is predicated on the fact that it provides a much closer correspondence (in a one-to-one mapping sense) than has been heretofore possible with other methods such as CMA. Moreover, the FD approach confers two advantages: (1) that with the advent of computers, the analysis is fairly simple and straightforward; and (2) under certain circumstances, it is possible to attach biological meaning to the Fourier coefficients (Lestrel, 1974).

With any quantitative approach that attempts to model the real world, limitations preclude its use in all situations, and FDs are no exception. Some morphological forms will be more amenable to analysis than others. In practical terms, limitations with conventional FDs include: (1) the tabulated function, $f(x)$, must be single-valued; (2) the data be set up in polar form; (3) the angles are made equal between vectors (to avoid the complications of a weighted analysis); and (4) integration is required for evaluation of the Fourier coefficients. The effect of these constraints is to limit analysis to morphological forms with relatively simple contours. Complex morphological forms that contain outlines that cross back onto themselves (producing intersections) for example, would be inadmissible.

Even without crossings, other morphologies produce problems. The presence of multiple values for a single vector cannot be satisfactorily handled with conventional FDs. An example drawn from human cranial base data in *Norma lateralis* (Lestrel and Sirianni, 1982; Lestrel and Roche, 1986) shows that vectors drawn from the centroid in the vicinity of the posterior clivus cross the cranial base at more than one place on the outline, resulting in a loss of data (Figure 5 in Lestrel and Roche, 1986). Such situations are difficult to handle without casting part of the data into an imaginary plane, an unattractive solution (Lestrel, 1989a). The use of equally spaced data is another constraint; and it is difficult to incorporate homologous landmarks into the FD if equal angles have to be maintained. Although most conventional FDs use equally-spaced points, this restriction can be overcome, at the price of considerably increased computational complexity. Nevertheless, few studies have attempted to do so. An alternative approach that largely circumvents these constraints is presented in the next section.

2.6 Elliptical Fourier functions

2.6.1 Two-dimensional outlines

The development of elliptical Fourier functions (EFFs), represents a parametric[8] formulation in the sense that the x- and y-directions are separately set up as functions of a third variable t (Kuhl and Giardina, 1982). This approach circumvents most of the constraints, discussed in the last section, that have characterized conventional FDs, although at the expense of an increase in numerical complexity. The requirement of equal intervals along the outline is now relaxed and multivalued functions are no longer a problem. Moreover, the parametric EFF coeffi-

---

[8] Parametric refers to a set of equations set up as functions of a common variable; for example, the circle $x^2 + y^2 = r^2$ can also be represented with parametric equations $x = r \cos \theta$ and $y = r \sin \theta$ in terms of the parameter θ.

cients are now generated using an algebraic approach, instead of the integral solutions required previously, which makes computation simpler and much faster. The relaxation of the constraints associated with conventional FDs confers a decided advantage in that the parametric EFFs allow for the numerical characterization of a much larger class of two-dimensional shapes than previously possible.

The Kuhl and Giardina parametric functions are defined in $x(t)$ as

$$x(t) = A_0 + \sum_{n=1}^{N} a_n \cos nt + \sum_{n=1}^{N} b_n \sin nt, \tag{2.42}$$

and in $y(t)$ as

$$y(t) = C_0 + \sum_{n=1}^{N} c_n \cos nt + \sum_{n=1}^{N} d_n \sin nt, \tag{2.43}$$

where n equals the harmonic number, N equals the maximum harmonic number, and the interval is over 2π as before.

If the sampled points along the polygon serving as the representation of the outline can be viewed as traveling at constant speed, then the first derivative, $dx_p/dt_p = f(t)$ or $dy_p/dt_p = g(t)$, is also piecewise constant and can be represented by a Fourier series. The derivatives of the segments or "deltas" between points, $\Delta x_p/\Delta t_p$ or $\Delta y_p/\Delta t_p$, are piecewise constant as well. These segments are associated with each division (since they are straight lines) along the polygon t_p to $t_p +$ 1 over the period 2π such that $\Delta t_p = [\Delta x_p^2 + \Delta y_p^2]^{1/2}$. Utilizing these relationships, Kuhl and Giardina derived estimates for the elliptical Fourier coefficients that do not require integrals, in contrast to conventional FDs. The Fourier coefficients for the x-projection are

$$a_n = \frac{1}{n^2\pi} \sum_{p=1}^{q} \frac{\Delta x_p}{\Delta t_p} [\cos(nt_p) - \cos(nt_{p-1})], \tag{2.44}$$

and

$$b_n = \frac{1}{n^2\pi} \sum_{p=1}^{q} \frac{\Delta x_p}{\Delta t_p} [\sin(nt_p) - \sin(nt_{p-1})], \tag{2.45}$$

where q is the total number of points along the polygon; as before, n is the harmonic number. Here t_p is the distance between point p and point $p + 1$ along the polygon, and x_p and y_p are the respective projections of the segment p to $p + 1$. The Fourier coefficients for the y-projections are

$$c_n = \frac{1}{n^2\pi} \sum_{p=1}^{q} \frac{\Delta y_p}{\Delta t_p} [\cos(nt_p) - \cos(nt_{p-1})], \tag{2.46}$$

and

$$d_n = \frac{1}{n^2\pi} \sum_{p=1}^{q} \frac{\Delta y_p}{\Delta t_p} [\sin(nt_p) - \sin(nt_{p-1})], \tag{2.47}$$

Besides the four coefficients a_n, b_n, c_n, and d_n that need to be evaluated, two constants, A_0 and C_0, also need to be estimated. The constants B_0 and D_0 are equal to zero which is analogous to conventional FDs. The constants A_0 and C_0 are computed from

$$A_0 = \frac{1}{2\pi} \sum_{p=1}^{q} \frac{\Delta x_p}{2\Delta t_p} [t_p^2 - t_{p-1}^2] + \alpha_p[t_p - t_{p-1}], \qquad (2.48)$$

and

$$C_0 = \frac{1}{2\pi} \sum_{p=1}^{q} \frac{\Delta y_p}{2\Delta t_p} [t_p^2 - t_{p-1}^2] + \beta_p[t_p - t_{p-1}], \qquad (2.49)$$

The α_p and β_p terms needed above are derived from

$$\alpha_p = \sum_{j=1}^{p-1} \Delta x_n - \left[\frac{\Delta x_p}{\Delta t_p} \sum_{j=1}^{p-1} \Delta t_j \right], \qquad (2.50)$$

and

$$\beta_p = \sum_{j=1}^{p-1} \Delta y_j - [\frac{\Delta y_p}{\Delta t_p} \sum_{j=1}^{p-1} \Delta t_j], \qquad (2.51)$$

where $\alpha_1 = \beta_1 = 0$. Reference should be made to Kuhl and Giardina (1982) for details of the above formulations.

A number of papers with a biological thrust have appeared since Kuhl and Giardina first published their algorithm. Two publications, one dealing with the shape of mosquito wings (Rohlf and Archie, 1984) and the other on the outline of *Mytilus edulis* shells (Ferson et al., 1985) represent the initial extension of EFFs in a zoological context. Rohlf and Archie (1984) compared the shape of mosquito wings using EFFs as well as the Zahn and Roskies formulation and conventional FDs. They found that EFFs produced the best results.

Subsequent applications have dealt with anatomical structures such as the maxilla, mandible, nasal bones, facial profile, and mastication (Lestrel, 1989a; 1989b; Lestrel et al., 1991; Lestrel and Kerr, 1993); pattern recognition (Lin and Hwang, 1987); botany (Kincaid and Schneider, 1983; White and Prentice, 1988; White et al., 1988); and cytology (Diaz et al., 1989; Diaz et al., 1990; Nafe et al., 1992).

A study of the comparison of plant leaf shape using conventional measures, Freeman chain codes, moment invariants, and EFFs, found that EFFs yielded the best discrimination between groups (White and Prentice, 1988; White et al., 1988). A study of the primate cranial base in *Norma lateralis* based on EFFs standardized for size yielded results consisting of a perfect (100%) discrimination of 122 specimens composed of five primate groups (Lestrel et al., 1988; Lestrel et al., 1996).

Recent investigations (Chapter 13) of both the cell boundary and the nucleus of human leukocytes utilizing EFFs produced characteristic differences in the shape of neutrophils, lymphocytes, and monocytes (Diaz et al., 1989). Finally, a recent study of the plasticity of mandibular response to clinical manipulation (functional appliance therapy) provides an example of the application of EFFs to characterize morphological shape changes in a dental setting (Chapter 15).

2.6.2 Extension to three dimensions

The extension of the EFF to 3-D outlines represents a straightforward approach with the inclusion of a third equation. Thus, any outline, whether in 2- or 3-D, can be approximated with this function. However, it must be reiterated that this 3-D extension only allows for the numerical description of *a curve in 3-space, not a solid*. Unfortunately, this parametric formulation precludes the analysis of solid (volume) 3-D morphologies.

If the boundary of a form can be modeled as a curve in 3-space, then the above equations can be extended without difficulty (Lestrel et al., 1993; Chapter 16). The Fourier series in $x(t)$ and $y(t)$ are given in 2.42 and 2.43, and the series in $z(t)$ is now

$$z(t) = E_0 + \sum_{n=1}^{N} e_n \cos nt + \sum_{n=1}^{N} f_n \sin nt. \tag{2.52}$$

As before, n equals the harmonic number, N equals the maximum harmonic number, and the interval is over 2π.

The extension to three-dimensional space can be viewed as an orthogonal system in three axes; that is, the polygon will be represented in a spatial domain which is sinusoidal in x, y, and z. The coefficients for the z-coordinate (Eq. 2.52) are evaluated similarly to the 2-D case; that is, the Fourier coefficients for the z-projection are:

$$e_n = \frac{1}{n^2 \pi} \sum_{p=1}^{q} \frac{\Delta z_p}{\Delta t_p} [\cos(nt_p) - \cos(nt_{p-1})], \tag{2.53}$$

and

$$f_n = \frac{1}{n^2 \pi} \sum_{p=1}^{q} \frac{\Delta z_p}{\Delta t_p} [\sin(nt_p) - \sin(nt_{p-1})], \tag{2.54}$$

There are now six coefficients, a_n, b_n, c_n, d_n, e_n, and f_n, and three constants, A_0, C_0, and E_0, that need to be evaluated. The constants B_0, D_0, and F_0, as before, are equal to zero. The constants A_0 and C_0 are calculated using Equations 2.48 and 2.49, whereas the E_0 term is

$$E_0 = \frac{1}{2\pi} \sum_{p=1}^{q} \frac{\Delta z_p}{2\Delta t_p} [t_p^2 - t_{p-1}^2] + \delta_p [t_p - t_{p-1}]. \tag{2.55}$$

The α_p and β_p terms are computed with Eqs. 2.50 and 2.51, and δ_p is calculated as

$$\delta_p = \sum_{j=1}^{p-1} \Delta z_j - \left[\frac{\Delta z_p}{\Delta t_p} \sum_{j=1}^{p-1} \Delta t_j \right], \qquad (2.56)$$

where $\alpha_1 = \beta_1 = \delta_1 = 0$.

If the boundary outline is planar (that is, two-dimensional), then the z-axis vanishes and Eqs. 2.52 to 2.56 become equal to zero. Once the coefficients in Eqs. 2.42, 2.43, and 2.52 have been computed, one can derive the usual estimates of amplitude and phase as discussed in Section 2.4.3. A study of the rabbit orbit margin represents the first application of the extension of EFFs to 3-D (Chapter 16).

2.7 Conclusions

The issue of how to precisely and completely describe the information that resides in all morphological forms remains a major challenge. Moreover, how the visual information present in the biological form is characterized by a mathematical representation has a direct bearing on subsequent explanations of biological processes. This chapter has briefly touched on some of the more formal aspects attendant upon the numerical description of the biological form, and provided an introduction to the mathematics underlying FDs.

In conclusion, it is argued that FDs, and especially EFFs, are powerful tools for eliciting boundary outline or contour information for a wide variety of complex 2-D and 3-D morphological forms of the type commonly encountered in the biological sciences. These methods need to be considered as initial steps toward the goal of extracting a significant percentage of the information that resides in all morphological forms.

References

Bliss, C. I. (1970). *Statistics in Biology.* Vol. 2. New York: McGraw-Hill.

Carslaw, H. S. (1950). *An Introduction to the Theory of Fourier's Series and Integrals.* (3d ed.) (reprint of original 1930 edition). New York: Dover Publications.

Davis, H. F. (1989). *Fourier Series and Orthogonal Functions* (reprint of original 1963 edition). New York: Dover Publications.

Davis, J. (1986). *Statistics and Data Analysis in Geology.* (2nd ed.) New York: John Wiley and Sons.

Diaz, G., Zuccarelli, A., Pelligra, I. & Ghiani, A. (1989). Elliptic Fourier analysis of cell and nuclear shapes. *Comp. Biomed. Res.* **22,** 405–14.

Diaz, G., Quacci, D. & Dell'Orbo, C. (1990). Recognition of cell surface modulation by elliptic Fourier analysis. *Comp. Meth. Prog. Biomed.* **31,** 57–62.

Ehrlich, R., Baxter Pharr Jr., R. & Healy-Williams, N. (1983). Comments on the validity of Fourier descriptors in systematics: A reply to Bookstein et al. *Sys. Zool.* **32,**

202–6.

Ferson, S., Rohlf, F. J. & Koehn, R. K. (1985) Measuring shape variation of two-dimensional outlines. *Syst. Zool.* **43**, 59–68.

Franklin, P. (1958). *An Introduction to Fourier Methods and the Laplace Transformation* (reprint of original 1949 edition). New York: Dover Publications.

Full, W. E. & Ehrlich, R. (1982). Some approaches for location of centroids of quartz grain outlines to increase homology between Fourier amplitude spectra. *Math. Geol.* **14**, 43–55.

Grattan-Guiness, I. (1972). *Joseph Fourier 1768–1830*. Cambridge: MIT Press.

Hamming, R. W. (1973). *Numerical Methods for Scientists and Engineers*, (2nd ed). New York: Dover Publications.

Harbaugh, J. W. & Merriam, D. F. (1968). *Computer Applications in Stratigraphic Analysis*. New York: John Wiley and Sons.

Harbaugh, J. W. & Preston, F. W. (1968). Fourier series analysis in geology. In *Spatial analysis: A Reader in Statistical Geography* ed. B. J. L. Berry & D. F. Marble. Englewood Cliffs, New Jersey: Prentice-Hall.

Herivel, J. (1975). *Joseph Fourier: The Man and the Physicist*. Oxford: Clarendon Press.

Hsu, H. P. (1967). *Fourier Analysis*. New York: Simon and Schuster.

Kieler, J. Skubis, K., Grzesik, W., Strojny. P., Wisniewski, J. & Dziedzic-Goclawska, A. (1989). Spreading of cells on various substrates evaluated by Fourier analysis of shape. *Histochem.* **92**, 141–8.

Kincaid, D. T. & Schneider, R. B. (1983). Quantification of leaf shape with a microcomputer and Fourier transform. *Can. J. Bot.* **61**, 2333–42.

Kuhl, F. P. & Giardina, C. R. (1982). Elliptic Fourier features of a closed contour. *Comp. Graph Imag. Proc.* **18**, 236–58.

Lestrel, P. E. (1974). Some problems in the assessment of morphological size and shape differences. *Yearbk. Phys. Anthropol.* **18**, 140–62.

Lestrel, P. E. (1980). A quantitative approach to skeletal morphology: Fourier analysis. *Soc. Phot. Inst. Engrs.* (SPIE) **166**, 80–93.

Lestrel, P. E. (1989a). Method for analyzing complex two-dimensional forms: Elliptical Fourier functions. *Am. J. Hum. Biol.* **1**, 149–64.

Lestrel, P. E. (1989b). Some approaches toward the mathematical modeling of the craniofacial complex. *J. Craniofac. Genet. Dev. Biol.* **9**, 77–91.

Lestrel, P. E. (in press). Morphometrics of Craniofacial Form. In *Fundamentals of Craniofacial Growth* eds. A. D. Dixon, D. A. N. Hoyte & O. Rönning. Boca Raton, Florida: CRC Press.

Lestrel, P. E. & Sirianni, J. E. (1982). The Cranial Case in *Macaca nemestrina*: Shape changes during adolescence. *Hum. Biol.* **54**, 7–21.

Lestrel, P. E. & Roche, A. F. (1986). Cranial base variation with age: A longitudinal study of shape using Fourier analysis. *Hum. Biol.* **58**, 527–40.

Lestrel, P. E., Stevenson, R. G. & Swindler, D. R. (1988). A comparative study of the primate cranial base: Elliptical Fourier functions. *Am. J. Phys. Anthrop.* **75**, 239.

Lestrel, P. E., Engstrom, C., Chaconas, S. J. & Bodt, A. (1991). A longitudinal study of the human nasal bone in *Norma lateralis*: Size and shape considerations. In *Fundamentals of Bone Growth: Methodology and Applications*. eds. A. D. Dixon, B. G. Sarnat & D. A. N. Hoyle. Boca Raton, Florida: CRC Press.

Lestrel, P. E. & Kerr, W. J. S. (1993). Quantification of functional regulator therapy using elliptical Fourier functions. *Europ. J. Orthod.* **15**, 481–91.

Lestrel, P. E., Bodt, A. & Swindler, D. R. (1993). Longitudinal study of cranial base changes in *Macaca nemestrina*. *Am. J. Phys. Anthrop.* **91**, 117–29.

Lestrel, P. E. & Swindler, D. R. (1996). The numerical characterization of the primate cranial base: A comparative study using Fourier descriptors. *Am. J. Phys. Anthrop.* Suppl. **22**, 148.

Lin, C. S. & Hwang, C. L. (1987). New forms of shape invariants from elliptic Fourier descriptors. *Pattern Recog.* **20**, 535–45.

Lu, K. H. (1965). Harmonic analysis of the human face. *Biometrics* **21**, 491–505.

Nafe, R., Kaloutsi, V., Choritz, H. & Georgii, A. (1992). Elliptic Fourier analysis of megakaryocyte nuclei in chronic myeloproliferative disorders. *Anal. Quant. Cytol. Histol.* **14**, 391–7.

Ostrowski, K., Dziedzic-Goclawaska, A., Strojny, P., Grzesik, W., Kieler, J., Christensen, B. & Mareel, M. (1986). Fourier analysis of the cell shape of paired human urothelial cell lines of the same origin but of different grades of transformation. *Histochem.* **84**, 323–8.

Parnell, J. N. & Lestrel, P. E. (1977). A Computer program for fitting irregular two-dimensional forms. *Comp. Prog. Biomed.* **7**, 145–61.

Persoon, E. & Fu, K. (1977). Shape discrimination using Fourier descriptors. *IEEE Trans. Pattern Anal. Mach. Int.* PAMI-**8**, 388–97.

Ravetz, J. R. & Grattan-Guiness, I. (1981). Fourier, Jean Baptiste Joseph. In *Dictionary of Scientific Biography* ed. C. C. Gillipsie. New York: Charles Scribner.

Rohlf, F. J. (1986). Relationships among eigenshape analysis, Fourier analysis and analysis of coordinates. *Math. Geol.* **18**, 845–54.

Rohlf, F. J. & Archie, J. W. (1984). A Comparison of Fourier methods for the description of wing shape in mosquitoes *(Diptera culicidae)*. *Syst. Zool.* **33**, 302–17.

Schwarcz, H. P. & Shane, K. C. (1969). Measurement of particle shape by Fourier analysis. *Sedimentology.* **13**, 213–31.

Spiegel, M. R. (1974). *Fourier Analysis.* New York: McGraw-Hill.

Strackee, J. & Nagelkerke, N. J. (1983). On closing the Fourier descriptor representation. *IEEE Trans. Pattern Anal. Mach. Int.* PAMI-**5**, 660–1.

Tolstov, G. P. (1962). *Fourier Series.* Englewood Cliffs, NJ: Prentice-Hall.

van Otterloo, P. J. (1991). *A Contour-Oriented Approach to Shape Analysis.* New York: Prentice-Hall.

Waters, J. A. (1977). Quantification of shape by the use of Fourier analysis: The Mississippian blastoid genus *Pentremites*. *Paleobiol.* **3**, 288–99.

White, R. J. & Prentice, H. C. (1988). Comparison of shape description methods for biological outlines. In *Classification and Related Methods in Data Analysis.* ed. H. H. Bock. Amsterdam: Elsevier Science Pub BV.

White, R. J., Prentice, H. C. & Verwijst, T. (1988). Automated image acquisition and morphometric description, *Can. J. Bot.* **66**, 450–9.

Zahn, C. T. & Roskies, R. Z. (1972). Fourier descriptors for plane closed curves. *IEEE Trans. Comp.* C-**21**, 269–81.

3

Growth and Form Revisited

DWIGHT W. READ

UCLA Department of Anthropology

3.1 Introduction

In his extraordinary work, *On Growth and Form*, D'Arcy Thompson (1961) set forth a framework for the analysis of morphological forms that still serves as inspiration for research on this topic. Thompson realized that complex forms may originate from simple principles—a theme that has returned with the now-familiar example of a fern leaf as a fractal image—that is, as geometric and topological aspects of form expressed through development. Some of his ideas led to incorporation of quantitative measurement in morphological studies, for example, the study of allometric relationships. It is striking, though, that his qualitative approach has served so well as a source of understanding of morphological form and as inspiration for new approaches to the study of form. This seeming timelessness suggests that we may do well to rethink how the study of the morphological form can be approached using Thompson's theme of seeing the underlying simplicity in what is an otherwise complex form.

This chapter begins with a brief review of the conceptual framework that Thompson set out for the study of growth and form. It then uses the review to introduce, in more detail, two themes in the study of form: representation of form, often in terms of the boundary of a form, and comparison of forms seen as a transformation of one form into the other. An example based on hominid teeth then serves to highlight a general problem not specifically addressed by these two themes; namely, the role of growth in the development of biological form. A synthesis between growth, taken as producing form, and the analytical study of form, including comparison and transformation of form, is presented and its implications explored. It is shown by example that using a biologically nonrelevant representation of the geometry of a form expressed in terms of homologous points leads to erroneous conclusions about shape versus size differences between forms. The need for a coordinate-free representation of the boundary of a form is discussed and such a representation, a curve based on inherent geometric properties, is outlined. The coordinate-free representation suggests a more qualitative approach to comparison of forms in terms of regions of maximum curvature, convexity versus concavity, and so on, in the boundary. The chapter concludes by returning to Thompson's concern with what constitutes legitimate comparison.

3.2 Conceptual foundations

Thompson saw the distinction between description and analysis as analogous to the difference between common speech and mathematical expression. Mathematical expression was not just quantification; rather, the simplicity and beauty of mathematical language provided for a deeper level of understanding. This deeper level Thompson saw as the beauty of geometry:

> The mathematical definition of a "form" . . . is expressed in few words or in still briefer symbols, and these words or symbols are so pregnant with meaning that thought itself is economized; we are brought by means of it in touch with Galileo's aphorism . . . that the Book of Nature is written in character of Geometry (1961:269)

From this perspective what needed to be understood was the geometry of form, not merely its quantitative measurement. Analysis of form was to be based not simply on quantitative measurement—as it subsequently became for many researchers (e.g., Reyment et al., 1984)—but required finding a mathematical expression that captures not only the particular form but also establishes it as an example in a class of forms.

Thompson also saw mathematics as providing what description alone—be it in words or in quantitative measurement—could not; namely, a way of going from the static to the dynamic:

> [A]nd this is the greatest gain of all, we pass quickly and easily from the mathematical concept of form in its statistical aspect to form in its dynamical relations: we rise from conception of form to the understanding of forces that give rise to it; and in the representation of form and in the comparison of kindred forms, we see in the one a diagram of forces in equilibrium, and in the other case we see the magnitude and direction of the forces which have sufficed to convert the one form into the other (1961:270).

At the same time Thompson realized that there are forms that did not lend themselves to this kind of geometrical expression, such as the outline of a skull, and accepted this limitation in his program of finding a mathematical representation of form.

For such forms Thompson turned to comparison and transformation based on coordinate systems. When Thompson was able to focus on geometric and topological properties of a developing form—such as the shape of a horn or the spirals of the nautilus shell—he arrived at comparison through the inherent geometric properties of a form, not properties determined by imposing an external reference framework, such as a Cartesian coordinate system, upon the form. However, for forms that did not lend themselves to a geometric interpretation, Thompson introduced the Cartesian coordinate system as a means to provide a basis for comparison. He recognized the difficulty in arbitrary comparison, and limited comparison to related forms where "the comparison shall be of a simple kind" (1961:273) and comparison would be between forms where "the form of the entire structure . . . should be found to vary in a more or less uniform manner, after the fashion of an approximately homogeneous and isotropic body" (1961:274).

However, this goal was violated in practice and Thompson (as have others) seemed to treat comparison of forms via a coordinate system as being on a solid foundation even when the comparison could not be given exposition through a simple transformation of a coordinate system. Where Thompson appeared to be most successful, such as in his comparison of ungulate limbs, the transformation of the coordinate system reflected an already-understood principle; for example, the change in diameter versus length as the load on the limb increased. Less successful (and not ignored by him) were comparisons that did not allow for a simple mathematical expression of the coordinate system transformation. At this juncture Thompson appears to have deviated from his general plan of relating form, and development of form, to mathematical expression and instead seems to have fallen back into the descriptive mode he had previously eschewed:

> . . . our co-ordinate systems may no longer be capable of strict mathematical analysis, they will still indicate *graphically* the relationship of the new co-ordinate system to the old, and conversely will furnish us with some guidance as to the "law of growth", or play of forces, by which the transformation has been effected (1961:285, italics in original).

With that statement Thompson, in effect, denied the earlier program that eschewed description for mathematical representation and now embraced description as if description alone were sufficient.

In a sense, though, the method of comparison through coordinate systems envisaged by Thompson is too powerful, in that it has no inherent constraint other than the requirement of comparison of forms of the same general type, for example, comparison of skulls. The latter allows for identification of homologous points, to serve as the basis for constructing a rectangular coordinate system for one form and as a curvilinear coordinate system for the other. Thompson generally restricted his comparisons to closely related genera, but he allowed himself the fanciful comparison of the dog skull with the human. This comparison showed that although there are obvious qualitative dissimilarities in local features such as the eye orbit, nonetheless there is a general suggestion as to the dimensions along which the compared skulls differ. Although suggestive, the example is also problematic, for without any constraint upon what constitutes legitimate comparison, it becomes possible to make any form comparisons, so long as there are homologous points for constructing the respective coordinate systems. Regardless of how the coordinate systems differ from one another, it is always possible to qualitatively discuss the apparent difference even if that comparison cannot be related to processes underlying the development of form. Here Thompson appears to have foregone his earlier program of discovering the geometry of form as it unfolds through development, for a more prosaic, and potentially less insightful, comparison of coordinate systems; thereby shifting the focus onto quantification, and away from geometry and topology, as the basis for comparison.

As is well known, his system of coordinate transformations has remained something of a curiosity; clearly something was there, but precisely what was always problematic. Subsequent examples published by others, such as the well-known hominid evolution sequence, suggested that global, rather than local, factors could account for changes, for example, in hominid skull shape. Yet even this descriptive insight has had little impact on practitioners in this area who tended to view anatomical detail in isolation, and use a cladistic analysis[1] of traits to arrange taxonomic classes with little concern for the insights to be obtained through consideration of the processes of growth and transformation envisaged by Thompson. Thus, description lost its foundations in the mathematical representation of form when it was replaced by quantitative measurement and statistical analysis of quantitative patterning with little concern for the dynamics of growth and development of the form. Geometric insight into form, and comparison of form through geometry, became comparison by embedding forms into a multidimensional measurement space, and transformation of forms was subsumed under distance measures such as Mahalanobis's D^2.

Yet even if one does accept the coordinate-system approach as the basis for showing how one form might be transformed into another—and descriptively, such a comparison is always possible and has been exploited cinemagraphically through morphing—more was needed for such a comparison than the verbal, qualitative comparisons provided by Thompson. Whereas transformation of coordinate systems could, for some examples, be expressed by a geometric or topological transformation of coordinate systems, this cannot be expected for all comparisons. This is particularly true when the only constraint imposed when

---

[1] Editor's note: Refer to Chapter 4 for a discussion of cladistics.

making the comparison is the presence of comparable homologous points in the forms being compared. Clearly, Thompson's vision needed a more thorough analytical basis, and this was ostensibly provided by the work of Bookstein (e.g., 1978, 1991) on the transformation of shapes.

3.3 Coordinate systems revisited

3.3.1 Transformation of shape

Bookstein's insight was to realize that the set of homologous points on a form, even if the form as a whole did not have a simple geometric description, could be reduced to triads of homologous points, whose geometry is captured by the relative location of the three points making up the triad. Representation of triads as triangles could be extended, quasi-geometrically, by constructing a truss system of adjacent triangles formed by connecting homologous points with straight lines (Strauss and Bookstein, 1982). Triangles can be compared via two orthogonal dimensions that express the amount of distortion needed to transform one triangle into the other. For example, a single, vertical, elongation distortion would change an equilateral triangle into an isosceles triangle; two orthogonal distortions would change an equilateral triangle into a triangle with unequal sides, and so forth. In this manner, the structure of homologous points can be viewed as a lattice-work of triangles. The local transformations needed to change a triangle in one lattice-work to a comparable triangle in the other can be determined and used to characterize how the geometry of one set of homologous points can be transformed into the geometry of the other.

In effect, Bookstein challenged the statistical/quantitative comparison of forms taken as static objects, and shifted the emphasis away from comparison of quantitative measurement to one of geometric transformation of triangles. This challenge was needed, as the statistical/quantitative approach that dominated analysis of form had created an artificial edifice, based, in part, upon invalid assumptions arising out of numerical taxonomy methods. The numerical taxonomy approach had begun by assuming sufficient comparison could be achieved by measuring an adequate set of characteristics with equal weighting (Sokal and Sneath, 1963:118–120), without ever clearly establishing what constituted an adequate set, other than relating it to the number of characteristics measured and the nature of the problem (see e.g., 1963:110–111). Methods such as principal component analysis were added to the original program of constructing units of comparison using clustering methods, which were based on similarity measures made directly over the measured characteristics. Principal component analysis held out an apparent solution to the age-old problem of separating out size and shape effects in the measurement and comparison of forms. When the first principal component, based on several measures made over a series of morphological forms, had positive loadings for all of the variables used to measure these forms, it was assumed that the it was a size dimension and the remaining components were shape dimensions (Marcus, 1990:87). However, that assumption was never given a theoretical foundation, as it depends upon unclearly specified definitions of what is meant by "size" and "shape" (Bookstein, 1989). Also, the unwarranted assumption that size and shape will be expressed in organisms as orthogonal quantities because principal components are orthogonal is contradicted in practice. For example, Read (1985) provided an example of a principal component analysis for which the first component, despite having positive loadings for all morphometric variables, clearly measured both size and shape differences.

3.3.2 Representation of a boundary

Whereas Bookstein challenged the statistical approach to morphometrics on the grounds that Thompson's theme of transformation of shape should be the primary focus of research, others have challenged the statistical approach to morphometrics as inadequate for even its purported goal of shape measurement. This contrasts with Bookstein's approach and its focus on transformation, rather than the representation, of forms. In his earlier work, Bookstein (1978) was concerned primarily with the transformation of the local geometry of one set of homologous points into the local geometry of another set of homologous points, ignoring the boundary connecting the points. Although the boundary is mentioned in his later work, it still remains an incidental aspect to methods for characterization of shape change based on homologous points (Bookstein 1990).

3.4 Local and global boundary representations

3.4.1 Local boundary representation

The statistical approach to the study of form assumed that adequacy of measurement was resolved by taking several measurements. Yet, whether the form so measured could be recreated accurately from the measurements was seldom, if ever, considered. This topic was investigated by Read (1982) with archaeological data (projectile points) rather than biological forms, but the principle is the same, and leads to a measurement system based on the geometry of the form as defined by its qualitatively distinguishable points (such as corners).

Read argued that the measurement system should be constructed to measure the form completely, that is, so that form can be reconstructed from the measurements. The system should be based on the geometry of the form (so that the measurements are independent of the coordinate system), and should lack redundancy (since neither the geometry nor the form need be preserved by procedures such as principal component analysis, nor is the assumed geometry for such procedures necessarily comparable to the geometry underlying the form in question). His method of measurement was extended in a subsequent article (Read and Lestrel, 1986) to more complex forms that take into account both the geometry of homologous points as well as the boundary of the form connecting homologous points. Briefly, Read used homologous points to define a local coordinate system within which the boundary connecting the homologous points could be given analytical expression. The form could then be represented by the geometry of the homologous points coupled with analytical expressions for the boundary segments.

3.4.2 *Global boundary representation*

In Read's approach, the boundary is seen as connected parts making up a whole, with mathematical representation having both geometric and analytical aspects. In contrast, other researchers have focused on Fourier methods to represent the boundary taken as a whole. Fourier methods have had a natural appeal due to their role in representation of periodic phenomena, and hence provide a means to represent a boundary viewed as a graph of a periodic function (Ehrlich and Weinberg, 1970; Anstey and Delmet, 1972; Lestrel, 1974; Chapters 6–12). Representation of the boundary of a form using elliptical Fourier functions, devised by Kuhl and Giardina (1982) and exemplified for a variety of biological forms (Lestrel, 1989; Lestrel et al; 1991; 1993; Chapters 14 and 16) and other forms (Rohlf and Archie, 1984; Diaz et al., 1989; Chapter 13), circumvents a problem that can arise when the Fourier representation of the boundary is expressed in polar coordinates. As is well known, an arbitrary boundary viewed as the graph of a function need not be single-valued. Kuhl and Giardina based their method on the fact that the boundary, even when it is a multi-valued function (if the boundary is considered to be the graph of a function defined using polar coordinates), can be given parametric representation using single-valued functions, making a Fourier series approximation a natural choice for representation of the parametric functions.

One starts by placing the boundary into a rectangular coordinate system. Let t be a parameter defined by measuring arc length along the boundary from some fixed, initial point, P, on the boundary. A rectangular coordinate location (x, y) on the boundary can then be given parametrically in the form $(x(t), y(t))$, $0 \leq t \leq T$, where $x = x(t)$, $y = y(t)$ and T is the total length of the boundary. The parametric functions $x(t)$ and $y(t)$ are continuous, single-valued, and periodic with period T so long as the boundary is a closed curve (and for non-closed curves, periodicity can be induced by retracing backwards along the boundary). Each parametric function can be represented by a Fourier series (see Chapter 2). The arbitrariness of the placement of the boundary of the form in a rectangular coordinate system for measuring the points *(x, y)*, and of the location of the initial point, can be obviated by a suitable translation and rotation of the coordinate system, and reassignment of the initial point. Kuhl and Giardina (1982) demonstrate that when the major axis of the first ellipse, defined by the Fourier series representation, is used to determine the initial point, the representation is coordinate-free up to the choice of a point of intersection of the boundary with the major axis of the ellipse. The latter indeterminacy can be removed by using a convention that defines which intersection will be used for the initial point.

Both Read's approach of using the geometry of the form to determine metric measurements, and Kuhl and Giardina's use of the elliptical Fourier series, place the emphasis on the representation of the boundary of a form. Homologous points are explicitly used by Read for the construction of local coordinate systems for the measurement of the boundary between homologous points. The elliptical Fourier procedure does not make use of homologous points, though homologous points can be explicitly included within the set of measurements upon which the Fourier function representation is based (see Chapter 16 for an example). In both instances, comparison of forms is made by comparisons of parameters inherent in these representations. The two approaches differ significantly, however, at the level of geometry. Read's approach takes into consideration local aspects of form and makes explicit use of homologous points for defining local coordinate systems. By its nature, a Fourier representation has coefficients that are affected by, but do not specifically measure, local shape properties. The elliptical Fourier method is primarily an analytical means to represent a curve as a function, and it makes use of the geometry of a form only indirectly via approximation of the overall shape as an ellipse used to orient the representation, which then allows for a rotation, translocation, and redefinition of the initial point to remove coordinate-system dependency.

3.5 Growth and boundary representation: a problem

These, and other approaches to measuring form, such as the use of indices, have been reviewed by Read (1990) and characterized in terms of their different assumptions about what aspect of a form is to be measured and/or compared. Common to approaches, however, is the use of adult forms as the basis of comparison, rather than comparison of differences arising out of growth and other structuring processes. This, it is now argued, is a fundamental problem that distorts the results of all of these approaches. Consider the following example based on mesiodistal (MD) and buccolingual (BL) dimensions of adult hominid teeth.

3.5.1 *Empirical example of the problem: hominid teeth*

For purposes of this example, assume that it is reasonable to simplify the boundary of a tooth crown into mesiodistal and buccolingual dimensions; that is, assume, for the purposes of reconstructing the shape, that the crown has a rectangular form (or at least a rectangle is a shape that can be reconstructed from these two measurements!). For each tooth in the tooth rows P3–P4 and M1–M3 (maxilla and mandible), fossil data show unidimensional change in tooth dimensions from early hominids (e.g., *Australopithecus afarensis*) to modern *Homo sapiens* (Read, 1975; 1984). The correlation between MD and BL is high (ranging from 0.80 to 0.93; Read, 1975) after occasional spurious fossil teeth have been removed. Some teeth exhibit isometry; others do not. For example, the (rectangular) shape of M_3 of modern *H. sapiens* differs from the shape of M_3 for *A. afarensis*. Using Read's system of shape representation, the change would be described as an elongation in the mesiodistal dimension and/or shortening of the buccolingual dimension. If the rectangle is divided into two triangles following Bookstein's method (and taking the corners of the rectangle as the homologous points), then the comparison would be based on a transformation of one set of triangles into the other. However, neither characterization of the evolutionary changes in the molar teeth is accurate, as both methods are based on comparing adult forms, and each ignores the growth process that (apparently) is involved.

3.5.2 *Data-driven model for tooth growth*

3.5.2.1 *A two-part growth process*

Read (1990) provided an argument for a two-part growth process in teeth. The model was derived theoretically to account for the fact that for some of the hominid molars and premolars, the mesiodistal and buccolingual dimensions are related by a straight line that does not go through the origin. A simple growth model might posit that teeth grow linearly from an initial bud to the adult size. For those teeth where the buccolingual and mesiodistal dimensions have changed linearly with respect to each other but the *y*-intercept is not zero, it follows that if teeth do grow linearly from a small tooth bud (i.e., from a point near the origin in a coordinate system) to their adult size, then each taxon must "know" how large the adult tooth should be, so that when there is a comparison between taxa, adult teeth will fall on a straight line (see Figure 3.1). This peculiar consequence appears implausible.

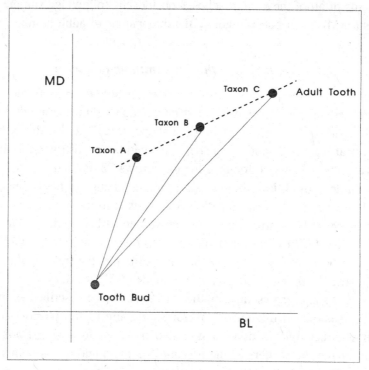

Fig. 3.1 Linear growth of a tooth from a bud to an adult form. The adult forms for three taxa are linearly related.

In place of this unrealistic model for tooth growth, Read (1990) showed that a two-part growth process, with the first part involving cell duplication of all cells within an initial tooth bud, and the second part involving cell duplication only of boundary cells, would account for the observed pattern (see Figure 3.2). Empirical support for this theoretical model is provided by Butler's work (1967, 1968) on tooth development. Butler notes that "there is an early proliferative phase when most of the cells are dividing " (1968:681), which would correspond to the first stage of the model, and "[t]hroughout its subsequent development the tooth adds to its stock of nondividing cells by growth confined to a zona cingularis" (1968:681), which would correspond to the second stage of tooth growth posited by the model. Further, under this theoretical model, the first growth sequence, that is, growth from an initial bud to a size where the growth process shifts to just the boundary cells being involved in further growth, is considered to be constant across taxa for a given tooth. What has changed, according to the model, is the extent of growth taking place in the second part of the process. By reducing the total amount of growth in the second phase, adult tooth size will become smaller, yet retain a linear relationship over time in the mesiodistal and buccolingual dimensions.

From the viewpoint of a single coordinate system, the model implies that tooth growth is initially isometric, and then becomes unequal in the mesiodistal and buccolingual dimensions, thus leading to a change in shape (see Figure 3.3). Empirical evidence for a difference in growth rates in these two dimensions can be found in a study of prenatal growth of M[1] (Butler, 1967). Butler also commented on the nonisometric growth pattern of deciduous M_2 and attributed this to lack of growth in the width of the tooth until a "threshold" length had been achieved (Butler 1968); a growth process consistent with the two-part growth process suggested here. It should be noted that Butler's (1967, 1968) data on M[1] and deciduous M_2 also suggest that the second stage of growth characterizes tooth

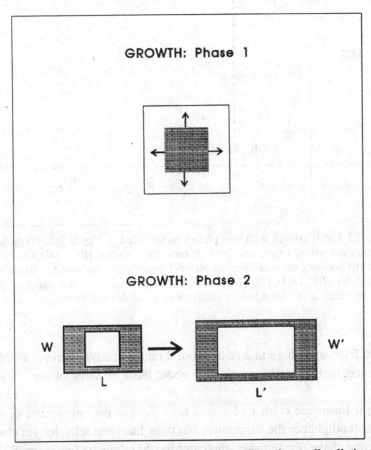

Fig. 3.2 Two-phase growth process. In the first phase, all cells in the cell mass (shaded area) duplicate and expand, isometrically, leading to the new, larger cell mass (unshaded rectangle). In the second phase, only the cell mass making up the boundary (shaded area, left figure) is involved in growth, possibly with different thickness in one dimension than in the other. The thickness of the cell mass remains constant as growth takes place, leading to a larger form (right figure) with a different shape: $L/W \neq L'/W'$.

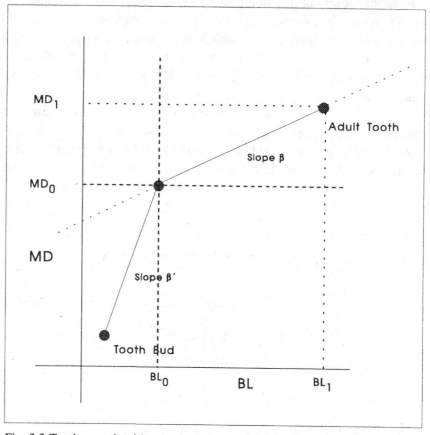

Fig. 3.3 Tooth growth with two phases as modeled in Figure 3.2. Tooth bud increases according to phase-1 growth until the location (BL_0, MD_0) is reached. Growth now occurs according to phase-2 growth and follows the linear trajectory from (BL_0, MD_0) to (BL_1, MD_1). The extension of this latter trajectory (dashed line) also characterizes differences in adult teeth between taxa.

development from a tooth germ size of about 2 mm in the two dimensions, to the adult tooth size, that is, tooth growth from about the 15th week of fetal development.

When adult forms are compared across taxa there is the appearance of shape change in the tooth; hence the suggestion that there has been selection for changes in both size and shape. Presumably, the latter would relate, had it occurred in this manner, to changes in kinds of food consumed (e.g., tougher, harder foods versus softer foods), which would require different tooth characteristics for efficient mastication. However, the apparent change in shape, it will now be suggested, is an artifact of using a single coordinate system for displaying the metric measurements of the teeth.

3.5.2.2 *Change in coordinate system*

Suppose that instead of a single coordinate system, the tooth is modeled with two coordinate systems—the first for the initial growth, with all cells dividing, and the second for subsequent growth with only boundary cells dividing. The result would be isometric growth from the perspective of each of the respective coordinate systems (see Figure 3.3). Further, the shift in coordinate system would, according to this model, correspond to a biological event; namely, the shift in the set of cells involved in tooth growth. For the second coordinate system, the slope of the line reflects the thickness of the one boundary in comparison to the other boundary, and from the viewpoint of this second coordinate system no shape change is involved. Instead, the single dimension that is varying is the total amount of growth taking place. The change in tooth shape, as measured with respect to the first coordinate system is, in this model, merely an artifact of using a measurement coordinate system that does not take into account the posited biological event of a shift in the set of cells involved in tooth growth. The apparent variation in the teeth, as measured in the first coordinate system (some appear to be isometric across taxa, others appear to have nonisometric variation) is also, according to the proposed model, simply the consequence of differences in the relative thickness of the mesiodistal and buccolingual boundaries.

Consequently, what the representation must account for in this model is *not* the apparent shape change as one goes from earlier to later hominids, but the shift from all cells duplicating to the duplication of only boundary cells, with tooth-specific differences in thicknesses in the buccolingual and mesiodistal dimensions. For a given tooth, the difference between early and late hominids is thus solely a difference in the rate of tooth growth over the same time period for tooth growth, and/or a change in the time period of tooth growth, and not a shape change, *per se*. Hence, analysis aimed at accounting for what appears to be change in shape is without basis as a biological event. Instead, the difference can be characterized by (1) a shift in coordinate system in the growth process of a tooth (with the timing of this shift unchanged as one goes from earlier to later hominids, and hence constant across taxa); and (2) a change in the rate and/or length of the time of growth during the second phase. The observed shape change, as seen from an anatomical perspective, thereby becomes an analytically derived, not a biologically primary, phenomenon.

3.5.3 *Mathematical model of a two-part growth process*

Mathematically, if the relationship between the mesiodistal and buccolingual dimensions is linear, then it may be discussed by the equation:

$$MD = \alpha + \beta BL \tag{3.1}$$

in a coordinate system with origin *(0,0)* corresponding to $MD = 0$ and $BL = 0$. The parameters α and β in Equation 3.1 have tooth-specific values with the empirically derived magnitude of β decreasing as one moves from anterior to posterior teeth (P3 to M3, mandible or maxilla) and, correspondingly, the magnitude of α increases (Read 1984). Equation 3.1 relates adult forms, and accounts for both intra- and interspecific comparisons of hominid teeth. Tooth shape, S, as measured by *MD/BL*, is thus given by:

$$S = (\alpha + \beta BL)/BL = \alpha/BL + \beta, \tag{3.2}$$

and for $\alpha \neq 0$, S varies with BL. The value of BL decreases with time and so the shape index, S, as defined by Equation 3.2, increases with time.

Now consider the two-part model of tooth growth. Initial growth proceeds according to the equation:

$$MD = \beta'BL. \tag{3.3}$$

During this growth phase, shape, S_1, is given by:

$$S_1 = MD/BL = \beta'BL/BL = \beta', \tag{3.4}$$

and shape is constant. Growth continues in this manner until the point (BL_0, MD_0) is reached, where $MD_0 = \beta'BL_0$ and $MD_0 = \alpha + \beta BL_0$ (see Figure 3.3). At this point it is hypothesized that a biological shift takes place in the growth process, with growth now taking place in the boundary cells. Hence, further growth follows the line $MD = \alpha + \beta BL$ until the adult size, (BL_1, MD_1) is reached (see Figure 3.3). For a coordinate system with origin at (BL_0, MD_0), the equation expressing growth in this second phase is:

$$MD' = \beta BL'. \tag{3.5}$$

The shape index, in terms of this new coordinate system, is given by:

$$S_2 = MD'/BL' = \beta BL'/BL' = \beta, \tag{3.6}$$

and S_2 is constant.

With respect to the original coordinate system, the model proposes that the correct measurement of growth during this second phase is $MD' = MD - MD_0$ and $BL' = BL - BL_0$ with the line of growth given by

$$(MD - MD_0) = \beta(BL - BL_0), \tag{3.7}$$

where $MD_0 = \alpha + \beta BL_0$. In other words, if we view the origin of the coordinate system as marking the point at which a growth phase begins, there has been a translocation of the origin of the original coordinate system by the amount (BL_0, MD_0) when the growth process has shifted from growth via all cells duplicating to growth via boundary cells duplicating.

3.5.4 Geometry of growth

This example suggests that comparison of adult forms is, by itself, inadequate. Rather, comparison needs to take into account, as Thompson discussed, the way in which the growth process produces the geometry of the adult form, and how changes in this growth process lead to divergence of adult forms from one another. As Thompson observed when considering variation in the shape of horns:

> The horn elongates by dint of continual growth within a narrow zone, or annulus, at its base. If the rate of growth be identical on all sides of this zone, the horn will grow straight; if it be greater on one side than on the other, the horn will become curved; and it probably will be greater on one side than on the other, because each single horn occupies an unsymmetrical field with reference to the plane of symmetry of the animal. If the maximal and minimal velocities of growth be precisely at opposite sides of the zone of growth, the resultant spiral *will* be a plane spiral; but if they be not precisely or diametrically opposite, then the spiral *will* be a gauche spiral in space" (1961:208, italics in original),

leading him to conclude:

> The distribution of forces which manifest themselves in the growth and configuration of a horn is no simple nor merely superficial matter. One thing is co-ordinated with another. To suppose that this or that size or shape of horn has been produced or altered, acquired or lost, by *Natural Selection*, whensoever one type rather than another proved serviceable for defense or attack or any other purpose, is an hypothesis harder to define and to substantiate than some imagine it to be (1961:213, italics in original).

Thompson was not rejecting natural selection as a driving force of evolution, but suggesting that the whole was more than the sum of the parts, each of which, supposedly, could have its own particular explanation in terms of natural selection. Instead, Thompson was arguing that the horn was a structural unit, with variation to be accounted for by differences in the shape of the growth zone, in growth velocity, and so on. From this perspective, natural selection could act only on these growth parameters, and hence act simultaneously on all parts that are under the control of these parameters, rather than on specific parts of the overall form that ensues.

3.6 Shape and growth: toward a synthesis

3.6.1 Generalized growth model

From the perspective of measurement and comparison of form, the implication is that direct comparison of adult forms is invalid. Rather, comparison should be of the growth processes producing the forms. Thus, the question should not be phrased in terms of how, for example, a straight horn might be transformed into a spiral horn via deformation of coordinate systems, but in terms of how there may have been changes in the shape of the growth zone and differences in growth rates around the boundary of that growth zone. The transformation that leads from a straight to a spiral horn would thus be expressed in terms of transformation of the shape of the growth zone (if the zones differ in shape), coupled with changes in growth rate as one traversed the boundary (if there were such changes). More formally, the shape of a horn should be derived from a representation of the boundary of the growth zone. In parametric form with arc length, s, as the parameter, this is given by:

$$x = x(s), \; y = y(s), \; 0 \le s \le S, \tag{3.8}$$

with the function $g = g(x, y)$ giving the growth rate, g, at each location on the boundary. The size of the horn is related to the length of time, T, that growth has taken place. By virtue of the parametric representation,

$$g(x, y) = g(x(s), y(s)) = G(s), \tag{3.9}$$

the growth function can be analytically separated from the boundary representation via arc length, s. Thus, it would be possible to vary the shape of the boundary, keeping fixed both the boundary curve, C, and the function $G(s)$, to determine the range of horn shapes that correspond to changes in the boundary shape. Alternatively, for a fixed boundary shape the function $G(s)$ could be varied to demonstrate the relationship between the adult horn shape and the growth function. In this way, what would otherwise involve transformation of 3-D coordinate systems for complex shapes (for example, transformation of the coordinate systems for the adult horn forms) becomes reduced to transformation of simpler shapes, possibly in two dimensions (the growth annulus for the horn). The latter comparison depends upon the geometry of the boundary, and thereby makes application of the method advocated by Bookstein for comparison of shapes problematic.

His method is based on a set of homologous points that determines the geometry of the shape; the boundary connecting these homologous points is of secondary concern. That approach would be flawed, however, when, as this horn example illustrates, the underlying biology appears to be driven by the characteristics of the boundary, and homologous points may even be absent (as would be true for shape boundaries with continuous first derivatives in a parametric function representation of that boundary). For such boundaries, the elliptical Fourier representation seems preferable by virtue of its reliance upon a parametric representation of a smooth boundary curve; that is, the biological form is congruent with the geometrical form in that the biological form has a direct relationship to the parameters of the mathematical representation.

3.6.2 *Geometry of homologous points and growth*

Even when the focus is on the geometry of the homologous points, without considering the boundary, the growth process is still central to determining a representation that can express the basis for change in the geometry of the homologous points. For example, Strauss and Bookstein (1982) exemplify the method of transformation of the geometry of homologous points by consideration of two species of the fish genus *Cottus*. The homologous points were defined by anatomical features on the outline of the body of the fish. They determined, when the two species were compared, that the geometry of one set of homologous points, expressed as a truss system, was changed into the geometry of the other through a change from a deep-bodied to a shallow-bodied form. In response, Read and Lestrel (1986) demonstrated that, for the same data and using the representation system of Read, in which the polygon determined by the homologous points is expressed as a sequence of angles and lengths, the change in the geometry can be characterized as (1) a change in a single angle in the facial area coupled with a necessary change in the length of one side of the polygon defined by the homologous points; and (2) an elongation and thinning of the tail portion of the fish as a separate event. Read and Lestrel argued that the two analyses presume different models of growth. The analysis based on change in a truss system essentially assumes that growth is radial (i.e., changes take place across the organism or form; for example, the initial growth stage in tooth growth in the model described in Section 3.5), and change in form represents differences induced through radially oriented growth. In contrast, the analysis given by Read and Lestrel assumes that growth is lateral with respect to the boundary (i.e., involves change in angles and/or change in distance between homologous points; for example, the second growth stage in tooth growth described in Section 3.5). This analysis presents change as a boundary phenomenon composed of local effects and the necessary consequences of local effects (e.g., a change in angle is coupled necessarily with an elongation of the distance between certain homologous points if the remaining geometry is to remain fixed, and the thinning of the tail section occurs as a separate event).

As discussed by Read and Lestrel (1986), the issue is not whether the geometry of homologous points, via triangles and their deformation considered within the framework of tensor analysis, is "more correct" than representation of a boundary and comparison of boundary representations. The issue is simply that any representation system and/or comparison of forms involves, at least implicitly if not explicitly, a model of biological processes of growth. In the above example, deciding whether the shape change of the fish species is better considered as a largely radial phenomenon, or as a modification of a boundary, requires an understanding of the biological process of growth as it is exemplified in this species of fish.

This brings us back to Thompson's fundamental thesis: that form should be understood via the process of growth, and differentiation into different forms arises out of growth processes.

3.6.3 Structure and sub-structure

Biological growth occurs directly via cell division, and form is a consequence of patterns of variable rates of cell division coupled with differentiation of the original cell mass into distinctive parts. Growth in some aspects of the organism, such as Thompson's horn cores, can be related to a growth site or region, with its geometry and differential growth rate leading to the adult form. Other aspects of the organism are not so easily characterized. The organism may well have an adult form that can be thought of as a splicing together of constituent forms, each produced via their own constraints and growth patterns, yet integrated within the overall form. For example, it has been argued that the supraorbital torus in *Homo erectus* is a consequence of the positioning of the brain case with respect to the face (Moss and Young, 1960; Hylander et al., 1991; Ravasa, 1991), or alternatively, that the supraorbital torus provides the buttressing needed to counteract forces generated through the masticatory apparatus (Endo, 1966; Russell, 1985). Should, then, the cranium as a whole be taken as a form, with identifiable landmarks, and analytical representation and comparison made of the whole form? Or should the whole form be decomposed into constituent subforms, each with its own growth trajectory driven by local properties of that subform, along with an architecture or geometry that accounts for the positioning of the constituent forms and the interlinkages this creates among them?

From an evolutionary perspective, and using the hominid cranium as an example, constituent parts may well be responding locally to selective pressures that only affect other constituent forms indirectly through the interconnections making up the whole structure. The sizes of the mandibular and maxillary structures relate to changes in tooth size, and these changes relate to mastication. Other uses of the teeth are distinct from changes in other constituent structures such as, for example, the brain case, where the volume presumably relates to behaviors leading to selection for larger brain size. It would be rather remarkable to find a single, all-encompassing geometry that would account for the totality of change in the cranium in the same way that horn shape variation was linked by Thompson to the geometry and the variability in growth rates across the growth site.

At the same time, it is evident that the positioning of the constituent parts, such as the masticatory apparatus, facial region, or brain case, is subject to constraints imposed by the transmission of forces, balance of the cranium on the foramen magnum, and so on. Triangulation of the overall structure via homologous points, whose relative positioning is affected both by local properties of constituent forms and by global properties relating to the position of these forms to one another, leads to a confounding of distinct structuring effects. Instead of considering how much of the geometry of one triangle is due to the relative positioning of constituent parts versus how much of that geometry is a consequence of the local geometry of the constituent part, comparison becomes submerged into a single, comprehensive statement: that is, of how the geometry induced by the combined effects in one instance can be transformed into the geometry of another instance.

The shift to comparison of form by comparison of changes in the geometry of forms, and away from comparison of forms by quantitative, multivariate methods has been both needed and enlightening. However, the simultaneous decrying of methods aimed at representation of form through methods such as elliptical Fourier functions (e.g., Bookstein 1990) assumes that (1) the geometry of homologous points is necessarily analytically prior to consideration of the boundary connecting homologous points; (2) what constitutes the homologous points that biologically define this geometry is known *a-priori*; and (3) that analysis can properly be framed in terms of comparison of the geometries of adult forms. In the fish example discussed in Section 6.3.2, homologous points were identified both as marking the location of features such as the eyes, and the intersection of appendages with an outline of the body of the fish, and so on. These defined points are comparable from one outline to another. But less evident is their collective biological significance as defining a single geometry whose deformation as one goes from one form to the other is the crucial step to be understood. The conclusions reached from the deformation argument are contingent upon the validity of the homologous-point deformation representing the biological reality characterized by the evolutionary pathway linking these two species of fish to a common ancestor. Only if one form is the same as a common ancestral form for both is it biologically meaningful to consider the one form as a deformation of the other.

3.6.4 Comparison of forms and structuring properties: an example

Deformation defined by comparing extant adult forms may be descriptively accurate but potentially limited from a biological perspective unless the deformation also represents the evolutionary pathway linking the two forms. Otherwise, it is but one choice out of several choices (as the alternative analysis by Read and Lestrel [1986] demonstrates) that could be made for describing how one geometry can be transformed into another geometry. Further, it is not self evident that the geometry of homologous points is biologically more meaningful (even leaving aside the problem of distinguishing which of all the potential homologous points should be used for defining this geometry). For example, the shape of primate mandibles is often modeled as a catenary curve (MacConaill and Scher 1949; Burdi and Lillie, 1966; Germane et al., 1992; but see Pepe, 1975, for a dissenting view). However, it should be noted that the curve ought to be defined by the central axis of the mandibular bone matrix and not necessarily the tooth row, as the tooth row has a curvature distinct from that of the mandibular bone for some primate species.

The choice of a catenary curve is motivated by viewing the distribution of forces on the mandible (MacConaill and Scher, 1949)—for example, forces in the form of the tongue providing, via outward pressure, a more or less equal distribution of forces on the lingual side of the mandibular bone and parallel to the axis of symmetry of the tongue. A triad of homologous points could be used for the location of the mandibular condyles and the front of the mandible at the point of maximum curvature; another triad might be the distal extremity of the third molars and the exterior side of the central incisors. Now compare two different mandibles. The catenary curve representation will define the difference between the forms in terms of two parameters, A and B, which measure the width and arc length of the catenary curve. If the homologous points are based on tooth locations, then the arc length relates to the summed buccolingual dimensions of the teeth and the catenary width relates (roughly) to the width of the face. Thus, change in the curve can be easily linked to potential sources of change in other parts of the masticatory apparatus and these two sources of change potentially can act independently of one another.

Because the catenary curve is geometrically definable, its parameters may be defined without reference to an externally imposed coordinate system. Its specific geometry within the class of catenary curves is determined by the parameters width and arc length. These parameters would also appear to be necessary for a biologically meaningful representation of the mandible, as they relate to (1) the width of the mandible, and hence the width of the facial region; and (2) the summed mesiodistal dimensions of the teeth in the mandible. Since width and arc length are not only mathematically independent parameters, but their biological counterparts (mandible width, summed mesiodistal lengths) can also vary independently, it follows that reduction of the mandible to a triad of points can give the appearance of shape change as the primary difference when forms are compared via homologous points, when in fact this change in the geometry of the points is derived from a size change. To illustrate, consider Figure 3.4a. Suppose we consider the distal ends of the tooth row and the point of maximum curvature as the homologous points. The three solid points mark the location of these three points on a catenary curve representing the mandible. The three points determine a triangle (solid lines). The two open points and the solid point show the changed lo-

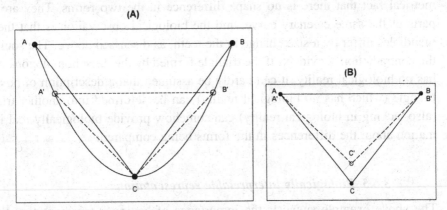

Fig. 3.4 (a) A catenary curve representing the curvature of a pongid mandible also showing evolutionary change in the overall length of the tooth row. Three homologous points are defined by the distal extreme of the tooth row (points A, A', B, and B') and the point of maximum curvature (points C and C', with C = C'). The points A, B, and C define one triangle (solid lines) and the points A', B', and C define a second triangle (dashed lines). (b) Comparison of the two triangles by superimposition with the homologous side of the one triangle (side connecting points A and B) and the homologous side of the other triangle (side connecting points A' and B') each rescaled to unit length. After rescaling and superimposition, the points C and C' are no longer coincidental, thus signifying a shape difference in the triangles.

cations of the homologous points after there has been size reduction in the teeth (hence the arc length of the catenary curve) and these form a second triangle (dashed lines).

With Bookstein's method of comparing forms using triangles formed by a triad of homologous points, the two triangles are superimposed using homologous sides, each rescaled to a unit length (see Figure 3.4b) in order to remove size differences: "In this step, size information is lost " (Bookstein, 1990:68). The remaining pair of points, one from each of the two triangles, will not be coterminous (see Figure 3.4b, points C and C') and the difference in the position of these two points contains information about the shape difference in the triangle: "All the information about the shapes of the triangles then inheres in the position of the third landmark" (Bookstein 1990:68). Since "[t]he triangle has no biological reality " (1990: 67), it must, apparently, be assumed that the triangles have captured the relevant shape information for comparison of the forms. Otherwise, the difference in the position of the third vertex of the two triangles would not be a measure of shape difference in the two forms. Thus, the conclusion reached from the different position of the third point in the triangles (see Figure 3.4b) is that there is a shape difference between the two mandibles. This, however, contradicts the geometrical fact that there is no shape difference in the two forms. They are both parts of the same catenary curve, and the biological observation is that the two mandibles differ by a size change in the teeth, as discussed above. The source of the contradiction is evident. If the triangle formed by the three homologous points has no biological reality, it can hardly be assumed that a description of how one triangle (which has no biological reality) can be deformed into another triangle (also lacking in biological reality) can somehow provide biologically real information about the differences in the forms being compared.

3.6.5 Biologically interpretable representations

The above example suggests the importance of having a representation that is close to the underlying biology of a form. Such a representation must ultimately be linked to genetically encoded information. The genetic linkage argues against using a representation or comparison based upon an externally defined reference system, such as a coordinate system, as it is difficult to imagine how an externally defined coordinate system could be genetically encoded. This being the case, it follows that when the goal of a boundary representation is to be biologically meaningful, at least it will need to be a coordinate-free representation.

3.6.6 Coordinate-free representations

A coordinate-free representation will be invariant under at least translation and rotation. Representations that are not initially coordinate-free can be made coordinate-free through a suitable normalization. Kuhl and Giardina (1982), for example, provided such a coordinate-free representation through a normalization indeterminate only in the choice of one or the other end of the major axis of an ellipse as a reference point (see van Otterloo, 1991:173–210, for an extensive discussion of several normalization methods that have been proposed). In effect, they removed the coordinate indeterminacy through use of the geometry of the first ellipse of the elliptical Fourier function fitted to the form.

For the problem addressed by them, namely, identification of a form for classificatory purposes, whether the fitted ellipse corresponds to any inherent geometry of the form was irrelevant. For biological applications the same would not be true in situations where one wants the coefficients of the Fourier representation to have biological significance. This is not to say that the coefficients cannot have such significance, only that the biological significance of the coefficients needs to be determined in an application, and not assumed.

For shape outlines, other representations are possible that are inherently coordinate-free. With planar curves (closed or not, simple or complex) a coordinate-free representation can be based upon arc length, s, where arc length is measured along the curve from some distinguished point, P. Let $\kappa(s)$ denote the curvature of the smooth curve, C, at distance s along the curve from P. Curvature $\kappa(s)$ is defined independent of a coordinate system by

$$\kappa(s) = |d\phi/ds|, \tag{3.10}$$

where ϕ is the angle of inclination of a line tangent to the curve at distance s from P.

Arc length and curvature completely characterize a planar curve and any representation of such a curve in a coordinate system can be changed to this coordinate-free representation. For example, if a planar curve specified in a coordinate system is given parametric representation in the form

$$(x, y) = (x(t), y(t)), \; a \le t \le b, \tag{3.11}$$

then curvature, κ, may be expressed in terms of the parameter t via

$$\kappa = |x'(t)y''(t) - x''(t)y'(t)| \, / \, [(x'(t))^2 + (y'(t))^2]^{3/2}. \tag{3.12}$$

For Euclidean 3-space (E^3), curvature is generalized from the planar case as

$$\kappa = |d\vec{T}/ds|, \tag{3.13}$$

where $T(s)$ is a unit tangent vector to the curve located at distance s from the point P.

If a smooth curve in E^3 specified in a coordinate system is given parametric representation in the form

$$(x, y, z) = (x(t), y(t), z(t)), \; a \le t \le b, \tag{3.14}$$

then

$$\kappa = |\vec{T}'(t)| \, /v(t), \tag{3.15}$$

where

$$\vec{v}(t) = (x'(t), y'(t), z'(t)), \; \vec{T}(t) = \vec{v}'(t) \, / \, v(t), \text{ and } v(t) = |\vec{v}(t)|.$$

In addition to curvature, the torsion τ, or amount of twist, of a curve must be specified as well:

$$\tau = -|d\vec{B}/ds| \, / \, |\vec{N}|$$
$$= -|d\vec{B}/dt| \, / \, (|\vec{N}|v), \tag{3.16}$$

where $\vec{B} = \vec{N} \times \vec{T}$, $\vec{N} = (\vec{a} - a_T\vec{T})/a_N$, $a_T = dv/dt$, and $a_N = |\vec{v} \times \vec{v}'| \, / \, v(t)$.

Note that $\tau = 0$ for planar curves. Thus, in E^3, a curve is parametrically specified via the parametric functions $\kappa(s)$ and $\tau(s)$, $0 \le s \le S$, S the total arc length of the curve, in a coordinate-free manner. That is, s, κ, and τ form a complete set of invariants for curves in E^2 or E^3 (Laugwitz, 1965:17).

The equations given above for computation of κ and τ in terms of a parametric representation of a coordinate-system-specified curve allows, for example, transformation from an elliptical Fourier representation to a coordinate-free representation. The transformation makes estimation of the functions κ and τ simpler as the numerical fitting of the elliptical Fourier function to measured data in the form of (x, y, z) coordinates is computationally straightforward and has been implemented in a computer program (Wolfe et al., 1995).

3.6.7 *Connections among coordinate-free representations*

If the coordinate-free representation via s and $\kappa(s)$ (for planar curves) is taken as a reference point, then the connections among several proposed methods for representing an outline become apparent. For example, Zahn and Roskies (1972) discuss representing an outline via s and the function

$$\theta^*(s) = \theta(s) - s, \; 0 \leq s \leq 2\pi, \tag{3.17}$$

where

$$\theta(s) = \phi(s) - \phi(0), \; 0 \leq s \leq 2\pi, \tag{3.18}$$

and then representing the function θ^* via a Fourier representation (i.e., the boundary is rescaled so that the total length is 2π and θ^* measures the difference in curvature between the curve and a unit circle). From the function θ^* (or from a Fourier representation of θ^*) the curvature function can be computed since $\theta^*(s) + s = \theta(s)$ and $d\phi(s)/ds = d\theta(s)/ds$. Hence a representation of the curve in terms of elliptical functions can be transformed into a representation via the θ^* function, or conversely the θ^* representation can be transformed into the curvature representation.

Rather than seeing these as competing representations (e.g., Rohlf, 1990:53; Lohmann and Schweitzer, 1990:151), their mathematical equivalence, in the sense of one being derivable from the other, shows their complementarity. Further, a pragmatic advantage of beginning with the elliptical Fourier representation arises because the elliptical Fourier function is numerically estimable from coordinate data, hence obviating the need for estimating the tangent lines required for direct measurement of the functions ϕ or θ.

Whereas curves may be given coordinate-free representation in E^2 or E^3, surfaces are not so easily treated. By their geometry, curves can be traced in only 2 ways, counterclockwise or clockwise: hence arc length provides a natural, unambiguous parametrization of a curve. In general, surfaces have no equivalent natural parametrization, given the indeterminacy in tracing a surface.

The extent to which it is necessary to represent a surface in addition to representing its boundary (in cases where the surface has a definite boundary) depends upon the relationship of the surface to the boundary. With some surfaces (e.g., those modeled by soap films), the surface is a minimum-tension surface, and completely characterized by the boundary (see Chapter 16). Although such surfaces may not be biologically common, they highlight the need to understand the relationship between a form with its surface and a growth region with its boundary. Characterization of the resulting surface may be complex to represent directly, but may be made simpler if it can be considered in terms of the boundary of the growth region and growth rate along that boundary as indicated in Equation 3.9.

Representation of a curve in terms of curvature allows for the location(s) of maximal and, perhaps less important for biological forms, minimal curvature to be computed. The elliptical Fourier representation is particularly helpful for such computations, as it provides a representation with continuous derivatives, and hence eliminates "corners" from the representation. With the elliptical Fourier representation, corners become curves of high curvature and can, thereby, be identified by constructing the curvature function, κ, from the Fourier representation of the parametric functions as indicated in Equation 3.12.

3.6.8 Qualitative production of form and evolution

If the locations of points of maximal curvature are biologically significant regions, it follows that understanding their production, in the sense of transformation of a curve segment with minimal curvature (i.e., almost straight) into a segment with a point of high curvature should be of biological interest. This leads to yet a different kind of representation of a shape based upon curvature, and comparison of shapes through transformation of regions of low curvature into regions of high curvature (see Leyton, 1988).

Consideration of the qualitative production of a complex shape draws attention to the fact that a complex shape *is* produced from simpler forms, and shape change has a qualitative, as well as a quantitative, aspect. Comparison of homologous-point structures takes the homologous points as given and accounts for neither their production nor their geometry. Instead, the homologous-point method focuses on a quantitative measure of the geometric changes and assumes the underlying geometry is unchanged, with only its quantitative aspects undergoing change. Reducing a set of homologous points to triads allows for a common geometrical representation via geometrically comparable triangles, but only so far as the compared forms have isomorphic sets of homologous points. However, change in a homologous point in an adult form need not signal noncomparability of form, but simply a qualitative change in an aspect of otherwise comparable forms. For example, the supraorbital torus of, say, *Homo erectus* may be given the status of a homologous "point" in a lateral cranium view. In *Homo sapiens*, that homologous "point" is lost by virtue of the repositioning of the cranium over the facial region. The repositioning has changed the underlying geometry/mechanics for the dispersal of stress throughout the facial region as induced by mastication and thereby has made the buttressed, supraorbital torus obsolete in *Homo sapiens*. The qualitative change—namely, loss of a homologous point—is not part of the homologous point comparison of shape change, but clearly is part of our understanding of changes in cranial morphology in the evolutionary sequence going from *Homo erectus* to *Homo sapiens*. The objection discussed above regarding deformation of the geometries of homologous points can be raised here as well, namely, that comparison is again made of adult forms as if one were transformed into the other evolutionarily, rather than both being derivative, to one degree or another, from a common ancestral form. Inclusion of the ancestral form as a reference form would lead to a comparison of forms in a manner akin to molecular trees. That is, a distance measure of extant species is based on the number of substitutions needed to arrive at the respective codon sequences in DNA for each of the extant species, with the number of substitutions measured from a common ancestral DNA molecule in the genome of the posited common ancestral species (see Leyton, 1988).

3.7 Conclusions

Thompson's insight that forms should be understood in terms of the interplay of the processes of growth, forces acting on forms, and the geometry of form has yet to be fully exploited. For Thompson, force was the basis for form: "The form . . . and the changes of form . . . may in all cases be described as due to the action of force" (1961:11), but force was not just physical force: "we shall manifestly be inclined to use the term growth in two senses . . . on the one hand as a *process*, and on the other as a *force*" (1961:13, italics in original), and so "growth . . . deserves to be studied in relation to form" (1961:10). Though selection is the ultimate arbitrator of form, selection does not act upon random variation in form but on variation in those forms that can be realized through growth. Comparison of forms, in this view, becomes comparison of forms seen as the consequence of alternative growth pathways rather than the comparison of the products of those growth pathways considered in isolation. The former leads, for example, to viewing hominid teeth as having had, under the effect of natural selection as an ultimate cause, unidimensional change driven by a decrease in total tooth growth. The latter leads to viewing hominid teeth as having changed due to both selection for shape change and for size change, with different teeth undergoing different amounts of shape change. The former derives from a plausible model of tooth growth; the latter utilizes a model of tooth growth with unrealistic implications, such as linearly related adult tooth sizes across hominid taxa with no mechanism for ensuring that linearity.

The contrast, comparison of form versus representation of form, with its implication that the one or the other somehow has greater priority, fails to bring out the need to embed the study of morphology into the production of form through growth and forces acting upon biological form. Without understanding these growth pathways it cannot be known how comparison should be made to ensure biological insight. As Burdi and Lillie comment with regard to embryogenic changes in the mandibular shape: "Change in the shape of the embryonic dental arch is not a haphazard affair but appears to be a carefully coordinated pattern of dimensional as well as directional growth" (1966:19). Similarly, it cannot be known *a-priori* if one form of representation or another provides biological insight in the absence of information on the implications of growth, seen both as a process and as a force. In the model of tooth growth described here, the process of tooth growth via cell division also serves as a force for change in tooth shape in that the growth process, through the posited shift from general growth via cell division to growth as a boundary process, is the basis for the change in shape across the several hominid taxa. D'Arcy Thompson saw clearly the danger in mere application of methods without consideration of evolutionary relationships:

> We should fall into deserved and inevitable confusion if, whether by the mathematical or any other method, we attempted to compare organisms [by the method of coordinates] separated far apart in Nature and in zoological classification. We are limited . . . by the whole nature of the case, to the comparison of organisms such as are manifestly related to one another and belong to the same zoological class" (1961:271).

It is the biological insight, and not the method, that is the ultimate arbiter of how form, and comparison of form, should be undertaken.

References

Anstey, R. L. & Delmet, D. A. (1972). Genetic meaning of zooecial chamber shapes in fossil bryozoans: Fourier analysis. *Science* **177**, 1000–2.

Bookstein, F. L. (1978). *The Measurement of Biological Shape and Shape Change. Lecture Notes in Biomathematics*, Vol. 24. Berlin: Springer.

Bookstein, F. L. (1989). "Size and shape": A comment on semantics. *Sys. Zool.* **38**, 173–80.

Bookstein, F. L. (1990). Introduction and overview: Geometry and biology. In *Proceedings of the Michigan Morphometrics Workshop* eds. R.J. Rohlf & F. L. Bookstein. Special Publication No. 2, Univ. of Michigan. 61–74.

Bookstein, F. L. (1991). *Morphometric Tools for Landmark Data.* Cambridge: Cambridge University Press.

Burdi, A. R. & Lillie, J. H. (1966). A catenary analysis of the maxillary dental arch during human embryogenesis. *Anat. Rec.* **154**, 13–20.

Butler, P. M. (1967). Relative growth within the human first upper permanent molar during the prenatal period. *Arch. Oral Biol.* **12**, 983–92.

Butler, P. M. (1968). Growth of the human second lower deciduous molar. *Arch. Oral Biol.* **13**, 671–82.

Diaz, G., Zuccarelli, A., Pelligra, I. & Ghiani, A. (1989). Elliptic Fourier analysis of cell and nuclear shapes. *Comp. Biomed. Res.* **22**, 405–14.

Ehrlich, R. & Weinberg, B. (1970). An exact method for the characterization of grain shape. *J. Sed. Petrol.* **40**, 205–12.

Endo, B. (1966). Experimental studies on the mechanical significance of the form of the human facial skeleton. *J. Fac. Sci. Univ. Tokyo (Sect. V, Anthrop.)* **3**, 1–106.

Germane, N., Staggers, J. A., Rubenstein, L. & Revere, J. T. (1992). Arch length considerations due to the curve of Spee: A mathematical model. *Am. J. Orthod. Dentofac.Orthop.* **102**, 251–5.

Hylander, W. L., Picq, P. G. & Johnson, K. R. (1991). Masticatory-stress hypothesis and the supraorbital region of primates. *Am J. Phys. Anthrop.* **86**, 1–36.

Kuhl, F. P. & Giardina, C. R. (1982). Elliptic Fourier features of a closed contour. *Comp. Graph. Image Proc.* **18**, 236–58.

Laugwitz, D. (1965). *Differential and Riemannian Geometry.* New York: Academic Press.

Lestrel, P. E. (1974). Some problems in the assessment of morphological size and shape differences. *Yearbk Phys Anthropol.* **18**, 140–62.

Lestrel, P. E. (1989). Some approaches toward the mathematical modeling of the craniofacial complex. *J Craniofac. Genet. Dev. Biol.* **9**, 77–91.

Lestrel, P. E., Engstrom, C., Chaconas, S. J. & Bodt, A. (1991). A longitudinal study of the human nasal bone in *Norma lateralis*: Size and shape considerations. In *Fundamentals of Bone Growth: Methodology and Applications.* eds. A. D. Dixon, B. G. Sarnat & D. A. N. Hoyle. Boca Raton, Florida: CRC Press.

Lestrel, P. E., Bodt, A. & Swindler, D.R. (1993). Longitudinal study of cranial base changes in *Macaca nemestrina*. *Am J. Phys Anthrop.* **91**, 117–29.

Leyton, M. (1988). A process grammar for shape. *Artificial Int.* **34**, 213–47.

Lohmann, G. P. & Schweitzer, P. N. (1990). On eigenshape analysis. In *Proceedings of the Michigan Morphometrics Workshop.* eds. R. J. Rohlf & F. L. Bookstein. Special Publication No. 2, Univ. of Michigan.147–66.

MacConaill, F. A. & Scher, E. A. (1949). Ideal form of the human dental arcade with some prosthetic application. *Dent. Rec.* **69**, 283–302.

Marcus, L. F. (1990). Traditional morphometrics. In *Proceedings of the Michigan Morphometrics Workshop* eds. R. J. Rohlf & F. L. Bookstein. Special Publication No. 2, Univ. of Michigan. 77–122.

Moss, M. L. & Young, R. W. (1960). A functional approach to craniology. *Am J. Phys. Anthrop.* **18**, 281–92.

van Otterloo, P. J. (1991). *A Contour-Oriented Approach to Shape Analysis.* New York: Prentice-Hall.

Pepe, S. H. (1975). Polynomial and catenary curve fits to human dental arches. *J. Dent. Res.* **54**, 1124–32.

Ravasa, M. J. (1991). Interspeciation perspectives on mechanical and nonmechanical models of primate circumorbital morphology. *Am. J. Phys. Anthrop.* **86**, 369–86.

Read, D. W. (1975). Hominid teeth and their relationship to hominid phylogeny. *Am. J. Phys. Anthrop.* **42**, 105–25.

Read, D. W. (1982). Towards a theory of archaeological classification. In *Essays on Archaeological Typology* eds. R. Whallon & J. A. Brown. Evanston: Center for American Archeology Press.

Read, D. W. (1984). From multivariate statistics to natural selection: a reanalysis of Plio/Pleistocene homnid dental material. In *Multivariate Statistics in Physical Anthropology* ed. G. van Vark. Amsterdam: D. Reidel Publishing Co.

Read, D. W. (1985). The substance of archeological analysis and the mold of statistical method: enlightenment out of discordance. In *For Concordance in Archeological Analysis* ed. C. Carr. Kansas City: Westport Press.

Read, D. W. (1990). From multivariate to qualitative measurement: representation of shape. *J. Hum. Evol.* **5,** 417–29.

Read, D. W. & Lestrel, P. E. (1986). Comment on uses of homologous point measures in systematics: a reply to Bookstein et al. *Sys. Zool.* **35,** 241–53.

Reyment, R. A., Blackith, R. E. & Campbell, N. A. (1984). *Multivariate Morphometrics*. London: Academic Press.

Rohlf, F. J. & Archie, J. W. (1984). A comparison of Fourier methods for the description of wing shape in mosquitos *(Diptera: culicidae)*. *Syst. Zool.* **33,** 302–17.

Rohlf, F. J. (1990). An overview of image processing and analysis techniques for morphometrics. In *Proceedings of the Michigan Morphometrics Workshop* eds. R. J. Rohlf & F. L. Bookstein. Special Publication No. 2, Univ. of Michigan.

Russell, M. D. (1985). The supraorbital torus: "A most remarkable peculiarity." *Current Anthrop.* **26,** 337–60.

Sokal, R. R. & Sneath, P. H. (1963). *Principles of Numerical Taxonomy*. San Francisco: W. H. Freeman and Co.

Strauss, R. E. & Bookstein, F. L. (1982). The truss: body form reconstruction in morphometrics. *Sys. Zool.* **31,** 113–35.

Thompson, D. W. (1961). *On Growth and Form*. Abridged Edition. ed. J. T. Bonner. Cambridge: Cambridge University Press.

Wolfe, C. A., Lestrel, P. E. & Read, D. W. (1995). *EFF23 2-D and 3-D Elliptical Fourier Functions. Software Description and User's Manual. PC/MS DOS Version 2.5.1.*

Zahn, C. T. & Roskies, R. Z. (1972). Fourier descriptors for plane closed curves. *IEEE Trans. Comput.* C-**21,** 269–81.

4

Methodological Issues in the Description of Forms

PAUL O'HIGGINS

University College, London

4.1 Introduction

There are many situations in which biologists wish to compare morphology. These include studies of normal and pathological variation, growth, and evolution. Each presents its own morphometric challenges. In the study of the shapes of populations of cells, for example, there may be a lack of unequivocally definable equivalent landmarks on which to base comparative measurements. In this case the investigator seeks morphometric methods that show little dependence on landmark identification. In contrast, studies of variation in skull shape might be based on landmarks that are equivalent between individuals in an evolutionary, developmental, or functional sense. Thus, comparative data may be based upon the relative locations of these landmarks rather than upon the shape or curvature of the outline itself.

In undertaking a morphometric study, one chooses, then, between methods that describe forms in terms of landmarks or interlandmark distances, and those that describe form with little or no reference to landmarks. There are, however, a large number of other issues involved; these are, to a degree, dependent on the problem at hand and the particular questions being addressed. For instance, in choosing landmarks, questions may arise concerning homology, the sampling of form, and the types of measurements to be taken. Many alternative strategies are available and choosing between them requires some knowledge, not only of the biology, but also of the morphometric issues concerning each approach.

Modern desktop computers, digitizing tablets, frame grabbers, and 3-D digitizing equipment allow many "new" types of data to be gathered. Therefore, it becomes important to consider how each of these data types may be compared between specimens, the limitations of each possible approach, and the extent to which the use of different types of data might affect the outcomes of morphometric investigations.

74

This chapter will review a number of approaches to the quantitative description of form, consider the differences between landmark-based and landmark-independent approaches, and examine the issue of homology in relation to landmarks and to boundary representations of forms with few or no landmarks. A large part of this review will be devoted to Fourier analysis, since this is the theme of this book. It is hoped that the discussion of Fourier analysis against a background of other morphometric approaches will be useful in providing context and perspective.

Landmark-dependent approaches are considered first, and methods which are less dependent on landmarks are discussed later. This is to some extent an artificial classification since landmark identification plays a part in the majority of strategies that can be applied to form description. A brief description of each technique is followed by a consideration of the theoretical and practical problems associated with it. This chapter will provide a critical overview of currently available tools for form description.

4.2 Landmark-based approaches to form analysis

4.2.1 Interlandmark distances

A classical application of landmarks to the study of biological forms comes from craniometry, in which landmarks may be of two kinds; anatomical (e.g., tips of prominences, sutural junctions) or extremal (e.g., the most dorsal or superior point). Examples of such landmarks, and measurements taken between them, are given by Brothwell and Trevor (1964) and Martin (1928). Interlandmark distances, and indices constructed from them, may be used in phenetics.[1] Classically, statistical analysis of interlandmark distances proceeds through univariate and multivariate analysis (Sneath and Sokal, 1973). Such distances, however, can be gap-coded (i.e., gaps in the statistical distribution of metrical characters between Operational Taxonomic Units (OTUs) are identified and coded as different character states) and used in cladistic[2] analyses.

4.2.1.1 Euclidean distance matrix analysis

Euclidean Distance Matrix Analysis (EDMA) refers to the comparison of forms based on matrices of Euclidean distances between equivalent landmarks (Lele and Richtsmeier, 1991). This procedure extends the classical approach to the com-

---

[1] Editor's note: Phenetics refers to a classification based on morphological similarity, without consideration of evolutionary relationships. An Operational Taxonomic Unit (OTU) has been defined as a collection of objects (specimens in biology), each member of which is described with a set of measurements, which becomes a dataset. Thus, OTUs are the lowest-ranking taxa in studies of variation. A set of OTUs is built in order to arrange the specimens in a hierarchical manner (see Sneath and Sokal, 1973).

[2] Editor's note: Cladistics implies an evolutionary classification based on the succession of splittings through which an organism has passed during its divergence from an ancestor.

parison of OTUs through the comparison of interlandmark distances. A series of equivalent landmarks is identified on each OTU to be compared. The distances between these landmarks are determined, and for each OTU a matrix of these interlandmark distances is produced (a form matrix). The geometric relations of all landmarks are preserved in the form matrix, since it contains all interlandmark distances. Form difference matrices (containing the ratios of corresponding distances between OTUs) are then calculated between single OTUs or between the average form matrices of populations of OTUs. The magnitudes of the ratios in the form distance matrix can be used to assess differences and to identify landmarks whose locations vary between forms. Statistically, the behavior of these ratios is complicated, but statistical inference can be approached through bootstrapping (Lele and Richtsmeier, 1991).

4.2.1.2 Criticisms of the use of interlandmark distances

The use of interlandmark distances or indices as the basis of morphological description has been criticized on several counts. First, the way in which interlandmark distances are commonly collected is such that no attempt is made to systematically describe the relative locations of landmarks, one to another. The result is a collection of measurements that may fail to describe the full 3-D disposition of landmarks as well as over-sample some regions at the expense of others. This is a criticism of study design rather than of the use of interlandmark distances *per se*; it does not apply when interlandmark distances are collected systematically, as in EDMA. Second, as Bookstein (1978) has pointed out, extremal landmarks (i.e., ones that occupy extreme limits of objects with respect to a particular line or plane; e.g., the most dorsal, most superior) are entirely orientation dependent. It is wise to avoid such landmarks. Third, no information relating to the curvature of form between landmarks is preserved; this criticism applies not only to interlandmark distances but to all landmark-based studies, and this issue will be considered later in this chapter.

A further criticism relates to the difficulties of visualization of shape differences. Typically, univariate and multivariate analyses are undertaken to investigate patterns of morphological variation. The results of such studies are often presented as plots of OTUs on canonical axes or principal components (PCs), or as a matrix of inter-OTU distances. These approaches lead to precise mathematical descriptions of patterns of covariance between (often disconnected) variables, but they do not, in themselves, produce a simple, readily interpretable, spatially integrated map of the size and shape differences under study. As such, their contribution to a ready understanding of the complex differences in size and shape between OTUs is limited.

However, it is possible to work backwards from eigenvectors to reconstruct the interlandmark distances of an OTU with any given set of PC scores. If these distances have been carefully measured, it is then possible to reconstruct the original landmarks in arbitrary registration and so to proceed to the visualization techniques that utilize coordinates, as outlined later in this chapter.

One advantage of the study of interlandmark distances over the study of landmark coordinates is that, in contrast to coordinates, these distances are independent of reference frame and registration (the way in which coordinates from different OTUs are "superimposed" on each other).

4.2.2 Co-ordinates and geometry

Use has also been made of Cartesian coordinate data in phenetic studies (e.g., Creel and Preuschoft, 1971; Corruccini, 1988). Rather than through interlandmark distances, specimens are described in terms of the x, y, and possibly z, coordinates of a set of landmarks. Coordinate data allow a description of morphology in which landmarks can be readily related one to another. As such, the description of form is complete (in the sense that all landmark locations are fully defined). Coordinates, like linear distances, are amenable to multivariate statistical analysis. Before carrying out such analyses, however, it is necessary to register ("superimpose") coordinates from different OTUs within the same reference frame.

With coordinate data, the particular patterns of between-OTU variation represented by a particular principal component or canonical axis will be entirely dependent on the way in which the OTUs have been registered (scaled, reflected, rotated, and translated to "register" or "superimpose" the data within the same reference frame) with respect to each other (see Bookstein, 1978). Thus, the perceived displacement of any particular landmark from one shape to another depends upon this registration. Different registrations will produce different impressions of the shape transformations, and regions close to registration points (if registration is undertaken using such points) will appear to change less than those more distant.

The methods of Procrustes analysis (reviewed by Rohlf, 1990a; Goodall, 1991) register forms by minimizing the "fit" (e.g., the mean square distance between landmarks on each OTU). Differences between objects after Procrustes fitting can be expressed in terms of Procrustes distances. Each OTU is represented as a point in Kendall's shape space, which is isometric with a sphere of radius 0.5 when an object is defined in terms of three landmarks in two dimensions (Kendall, 1984). When more landmarks or dimensions are used the shape space becomes more complicated. The Procrustes distance coefficient between any two OTUs is non-Euclidean and can be thought of as the closest great circle distance between them (Kendall, 1984). For practical purposes, statistical analyses (such as principal components) of Procrustes-fitted data are generally carried out in the tangent plane to shape space (which, as long as variations are small, adequately approximates the curving surface of Kendall's shape space). In this case, the distances between OTUs are treated as if they are Euclidean, and normal statistical assumptions are made.

It is worth reiterating that statistical studies of coordinate data rely on registration and scaling, whereas studies of interlandmark distances are independent of registration. This difference in approach allows the investigator to confirm or modify the conclusions drawn by using one set of techniques (e.g., Procrustes analysis) in the light of studies using the other (e.g., EDMA). It can be argued that such confirmatory analyses should form an important part of all morphometric studies since all techniques suffer to a certain extent from their own peculiar constraints and limitations.

4.2.3 Visualization and graphical representation of shape differences using landmark data

4.2.3.1 Transformation grids

In contrast to multivariate analysis, an alternative strategy for comparing coordinate representations of form is to describe "shape changes" or "shape differences" (both are commonly used) as a deformation that smoothly rearranges the configuration of landmarks as a whole. The best-known representation of such a deformation is in the form of a "transformation grid" (Thompson, 1917) in which the differences in morphology between OTUs are described through distortions of a rectangular grid.

There have been several attempts to produce mathematically defined, reproducible visualizations of shape transformations. For practical reasons these are commonly restricted to 2-D (x and y coordinates). Earlier attempts (e.g., De Coster, 1939; Moorees and Lebret, 1962) tended to suffer from problems associated with the registration of one form on another, and with the extrapolation of the shape differences indicated by differences in landmark location to the spaces between them (Bookstein, 1978; Sneath, 1967).

One approach (Sneath, 1967) is to register landmark configurations from two forms using least squares and to model the displacements of landmarks between the first (base form) and second (target form) in both the x and y directions using pairs (one for x and one for y) of linear, quadratic, and cubic power surfaces (trend analysis). These surfaces are used to displace the nodes of a square grid in both x and y (see below; thin plate spline) producing a distorted (transformation) grid reminiscent of those hand drawn by Thompson (1917).

A very similar approach was introduced to morphometrics by Bookstein (1989). A pair of surfaces is defined as in Figure 4.1a. The x and y coordinates of the base form are represented in x and y, while the x coordinate, and then the y co-

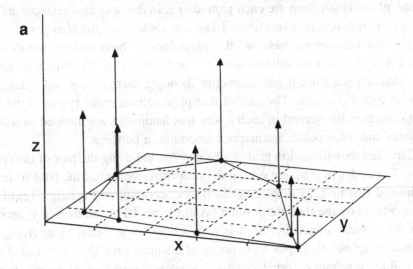

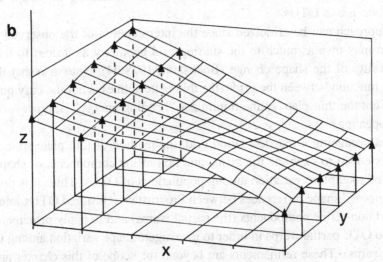

Fig. 4.1 The construction of a Cartesian transformation grid using thin plate splines. In (a) the coordinates of landmarks taken from an ape skull are plotted in x and y. The z axis represents the x coordinates of corresponding landmarks on the target form (not shown). In (b) a thin plate spline is used to fit a surface to the z coordinates shown in (a). A grid drawn in the coordinate system of the original form is shown in the plane of the x and y axes. Arrows connecting the nodes of the grid to the surface indicate the degree to which the nodes of the square starting grid are displaced in x in the transformation grid. For clarity, only some of the arrows are shown connecting the nodes with the surface. Displacements in y are treated in the same way; thus, a pair of thin plate splines (one in x and one in y) are used to draw a Cartesian transformation grid.

ordinate, of the target form are each plotted in z. In this way two surfaces are defined (i.e., two diagrams like Figure 4.1a); one illustrating the changes in x and the other in y, between the base and the target forms. These surfaces are then fitted by a pair of thin plate splines (one for x and one for y). The thin plate spline is a sensible choice since it minimizes the "bending energy" required to take the first form into the second. The pair of thin plate splines sends points in the first form to points in the second in such a way that landmarks are mapped exactly to landmarks and other points are mapped smoothly in between.

A Cartesian transformation grid can be constructed using the pair of thin plate splines. The nodes of a square grid in the coordinate system of the base form are repositioned first in x, by displacing them in x according to the height (z) of the surface defined by the thin plate spline (as in Figure 4.1b) and then in y, according to the surface defined by the thin plate spline for y. Thus, by applying the same pair of splines to displace the nodes of a square grid, they are shifted in x and y and the resultant deformed grid is known as a transformation grid (Figure 4.2a) that is visually very close to those derived by Thompson (1917). The perceived mapping does not depend on the particular coordinate systems of the figures, making this is a registration-free method for visualizing the shape differences between two OTUs.

This approach can be criticized since the interpretation of the observed transformation may owe as much to the starting grid geometry as it does to the biological reality of the shape change (Bookstein, 1978:94). Note also that the interpolant function between the grids (the thin plate spline) is not the only possible choice. That the thin plate spline minimizes "bending energy" is, however, intuitively appealing.

Besides producing a transformation grid, the method of thin plate splines can also be extended to examine the affine and non-affine components of shape difference and to explore variation among populations of OTUs. Thus, it is possible to "decompose" shape differences between Procrustes- registered OTUs into their affine and non-affine components (the partial warps) and to apply principal components to OTU partial warps in order to investigate shape variation among OTUs (relative warps). These refinements are beyond the scope of this chapter and the interested reader is referred to Rohlf and Bookstein (1990) and Reyment (1991).

4.2.3.2 Finite element analysis

In biology, finite element analysis, originally devised for describing the effects of stresses on engineering materials, has been adapted to the task of characterizing shape changes. In this approach, shape differences are described in terms of the directions and magnitudes of the principal strains in the transformation of one form to another. These methods are also "registration free" since they provide information about the "stretching" of elements rather than the movement of land-marks relative to the coordinate system.

"Homogeneous" finite element methods (e.g., Bookstein, 1978; Moss et al., 1987) work under the assumption that shape changes (strains) are uniformly dis-tributed throughout each element (are homogeneous). This is not necessarily true of biological forms in which an element may span diverse tissues; consequently, this simplifying assumption of homogeneity may have an effect on the biologi-cal interpretation of results. Elements may be of different shapes but the simplest possible ones are triangles whose apexes are equivalent landmarks between two forms. The shape transformation between homologous elements can be described by the major and minor axes of the ellipse obtained by deforming one triangle, together with its inscribed circle, into the other (Bookstein, 1978). The directions of these axes (principal strains) indicate the directions of maximum and minimum shape change, and their magnitudes indicate the relative measures of these changes (Figure 4.2b). It is noteworthy that from the biometrician's perspective, the prin-cipal strains are invariant to element registration and that they relate the uniform deformation of each element in the base form to the equivalent element in the tar-get form.

"Nonhomogeneous" finite element methods, on the other hand, do not make the assumption of homogeneity; they use more complex elements (i.e., cubes rather than triangles), and allow the computation of local deformations around land-marks. The Finite Element Scaling Method (FESA; e.g., Lew and Lewis, 1977) is nonhomogeneous and has been widely applied in studies of craniofacial growth and sexual dimorphism (Cheverud and Richtsmeier, 1986; Richtsmeier, 1986; 1989).

Although the selection of landmarks and finite elements is largely arbitrary, the interpretation of shape changes in particular anatomical regions may differ ac-cording to element design (Zienkiewicz, 1971; Cheverud and Richtsmeier, 1986). For instance, thin triangles will tend to "amplify" small shape changes. In order to minimize these effects, O'Higgins and Dryden (1993) have recently proposed the use of the Delauney triangulation (see Green and Sibson, 1977).

4.2.3.3 Biorthogonal grids

Bookstein's (1978) solution to the problem of element design is to compute the deformation of many elements interpolated over the interior of the forms under study and to derive a smooth map of shape changes over these elements. He calls this method "biorthogonal grids." O'Higgins and Dryden (1993) constructed such grids using the thin plate spline as the interpolant. A small triangle was trans-formed between OTUs using the pair of thin plate splines, and its principal strains were calculated. Moving off a short distance in the direction of the first principal

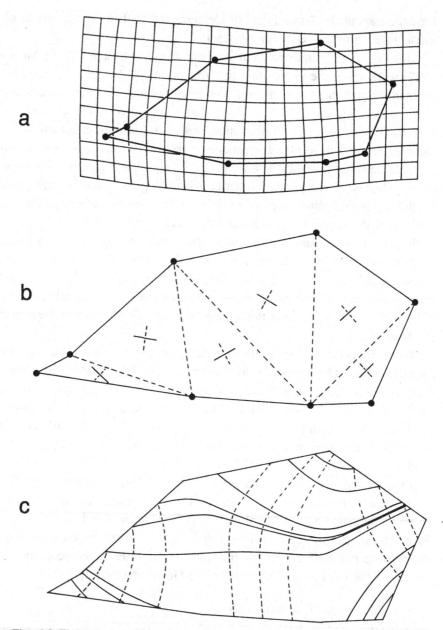

Fig. 4.2 The transformation from female to male gorilla crania: (a) Cartesian transformation grid; (b) finite elements analysis; the crosses indicate the directions of the major (————) and minor (——) principal strains; the limbs of the crosses are drawn to scale indicating the magnitudes of these strains; (c) biorthogonal grid; curves showing major (————) and minor (——) principal strains. Magnitudes are omitted for clarity.

strain, a second triangle was transformed and its strains calculated. The strains were smoothly joined in a diagram and the process was repeated for the second (minor) principal strain. The result is a biorthogonal grid (see Figure 4.2c).

Using this procedure, the vagaries of element design are largely side-stepped since the whole interior of the forms is taken to smoothly deform between them and the matching of internal and boundary "homologies" is taken to conform to a smooth mapping of the homologous landmarks. Note, however, that the choice of homology mapping function (e.g., the thin plate spline) is not unique.

These approaches (Cartesian transformation grids, finite elements, and biorthogonal grids) for the visualization and geometric study of landmark data offer interesting possibilities for the spatially integrated, graphical analysis of form differences. All of these methods should be employed, however, with due regard to their limitations. Cartesian transformation grids derived using thin plate splines (Bookstein, 1989) suffer from a potential problem in that starting grid geometry may influence the interpretation of shape transformation (Bookstein, 1978). Additionally, it should be noted that a different interpolant applied to the same data would produce different transformation grids. These considerations aside, Cartesian transformation grids, like finite element methods and biorthogonal grids, result in reproducible, mathematically defined graphical descriptions of shape change independent of registration. All of these approaches may, however, produce different results when different landmarks are selected (a little-studied issue) and element design will influence the outcome of finite element analyses.

It is important to appreciate that these methods do not attempt to model the biological mechanisms of shape transformation (e.g., growth processes); rather, they produce geometric or graphical descriptions of shape differences and transformations. As such they allow conclusions to be drawn about patterns, but not about the mechanisms that underlie the transformation or difference. Moreover, in the study of these patterns, the limitations of each method should be clearly borne in mind, and it would seem useful to consider the results from several approaches simultaneously to provide a check on interpretation.

4.2.4 General issues relating to the use of landmark-based data

Recent developments in methods for the study of landmark-based data offer interesting possibilities for the analysis of form differences. There are, however, a number of issues relating to the use of landmark-based data that deserve consideration.

If landmark data are to provide the basis for comparison of forms, it is important that the landmarks be, in some way, equivalent between OTUs. The term "homology" is often applied in this context, although the usual usage of this term is to refer to components (organs, parts, characters) of organisms rather than landmarks on these structures.

Pre-Darwinian homology was determined through the correspondence of parts in their relative position (Owen, 1847). A homologue was seen as the same organ in different animals under every variety of form and function. This correspondence was taken to reflect the *Bauplan* of ideal types and does not refer to evolution at all. Homologous structures in this sense were defined on the basis of their correspondences and relations.

After Darwin (1859), homology was reinterpreted as reflecting the structure of ancestral types; evolution was taken as the explanation of sameness. Indeed, to Darwin, homology was another component of the evidence for evolution; homologous structures between species are such because they derive from a common structure in a common ancestor.

Sneath and Sokal (1973) have pointed out the inherent circularity in the identification of homology (using its post-Darwinian definition), since this identification depends on knowledge of evolutionary history, and knowledge of evolutionary history depends on the study of homologous features. In order to develop a pragmatic solution to this "homology problem," Sneath and Sokal (1973:77–82) loosely describe homology as composed of "compositional correspondence" (implying a qualitative resemblance) and "structural correspondence" (referring to a spatial arrangement of parts, as in Owen's definition). This leads them to the approach they term "operational homology." Thus, two characters are operationally homologous if they are "very much alike in general and particular." In so doing, they posit homology based on criteria of similarity, and leave the testing of homologies to repeated finer detail analyses. It was hoped that further analyses will lead to further support or refutation of operational homologies. This operational approach allows the taxonomist to "get started" in a study, since homologies are identified without reference to phylogenetic reconstruction.

Cladistic approaches to phylogenetic reconstruction have pointed to a method for the testing of operational homologies in an evolutionary sense. Eldredge and Cracraft (1980:36) in viewing homology from a cladistic perspective indicate that the solution to the "homology problem" rests with the concept of synapomorphy (the sharing of derived character states). "*Homologous similarities are inferred inherited similarities that define subsets of organisms at some hierarchical level within a universal set of organisms*" (italics in original). Thus, they suggest that the "test" for homology is not similarity, but the congruence of other hypothesized synapomorphies in defining sets of a cladogram. Postulated (operational) homologies are used to construct a cladogram in which (if it is taken to be a true reflection of phylogeny) congruent characters are attributed to synapomorphy or "true homology." Noncongruent characters (homoplasies) are taken as reflecting convergence and parallelism. One problem with this cladistic approach is that, in practice, a cladistic analysis may result in the identification of several equally parsimonious cladograms. In this case some operational homologies are supported by some analyses and not by others, and so it is not possible to unequivocally confirm evolutionary homology.

This cladistic approach to homology, like the approach of "operational homology" in phenetics, relies on the testing of hypothesized homology through iterative analysis and reanalysis. However, unlike the operational approach, the cladistic testing of homology utilizes hypothesized evolutionary relationships and, as such, is truer to Darwin's, rather than Owen's, definition. This is not to say that an operational approach to the identification of homology is unnecessary for the practice of cladistics; rather, it is "useful, even necessary, to organize data as putative homologies, which are either corroborated or refuted by the cladogram that best fits the data" (Nelson, 1994).

The question arises as to how the concept of evolutionary homology can be applied to landmarks as opposed to characters or organs. Landmarks might be identified as operationally homologous through the identification of corresponding local relations (after Owen). Such operationally homologous landmarks can and do adequately serve as the basis for phenetic studies, and may form the basis of phylogenetic reconstructions using characters derived from sets of landmarks. Thus, individual landmarks serve as the basis for the quantification of characters whose homology might be tested through subsequent phylogenetic analyses. In practice, an operational approach to landmark homology is adopted in the early stages of both phenetic and cladistic analyses, and the problem of Darwinian homology appears to have little impact; operationally homologous landmarks are, in general, readily recognized and adequately serve as the basis for evolutionary studies (of variation, phylogeny, and biogeography) among closely related OTUs.

Another problem relating to landmark equivalence may be encountered in developmental studies. A landmark defined as the junction between three bones in the skull may be taken to be equivalent to a similarly defined landmark on another skull or on the same skull at a different time (e.g., a radiographic study). Developmental variation may, however, result in differences in the derivation of the exact parts of the three defining bones that meet to form the landmark. Likewise, local growth phenomena (e.g., bony remodeling, shifting muscle insertions) influence the derivation of landmarks at tips of prominences or in pits. As such, landmarks that *appear* equivalent in terms of their local relations need not necessarily reflect the locations of homologous material. Thus, in what sense, if any, can such landmarks be considered homologous?

Wagner (1994) has recently addressed this issue. He notes that, despite the fact that during growth bony material is likely to be completely replaced, structural identity is maintained. This maintenance of identity requires the action of "morphostatic" mechanisms, and although landmarks may not be equivalent in the sense of being located on homologous material, they may be equivalent in terms of the continuity of these morphostatic mechanisms. Developmental equivalence between landmarks may therefore be considered to equate to homology in the sense of "correspondence caused by continuity of information" (van Valen, 1982).

These considerations open up a possible role for landmark-based descriptions of form in understanding ontogenetic processes. In the case of the skull, for instance, displacements of landmarks during ontogeny result from underlying processes such as sutural growth or bony remodeling. Therefore, the combination of morphometric data with data on these processes (e.g., remodeling activity) might offer new insights into the ontogeny of shape transformation. An example of such a study is provided by the work of O'Higgins and Dryden (1992), in which cortical remodeling maps are combined with transformation grids in an attempt to examine the integration of facial bone displacements with cortical remodeling in the mangabey.

The issue of equivalence between landmarks arises in yet another circumstance. For example, in biomechanical studies, functional equivalence may be more important than either evolutionary or developmental homology. In that case, the ontogenetic or phylogenetic equivalence of landmarks on bat and bird wings is of less concern than their functional equivalence in considering the biomechanical basis of flight in these species. In comparing the functional morphology of such structures, landmarks defining, for instance, the extremes of lever arms or the locations of muscle insertions, may be selected on the basis of their functional equivalence.

Besides the problems inherent in identifying equivalent landmarks on OTUs, there are a number of other issues that surround their use. Landmark-based methods leave the form between landmarks unsampled. A problem is presented where no landmarks can be readily identified in a particular anatomical region because of a lack of surface features (e.g., on the smooth bones of the vault). In this case it is possible to interpolate landmarks according to the locations of observed (operationally homologous) landmarks and surface curvature. It is doubtful, however, that such landmarks (more appropriately termed pseudo landmarks) can be considered homologous between OTUs, in either a developmental or evolutionary sense, since their location relates to mathematical, as well as biological, constraints. This consideration is important in studies that may use several constructed landmarks (see Section 4.2).

A further issue arises in the context of different types of landmarks, since, by their nature, some can be readily located (e.g., a sutural junction), whereas others can only be approximately identified (e.g., the tip of a prominence). Practical considerations, therefore, play a role in limiting the number of landmarks that can be usefully included in a morphometric study.

Thus, a number of issues surround the choice and use of landmark data as the basis of form description. Despite some of these constraints, landmarks continue to provide an important basis for the analysis of form and offer one important advantage over morphometric approaches that use few or no landmarks; it is possible to investigate variations between OTUs in terms of "homologous" regions. For example, it would be impossible to consider differences in the facial skeleton between two apes unless the location of the face relative to rest of the skull were defined on each, for which, some landmarks defining the locations of skull components are required.

Comparison of the disposition of equivalent landmarks between OTUs is a good way of describing changes in the "homology map." In some studies, however, there may be difficulties in defining equivalent, or indeed, any landmarks (e.g., cell shape, shell shape). In these circumstances, and in the case where comparisons are sought based on the form of outlines between definable landmarks, it may be necessary to turn to alternative morphometric strategies.

4.3 Forms with reduced landmark dependency

In certain circumstances the biometrician may be faced with the challenge of examining shape variations between OTUs lacking sufficient readily identifiable, equivalent, landmarks. Examples are found in studies of cells, leaves, insect wings, ostracods, and so on. Alternatively, although landmarks may be readily identified, it is possible that outlines of regions between these landmarks are the focus of study. Such an example is presented by the cranial vault and the curvature of vault bones between landmarks.

In each of these situations the investigator may justifiably seek morphometric strategies that show little or no dependence on landmark identification. Most such work has been restricted to the analysis of 2-D outlines, although many of the available approaches are extensible to 3-D.

The outlines of objects can be traced from photographs, or projected using a digitizing tablet, or derived from video images using readily available image analysis software. Details of such an apparatus are given in Johnson et al., (1985) and an algorithm for tracing the outline of an object is given by Rohlf (1990b). Points are sequentially read from the outline at determined intervals and stored as a series of x and y coordinates. If general measurements such as the perimeter (or enclosed area) are sought, these can be derived directly from raw digitizer output using standard software.

4.3.1 Outlines and the enclosed area within them

There are several landmark-independent methods available for the description of forms in terms of their outlines and the area enclosed within them.

4.3.1.1 Shape factors

Shape descriptors that are invariant to differences in OTU position and orientation (i.e., non-registration-dependent quantities) are desireable. Examples are area, perimeter, maximum length, and so on. A very simple measure of shape is given by the aspect ratio:

$$F_1 = \frac{\text{max } length}{\text{max } breadth},$$ (4.1)

where the *max breadth* is 90° to the *max length*. Given the area, A, and the perimeter, P, of an object, two further quantities can be readily calculated:

$$F_2 = \frac{4\pi A}{P^2},$$ (4.2)

$$F_3 = \frac{P - \sqrt{P^2 - 4\pi A}}{P + \sqrt{P^2 - 4\pi A}}.$$ (4.3)

F_1 provides a measure of elongation and F_2 and F_3 are measures of the undulation of the outline relative to a circle (in which $P^2 = 4\pi A$).

Note that quite different outlines can have similar values for one or more, of these simple measures, and it is advisable to consider all three simultaneously. Examples of their use in biology are studies of cell shape (e.g., Young et al., 1974) and cranial form in the primates (O'Higgins, 1989).

4.3.1.2 Moments

Sometimes a form may be specified as a collection of interior points. In the case of digitized images, the positions of interior points are specified by the x and y coordinates of pixels, the distribution of which can be used to describe a form.

For a single variable, for example, the x locations of pixels, m_p, the pth moment of x is given by:

$$m_p = \Sigma(x^p) = \int_{-\infty}^{\infty} x^p f(x) dx.$$ (4.4)

Thus, in a binary image, the zero order moment is the number of pixels enclosed by the outline. The first moment is the mean of x, the second its variance, and so on.

For a 2-D distribution along arbitrary axes x and y, the moment of the order $(p+q)$ is defined by:

$$m_{pq} = \int_{-\infty}^{\infty} \int_{-\infty}^{\infty} x^p y^q f(x,y) dx \, dy. \tag{4.5}$$

This series of moments uniquely describes an image and can, therefore, be used to reconstruct it.

As described above, the moments are dependent on position and orientation and, as such, are of little use in taking measures of shape that will allow comparisons between forms differing in registration. Central moments, which are translation-independent, can be calculated by referring the xs and ys to the centroid. Hu (1962) has further described 2-D moment invariants, which show size, rotation, and contrast invariance.

Moments represent a "landmark-free" method of form analysis and as such have been applied in situations where landmark identification is difficult (e.g., cell biology; Dunn and Brown, 1986). Rohlf (1990b) comments that there is evidence to indicate that moments perform well in classification but that, in his own experience, there have been problems in their use due to lack of statistical independence between moment invariants, and their sensitivity to rounding errors.

4.3.1.3 Skeletons and medial axis transforms

Blum (1967) has introduced a very different approach to describing shapes. The shape is defined by a symmetric axis or skeleton that consists of all points within a form that do not have a unique nearest boundary point upon the shape. Associated with each point on the symmetric axis is a width function defining the distance to any of the set of equally distant nearest boundary points. The "grassfire model" (Blum, 1973) makes comprehension easier. The shape is characterized as an area of dry grass. If it is fired simultaneously all around the edge, it will burn toward the interior. If an even rate of burning is assumed, the points at which the fire meets itself comprise the points defining the skeleton; the time taken to reach these points is the function.[3] The skeletal pair (axis and function) exhaustively describe as well as allow for the complete reconstruction of the form, independent of landmarks.

Straney (1990) considers Blum's and alternative approaches to skeletonization including Bookstein's (1979) variant; the line skeleton. Bookstein's differs from the symmetric axis by being composed of line segments rather than line segments

---

[3] Editor's note: The medial axis transform can also be visualized as a series of concentrically-overlapping circles that touch the outline in a tangential or orthogonal fashion. The skeleton is then defined as the locus of all of the centers of these circles, which are equidistant from all borders of the outline.

and parabolic arcs. Additionally, the width function associated with Bookstein's (1979) line skeleton is not necessarily symmetrically located within the form, and so is not single-valued as that of Blum (1973).

Line skeletons and symmetric axes have been applied to studies of mandibular growth (Bookstein, 1979; Webber and Blum, 1979). The results of these studies indicate that the branch points and angles between skeletal branches may be similar between OTUs. It has therefore been suggested that the branch points may serve as useful landmarks for morphometric analysis. Such an approach, in which branch points are taken as equivalent landmarks between OTUs, was followed by Straney (1990) in a study of the evolution of the baculum of rats.

The use of skeletons of images as the basis for the identification of operationally homologous landmarks, therefore, represents a strategy for the comparison of forms with limited external landmarks. It should be noted, however, that different skeletonization algorithms and subtle differences in outline form may result in skeletons with quite different topologies. Also, the identification of such operational homologies may not be necessarily supported from a developmental or evolutionary perspective.

4.3.2 Boundaries

The three approaches outlined above, shape factors, moments, and skeletons, characterize form in terms of an outline and the area enclosed within it (note: gray scale extensions of these methods are possible with pixel data). More commonly, forms are studied in terms of the boundary alone.

Raw data from a digitizing tablet, or a video digitizer, generally consist of a stream of unevenly spaced x and y coordinates describing the boundary of each OTU. These cannot be directly compared between OTUs because of differences in landmark number and spacing, and in the registration of objects.

4.3.2.1 Pseudo landmarks

Several strategies exist for comparing OTU outlines. One approach is to divide the outline of each OTU into segments, each of which can be imagined as being delimited by a *pseudo* landmark. Such pseudo landmarks are operationally, but not necessarily biologically, equivalent.

The matching of pseudo landmarks between OTUs is relatively simple if one biologically equivalent landmark can be identified on each OTU, since all others can be counted sequentially from it. If no biologically equivalent landmarks can be identified, then matching can be achieved through Procrustes analysis. Such a maneuver will simultaneously "match" pseudo landmarks and register outlines with respect to each other.

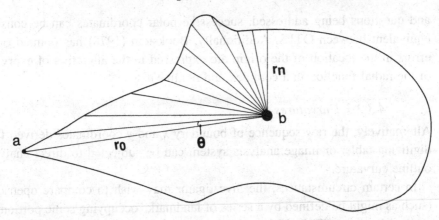

Fig. 4.3 Polar coordinates (r_0-r_n) of boundary points calculated from starting point (a) and centroid (b).

Pseudo landmarks, once identified, may form the basis of morphometric studies using any of the methods for the study of landmark data outlined earlier in this chapter. It must be borne in mind, however, that pseudo landmarks are unlikely to be equivalent in either an evolutionary or developmental sense.

4.3.2.2 Polar coordinates

Rather than divide the outline into equally spaced segments with pseudo landmarks at each segmental junction, it is possible to transform the outline data into polar coordinates centered on the objects themselves. The boundary of a convex shape is reexpressed in terms of the lengths of radii spaced at equiangular intervals. On each shape an origin (center) is defined for the polar series together with a starting point on the outline from which the series will be deemed to begin (Figure 4.3).

As an example, Yasui (1986) used polar coordinates to study shape variation in Japanese crania. He registered polar representations of outlines with respect to each other by rotating them about their centroid and determining a criterion of best fit.[4] This is similar to the approach of O'Higgins et al., (1986) and is a form of Procrustes superimposition (Goodall and Bose, 1987).

The line connecting the origin (center) to the starting point is the line to which all other polar coordinates are referred. If the origin and starting points are "homologous landmarks" (e.g., Lestrel, 1982), then all comparisons between polar coordinate pairs from outlines are, in effect, referred to these homologies. Again, it is for the investigator to decide whether, in the context of the particular study

---

[4] Editor's note: This is, in effect, a crosscorrelation procedure (Parnell and Lestrel, 1977).

and questions being addressed, successive polar coordinates can be considered equivalent between OTUs. Additionally, Bookstein (1978) has pointed out that errors in the location of the origin are expressed in the alteration of every value of the radial function in a complex and nonlinear way.[5]

4.3.2.3 Curvatures

Alternatively, the raw sequence of boundary x and y coordinates derived from a digitizing tablet or image analysis system can be subjected to direct analysis of outline curvature.

In certain circumstances, the investigator may wish to compare open curves (such as might be defined by a series of landmarks occupying some portion of an outline) between OTUs. If these can be expressed such that one coordinate (e.g., y) is a single-valued function of the other (x), then it is possible to fit a polynomial, cubic spline, or other function, to the sequence of y coordinates (see Rohlf, 1990c).

Given a closed outline, that is, one in which the y coordinates cannot be expressed as a single-valued function of x, different strategies need to be considered. One approach is to describe the outline in terms of the tangent angle at the points on the outline. These points may represent operational homologies, they may be spaced equidistantly around an outline, or they may represent the nodes of outlines divided into equal numbers of segments.

In the case in which the points are not operational homologies, tangent angles may be matched between outlines by relating them sequentially to an operationally homologous starting point on each form, or through a Procrustes superimposition.

A comparison of raw tangent angles is, however, sensitive to the orientation of forms, since the tangent angle is measured relative to some arbitrary line. This orientation-dependency can be eliminated by relating all tangent angles to the starting point. This is accomplished by calculating the difference in tangent angle between the start point, $\theta(0)$, and each outline point, $\theta(t)$, that is:

$$\phi(t) = \theta(t) - \theta(0), \tag{4.6}$$

in which $\phi(t)$ expresses the angular change between the tangent at a particular point on the outline and the starting point.

If the outline perimeter is scaled to a length of 2π, and the distance along the outline between the starting point and the current point is t, then the angles, measured in radians, can be derived from a new function as:

$$\phi^*(t) = \phi(t) - t, \tag{4.7}$$

---

[5] Editor's note: This particular problem can be ameliorated by shifting the location of the center of the radii to the centroid and recomputing equiangular intervals between the radii.

which is equal to *0* for every point on the outline of a *circle* (for details see Zahn and Roskies, 1972). Thus, when applied to a plane closed curve, the magnitude and sign of $\phi^*(t)$ is related to the difference between the actual curvature and that of a circle. Note that $\phi^*(t)$ is invariant under translations, rotations, and changes in perimeter.

Rohlf and Archie (1984) applied the $\phi^*(t)$ function to the description of the shape of mosquito wings (Figure 4.4). They calculated $\phi^*(t)$ for 100 equally spaced values of *t*, subjected the $\phi^*(t)$s to Fourier analysis, and used the resulting harmonics as the basis for multivariate phenetic analyses. Lohmann has also used the $\phi^*(t)$ functions, in combination with principal components analysis, as the basis of his method of "eigenshape analysis" (reviewed in Lohmann and Schweizer, 1990).

Young et al., (1974) describe a related measure of shape based on the notion of "bending energy." A 2-D outline made out of a homogeneous material, if allowed to adopt its "free" form would assume the shape of a circle because this is the shape which minimizes the stored energy. To make more convoluted outlines requires the expenditure of work in the form of bending energy. The measure they describe creates equivalence classes of figures with equal "stored energy." The shape is divided into small regions and for each region the curvature, K_n, is defined as the change in direction per unit length. The total "bending energy" is given by the sum of K_n^2s over the whole outline.

They also describe a simple approach to the calculation of bending energy, directly computed from the chain code (directions from pixel to pixel) of an outline through difference codes (changes in direction from pixel to pixel). The calculated value of bending energy is invariant with respect to position and rotation but is affected by size as well as shape differences. This accords with the intuitive feeling that it takes more energy to bend a short length of material into a circle than a long one.

4.3.2.4 Complexity

In one sense, bending energy relates to the complexity of an outline, since highly convoluted outlines have larger values of bending energy than less-convoluted outlines of the same length.

Another approach to describing the complexity of form typical of biological materials is based upon the concept of fractal dimension (Mandelbrot, 1983). As a curve in a plane becomes increasingly more convoluted, it fills up more and more of that plane. A simple line has a dimension of one. A complex curve, to some degree, fills a plane, and so it can be considered to have a dimension greater than one; this is its fractal dimension. The fractal dimension can be employed as a summary measure of complexity.

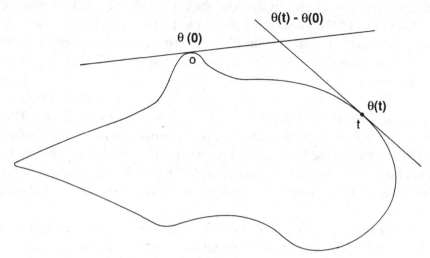

Fig. 4.4 Calculation of the tangent angle function from a skull outline. The boundary is scaled to length 2π. $\phi^*(t) = \theta(t) - \theta(0) - t$ (see text).

If a line of length 1 is divided into N equal parts, each equal in length, r, where $r = l/N$, then:

$$Nr = 1. \tag{4.8}$$

For two dimensions:

$$Nr^2 = 1, \tag{4.9}$$

where r is being expressed in terms of a one-dimensional characteristic of area, that is, its "linear scale"; for example, the diameter of a circle, or the length of a side for a square. This leads to the general equation, which is:

$$Nr^D = 1, \tag{4.10}$$

where D is the fractal dimension.

D can be estimated by the relationship between the estimated length of an outline and the scale of measurement. Examples of the use of fractal dimension as a means of summarizing complexity in biological forms are provided by Reyment (1991:152) and by Katz and George (1985).

4.3.2.5 Fourier analysis

The methods considered above, in the context of the analysis of forms with few or no landmarks, can be divided into two groups. The first group of methods produces summary measures of form (e.g., shape factors, bending energy, fractal dimension), which can be compared directly. The second results in a re-expression of the boundary information present in the original x and y coordinates (e.g., polar coordinates, tangent angle function). The latter ones offer one significant advantage; they provide largely registration- independent data (although they are dependent on the starting point and, in the case of polar coordinates, the centroid location). These new data may be compared using the techniques of multivariate analysis, or they may be subjected to a further transformation such as Fourier analysis; a transformation of data from the spatial to the frequency domain. This may be useful in understanding periodicity (but which rarely has a basis in the biological determinants of morphology), or in summarizing large datasets.

Fourier analysis results in the decomposition of a periodic function (e.g., polar coordinates, $\phi^*(t)$) into a series of sinusoidal waves of differing frequencies, composed of phases and amplitudes which, when summed, can reproduce the original form. Fourier analysis has been applied to the measurement of biological shapes by a number of workers (e.g., Lu, 1965; Kaesler and Waters, 1972; Lestrel, 1982; Rohlf and Archie, 1984; Ferson et al., 1985). Briefly, a periodic function, f(t), can be approximated by:

$$F(t) = a_0 + \sum_{n=1}^{k} a_n \cos(nt) + \sum_{n=1}^{k} b_n \sin(nt), \qquad (4.11)$$

where the a_n are the cosine components, and the b_n are the sine components. They describe the cosine and sine waves at a particular frequency, n (k is the maximum harmonic order of the calculated series). The single Fourier series will provide a fit to any smooth single-valued periodic function. It can be applied to both polar and curvature representations of an outline, and the resulting Fourier coefficients can be used to reconstruct that outline. The polar representation of the Fourier series can be written as:

$$r = F(\theta) = a_0 + \sum_{n=1}^{k} a_n \cos(n\theta) + \sum_{n=1}^{k} b_n \sin(n\theta). \qquad (4.12)$$

In this form, the function, $F(\theta)$, describes the magnitudes of successive radii, r, at successive angular displacements, θ.

Polar coordinates are only amenable to Fourier analysis when each radius intersects the outline at only one point. When the outline is more complex, it may be possible to calculate the Fourier series from x and y coordinates through the tangent angle formulation (Zahn and Roskies, 1972).

Alternatively, the x and y coordinates from an outline may be submitted to elliptic Fourier analysis (Kuhl and Giardina, 1982; Lestrel, 1989) in which Fourier series are separately fitted to Δx and Δy expressed as functions of cumulative chordal distance (see Chapter 2), or to dual-axis Fourier analysis in which the x and y coordinates are fitted directly (Moellering and Rayner, 1981).

There is also an alternative representation of the Fourier series; the amplitude-phase-lag representation:

$$r = F(\theta) = R_0 + \sum_{n=1}^{k} R_n \cos(n\theta + \phi_n). \qquad (4.13)$$

The ϕs are known as the phase lag components; they contain all the rotational information (i.e., about "starting point"). As such, they register the waves of different frequencies with respect to each other in a way that allows reconstruction of the original outline. The phase lag components are readily calculated from the sine and cosine components considered earlier (Eq. 4.12) :

$$\phi_n = \tan^{-1}\left(\frac{b_n}{a_n}\right). \qquad (4.14)$$

The amplitude components, R_n, are a measure of the contribution of each harmonic to the whole form. They contain no phase information, and so are independent of the boundary landmark chosen as the start of the polar series. They, too, are readily calculated from:

$$R_n = \sqrt{a_n^2 + b_n^2}. \qquad (4.15)$$

This representation offers some advantages. In situations where dependency on the starting point definition is considered problematical, the amplitude components alone can be compared between shapes. It should be noted, however, that amplitude components alone do not uniquely specify a shape; different shapes may share the same amplitude components. In a biological situation it seems unlikely, however, that OTUs will differ in phase components alone. Thus, O'Higgins and Williams (1987) and O'Higgins (1989) have shown, in studies of cranial form in primates, that using amplitude components alone gives a similar pattern of between-species discrimination when compared to the combined amplitude/phase-lag spectrum. Nevertheless, the degree of between-OTU discrimination in the former analysis was reduced relative to that in the latter because of the lack of phase information.

4.4 Analysis of data and the reconstruction of form

In Section 4.3 a number of methods were considered by which Cartesian coordinate data representing a boundary can be used to provide measures of form. Some of these approaches (e.g., shape factors, bending energy, fractal dimension) result in simple summary measures of shape that can be readily compared between OTUs. Others result in a reexpression of the information contained within the original Cartesian representation in a way that is, to a greater or lesser degree, independent of the original registration. Thus, moments, medial axis transforms, polar coordinates, curvature functions, and Fourier series exhaustively describe, and can be used to reconstruct, an OTU. In this section some general aspects dealing with the statistical analysis and reconstruction of form will be considered, with an emphasis on the role of Fourier analysis. The section ends with some broad conclusions concerning shape analysis.

4.4.1 Size

Many of the methods for shape description considered here result in data which are invariant with respect to size. This arises because the data are ratios, or angles, or are standardized with respect to some "size variable." Other methods, for example, landmarks, linear measurements, polar coordinates, and medial axes, preserve information about scale. It is important, therefore, to consider how to account for size differences and their consequences. The literature on scaling is large, and the reader is referred there for details of methods and approaches to scaling (e.g., Jungers, 1985; Schmidt-Nielsen, 1984). It is, however, worth raising some general points here.

Two specimens may differ not only in shape but also in size. At first this seems obvious and clear-cut, but there are semantic and mathematical difficulties in discussing size independent of shape in most circumstances. In the comparison between two objects of identical shape, the difference between any pair of homologous measurements, one from each OTU (e.g., lengths, widths, heights), will indicate the scaling required to make them identical. In most biological situations, however, OTUs will differ in shape. Consequently, intuition comes into play and the concept of size becomes less well defined. Sneath and Sokal (1973) ask, "which is bigger, a snake or a turtle?" The term "size" in this circumstance relates to the differences in scale over whole objects. As such, "size" might be best thought of as a vague term relating to the differences in magnitude of many dimensions.

Many different approaches have been taken in determining "size" differences between OTUs and for choosing suitable scaling variables between differently shaped OTUs. The choice of methods depends on the questions being addressed and on the investigator's concept of size. In biomechanical studies, a quantity such as body mass might be appropriate to scale measurements (e.g., Alexander, 1991). Alternatively, the length of a lever arm might be chosen, the choice of scaling variable being justified from an engineering perspective. In phenetic or cladistic studies, however, the problem is more difficult: is body mass the most appropriate choice? Different workers have used different "size variables." Some of these are external to the object under study. For instance, Wood (1976) used femur length as an estimate of body size, and determined the allometric relationship between this and a number of cranial and other dimensions. Other workers have used "size variables" that are derived from the object itself. For instance Albrecht (1978) used three different measures of size: the greatest length of the skull, the geometric mean of the log transformed cranial variables, and an estimate of cranial volume, in a study of the craniofacial morphology of the Sulawesi macaques.

Considering the difficulties in providing an unequivocal size measure for differently shaped forms, it seems sensible to use some measure that describes the magnitudes of many variables. Examples of such measures include centroid size (Bookstein, 1978)—a measure of the deviation of landmarks from the centroid of a shape—and area (which relates to all boundary points).

Turning to Fourier analysis, it is often stated that the constant or zero-order cosine components (a_0s) are suitable size measures with which to scale OTUs described in terms of polar coordinates, since the a_0s are closely related to the area, and form a natural part of the Fourier series. Although scaling by the a_0 term is usually sensible and appropriate, it seems, from the considerations above, that in some circumstances another "size measure" may be more suitably applied to the polar coordinates themselves *prior* to Fourier analysis. In any case, the choice of a "size measure" in any one study needs to be undertaken in the knowledge that each may be different and should be based on the particular biomechanical, ontogenetic, or phylogenetic issue that is being addressed.

4.4.2 Data reduction and shape reconstruction

Several of the methods (e.g., moments, medial axis transforms, polar coordinates, curvature functions, and Fourier series) described earlier result in exhaustive descriptions of individual OTUs and generate a large number of variables. As such, statistical comparisons are often best achieved through the use of the techniques of multivariate analysis. Principal components analysis (PCA) of the covariance or correlation matrices between OTUs may be carried out to investigate patterns of variation. In such analyses, each OTU may be described in terms of registered (Procrustes) Cartesian coordinates of outline points, polar coordinates of outline points, curvature functions, moments, or Fourier coefficients.

Alternatively, analyses might be undertaken to investigate patterns of variation between groups of organisms through Mahalanobis's distances, canonical axes or discriminant functions. In these circumstances the number of measurements (coordinates, moments, $\phi^*(t)$s, etc.) must be considerably less than the number of individuals included in the analysis. If large numbers of specimens are not available it becomes necessary to attempt to reduce the quantity of data from each. Many strategies are available to achieve this aim. These include the selection of fewer data on the basis of what appears "sensible" from the biologist's perspective (a highly subjective exercise), selection on the basis of some measure of likely discriminating value (e.g., F-ratios), and data reduction by way of PCA.

It is also possible to apply the technique of Fourier analysis to the task of data reduction for statistical analysis, since good approximations of the original form can be achieved with relatively few Fourier components. One approach to selecting the number of Fourier components to be used is by means of the harmonic amplitudes (R_ns). These can be plotted against harmonic order to produce a power spectrum (Figure 4.6) that allows a rapid, objective, quantitative assessment of the contribution to the overall form of components of successive frequencies. Figure 4.5 illustrates a chimpanzee cranium reconstructed from increasing numbers of Fourier components via polar coordinates; a good approximation is achieved by relatively few of the lower-order harmonics in the series. This is because the form is generally globular and smooth; higher-order Fourier terms are required to describe finer, more jerky aspects of outline. Note that different boundary forms will be better summarized by different combinations of Fourier components.

The selection of the lower-order Fourier components for statistical analysis of their harmonic amplitudes does not necessarily ensure optimal discrimination between OTUs. This is because information describing aspects of form that differ

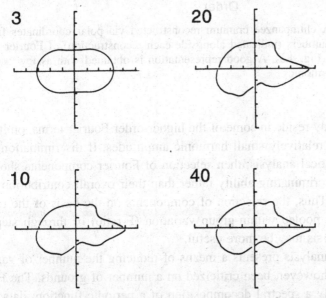

Fig. 4.5 A schematic illustrating the production of a power spectrum (see Fig. 4.6) for relatively simple, gently undulating forms. Most of the shape information is represented by the first few, low-order Fourier components.

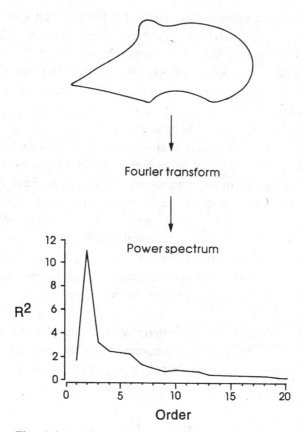

Fourier transform

Power spectrum

Fig. 4.6 A chimpanzee cranium reconstructed via polar coordinates from increasing numbers (indicated alongside each reconstruction) of Fourier components (see Fig. 4.5). A good representation is obtained with as few as twenty Fourier harmonics.

between OTUs may reside in some of the higher-order Fourier terms, omitted on the basis of their relatively small harmonic amplitudes. If discrimination is the objective of statistical analysis, then selection of Fourier components should be based on their discriminating ability rather than their overall contribution to the power spectrum. Thus, the selection of components on the basis of the ratio of between-group to pooled within-group variation (F-ratio) or through step-wise discriminant analysis may be more useful.

Thus, Fourier analysis presents a means of reducing the number of variables per OTU. It has, however, been criticized on a number of grounds. The Fourier series is essentially a spectral decomposition of a periodic function; data in the spatial domain are transformed into the frequency domain. In the process, all information regarding the relative locations of boundary points is referred to the

starting point, and even if this is taken to be homologous between forms, it is unlikely that individual Fourier coefficients can be considered homologous in an
evolutionary or developmental sense. As such, the application of Fourier analysis in the study of biological variation has been the subject of debate.

Bookstein (1978) and Bookstein et al., (1982) have criticized Fourier analysis
because the Fourier decomposition of a curvature function around an outline
allows for one landmark only, the starting point. If the aim of the study is to examine differences in homology relationships, then homologies will not be discernible from the Fourier coefficients.[6] The description of the pattern of disposition of homologies is confused, not aided, by Fourier analysis.

Ehrlich et al., (1983) have replied to these criticisms. They rightly state that
"examination of homologous skeletal features is only one of many approaches"
to biomorphological studies. Their work on the foraminiferan *Globorotalia tru-
catulinoides* indicates that there is a "consistent angular relationship between the
orientation of the second harmonic and the spiral side keel." With this noted, they
indicate that it could be suggested that "the very fact that the radial Fourier series is locking onto homologous points is a good reason not to use the Fourier series; that is, given the relationship, why go through the complex calculations?"
The justification they give is that "the possibility always exists, however, that the
additional data needed to fully reconstruct the profile between homologous points
may contain biologically interesting information."

Read and Lestrel (1986) provide an example in which measurements taken between homologous points fail to describe significant differences in morphology
between structures because the measurements omit the boundary connecting the
landmarks. They express agreement with the observation made by Bookstein et
al., (1985:3) that "although no morphometric method can be wrong in all contexts, neither is any method universally applicable."

4.5 Concluding remarks

This chapter has been concerned with a survey of methods available for the description of biological forms. A number of issues have been raised concerning
their use. These issues impact every aspect of shape analysis: the definition of
landmarks, the identification of homologies, the strategies available when few homologies can be identified, the ways in which data might be compared between
forms, and the consequences of data collection and transformation strategies on
the perception of shape differences.

---

[6]Editor's note: Chapters 14 and 16 present a procedure that preserves homology withtin a boundary-outline
context.

Clearly, there is no one approach or collection of approaches to shape analysis that is ideal in all circumstances. Fourier analysis, the subject of this volume, is just one of many techniques for shape analysis available to the modern investigator. In most biological circumstances, comparisons of individual harmonics between shapes will rarely provide readily interpretable information about basic biological processes such as ontogeny or phylogeny. This is because Fourier components are not necessarily homologous between OTUs in terms of the features they describe. In combination, however, the shape summaries that can be produced through Fourier analysis may be useful in the multivariate assessment of overall morphological differences arising because of underlying biological processes.

The morphometrician must choose strategies for shape description and comparison that are appropriate to the question at hand. Where several approaches appear equally applicable it seems sensible to apply a few in order to compare outcomes. Such a broad approach not only serves to provide a comparison of techniques, but also ensures that interpretations of results are widely based and so are less likely to be influenced by the foibles of one single technique. The morphometrician must be aware of the theoretical and practical limitations of each method and must apply this knowledge to each situation.

4.6 Acknowledgments

I should like to thank Pete Lestrel for organizing this volume and for inviting me to contribute to it. Ian Dryden, Lisa Wiffen, and John Kent of the Department of Statistics, University of Leeds have provided me with invaluable advice and help in understanding some of the methods outlined in this chapter. Martin Thompson of the Department of Anatomy and Human Biology, the University of Western Australia, kindly prepared the illustrations for this chapter.

References

Albrecht, G. H. (1978). The craniofacial morphology of the Sulawesi macaques. In *Contributions to primatology 13* ed. F. Szalay. Basel: S. Karger.

Alexander, R. M. (1991). How dinosaurs ran. *Scientific American* **264,** 62–8.

Blum, H. (1967). A transformation for extracting new descriptors of shape. In *Models for the Perception of Speech and Visual Form* ed. W. Whaten-Dunn. Cambridge: MIT press.

Blum, H. (1973). Biological shape and visual science. *J. Theor. Biol.* **38,** 205.

Bookstein, F. L. (1978). *The Measurement of Biological Shape and Shape Change. Lecture notes in Biomathematics* Vol. 24. Berlin: Springer.

Bookstein, F. L. (1979). The line skeleton. *Comp. Graph. Image Proc.* **11,** 123–37.

Bookstein, F. L. (1989). Principal warps: Thin plate splines and the decomposition of deformations. *IEEE Trans. Pat. Anal. Mach. Intel.* **11,** 567–85.

Bookstein, F. L. (1991). *Morphometric Tools for Landmark Data: Geometry and Biology*. Cambridge: Cambridge University Press.

Bookstein, F. L., Chernoff, R., Elder, J., Humphries, J., Smith, G. & Strauss, R. (1982). A comment on the uses of Fourier analysis in systematics. *Systematic Zool.* **31**, 85–92.

Bookstein, F. L., Chernoff, R., Elder, J., Humphries, J., Smith, G. & Strauss, R. (1985). *Morphometrics in Evolutionary Biology*. Special publication 15. Philadelphia: The Academy of Natural Science.

Brothwell, D. & Trevor, J. (1964). Craniometry. *Chambers Encyclopaedia*. Vol. I. London: George Newnes Ltd.

Cheverud, J. M. & Richtsmeier, J. T. (1986). Finite element scaling applied to sexual dimorphism in rhesus macaque (*Macaca mulatta*) facial growth. *Syst. Zool.* **35**, 381–99.

Creel, N. & Preuschoft, H. (1971). Hominoid taxonomy: A canonical analysis of cranial dimensions. *Proc. 3rd. Int. Congr. Primat. Zurich 1970.* **1**, 36–43.

Corruccini, R. S. (1988). Morphometric replicability using chords and Cartesian co-ordinates of the same landmarks. *J. Zool., Lond.* **215**, 389–94.

Darwin, C. (1859). *On the Origin of Species by Means of Natural Selection, or The Preservation of Favoured Races in the Struggle for Life*. London: John Murray.

De Coster, L. (1939). The network method of orthodontic diagnosis. *Angle Orthod.* **9**, 3–10.

Dunn, G. A. & Brown, A. F. (1986). Alignment of fibroblasts on grooved surfaces described by a simple geometric transformation. *J. Cell. Sci.* **83**, 313.

Eldredge, N. & Cracraft, J. (1980). *Phylogenetic Patterns and the Evolutionary Process*. New York: Columbia University Press.

Ehrlich, R., Baxter Pharr, R. & Healy-Williams, N. (1983). Comments on the validity of Fourier descriptors in systematics: A reply to Bookstein et al. *Syst. Zool.* **32**, 202–4.

Ferson, S., Rohlf, F. J. & Koehn, R. K. (1985). Measuring shape variation of two-dimensional outlines. *Syst. Zool.* **34**, 59.

Goodall, C. R. (1991). Procrustes methods in the statistical analysis of shape. *J. Roy. Stat. Soc. B.* **53**, 285–339.

Goodall, C. R. & Bose, A. (1987). Models and Procrustes methods for the analysis of shape differences. *Proc. 19th Symp. of the Interface between Computer Science and Statistics*.

Green, P. J. & Sibson, R. (1977). Computing Dirichlet tessellations in the plane. *Comp. J.* **21**, 168–73.

Hu, M. K. (1962). Visual pattern recognition by moment invariants. *IRE Trans. on Information Theory.* **8**, 179–87.

Johnson, D. R., O'Higgins, P., McAndrew, T. J., Adams, L. M. & Flinn, R. M. (1985). Measurement of biological shape: A general method applied to mouse vertebrae. *J. Embryol. Exp. Morph.* **90**, 363–77.

Jungers, W. L. (1985) *Size and Scaling in Primate Biology*. New York: Plenum Press.

Kaesler, R. L. & Waters, J. A. (1972). Fourier analysis of the Ostracod margin. *Geol. Soc. Am. Bull.* **83**, 1169.

Katz, M. J. & George, E. D. (1985). Fractals and the analysis of growth paths. *Bull. Math. Biol.* **47**, 273–86.

Kendall, D. G. (1984). Shape manifolds, Procrustean metrics and complex projective spaces. *Bull. Lond. Math. Soc.* **16**, 81–121.

Kuhl, F. P. & Giardina, C. R. (1982). Elliptic Fourier features of a closed contour. *Comp. Graph. Image Proc.* **18**, 236–58.

Lele, S. & Richtsmeier, J. T. (1991). Euclidean Distance Matrix Analysis: A co-ordinate free approach for comparing biological shapes using landmark data. *Am. J. Phys. Anthrop.* **86,** 415–28.

Lestrel, P. E. (1982). A Fourier analytic procedure to describe complex morphological shapes. In *Factors and Mechanisms Influencing Bone Growth* eds. A. D. Dixon & B. G. Sarnat. New York: Alan R. Liss, Inc.

Lestrel, P. E. (1989). Method for analysing complex two-dimensional forms: Elliptical Fourier functions. *Am. J. Hum. Biol.* **1,** 149–64.

Lew, W. D. & Lewis, J. L. (1977). A nonhomogenous anthropometric scaling method based on finite element principles. *J. Biomech.* **13,** 815–24.

Lu, K. H. (1965). Harmonic analysis of the Human face. *Biometrics.* **21,** 491.

Lohmann, G. P. & Schweizer, P. N. (1990). On eigenshape analysis. In *Proceedings of the Michigan Morphometrics Workshop.* eds. F. J. Rohlf & F. L. Bookstein. Special Publication Number 2. Ann Arbor, Michigan: The University of Michigan Museum of Zoology.

Mandelbrot, B. B. (1983). *The Fractal Geometry of Nature.* New York: W.H. Freeman.

Martin, R. (1928). *Lehrbuch der Anthropologie.* (2nd ed.). Vols. 1–3. Jena: Gustav Fischer.

Moellering, H. & Rayner, J. N. (1981). The harmonic analysis of spatial shapes using dual axis Fourier shape analysis (DAFSA). *Geographical Anal.* **13,** 64–77.

Moorees, C. F. A. & Lebret, L. (1962). The mesh diagram and cephalometrics. *Angle Orthod.* **32,** 214–24.

Moss, M. L., Vilman, H., Moss-Salentijn, L., Sen, K., Pucciarelli, H. M. & Skalak, R. (1987). Studies on orthocephalization: Growth behavior of the rat skull in the period 13–19 days as described by the finite element method. *Am. J. Phys. Anthrop.* **72,** 323–42.

Nelson, G. (1994). Homology and systematics. In *Homology: The Hierarchical Basis of Comparative Biology.* ed. B. K. Hall. San Diego: Academic Press.

O'Higgins, P. (1989). *A morphometric study of cranial shape in the hominoidea.* Ph.D. thesis, University of Leeds.

O'Higgins, P., Johnson, D. R. & McAndrew, T. J. (1986). The clonal model of vertebral column development: A reinvestigation of vertebral shape using Fourier analysis. *J. Embryol. Exp. Morph.* **96,** 171–82.

O'Higgins, P. & Williams, N. W. (1987). An investigation into the use of Fourier coefficients in characterizing cranial shape in primates. *J. Zool. Lond.* **211,** 409–30.

O'Higgins, P. & Dryden, I. (1992). Studies of craniofacial development and evolution. *Archeol Oceania 27/ Persp. Hum. Biol.* **2,** 105–12

O'Higgins, P. & Dryden, I. (1993). Sexual dimorphism in hominoids: Further studies of cranial "shape change" in *Pan, Gorilla,* and *Pongo. J. Hum. Evol.* **24,** 183–205.

Owen, R. (1847). Report on the archetype and homologies of the vertebrate skeleton. *Rep. Br. Ass. Advmt. Sci.* (1846). **16,** 169–340.

Parnell, J. N. & Lestrel, P. E. (1977). A computer program for fitting irregular two-dimensional forms. *Comp. Prog. Biomed.* **7,** 145–61.

Read, D. W. & Lestrel, P. E. (1986). Comment on uses of homologous-point measures in systematics: A reply to Bookstein et al. *Syst. Zool.* **35,** 241–53.

Reyment, R. A. (1991). *Multidimensional Palaeobiology.* Oxford: Pergamon Press.

Richtsmeier, J. T. (1986). Finite element scaling analysis of human craniofacial growth. *J. Craniofac. Genet. Dev. Biol.* **6,** 289–323.

Richtsmeier, J. T. (1989). Applications of finite element scaling in primatology. *Folia Primatol.* **53,** 50–64

Rohlf, F. J. (1990a). Rotational fit (Procrustes) methods. In *Proceedings of the Michigan Morphometrics Workshop.* eds. F. J. Rohlf & F. L. Bookstein. Special Publication Number 2. Ann Arbor, Michigan: The University of Michigan Museum of Zoology.

Rohlf, F. J. (1990b). An overview of image processing and analysis techniques for morphometrics. In *Proceedings of the Michigan Morphometrics Workshop.* eds. F. J. Rohlf & F. L. Bookstein. Special Publication Number 2. Ann Arbor, Michigan: The University of Michigan Museum of Zoology.

Rohlf, F. J. (1990c). Fitting curves to outlines. In *Proceedings of the Michigan Morphometrics Workshop.* eds. F. J. Rohlf & F. L. Bookstein. Special Publication Number 2. Ann Arbor, Michigan: The University of Michigan Museum of Zoology.

Rohlf, F. J. & Archie, J. W. (1984). A comparison of Fourier methods for the description of wing shapes in mosquitos. *Syst. Zool.* **33,** 302–17.

Rohlf, F. J. & Bookstein, F. L. (1990). *Proceedings of the Michigan Morphometrics Workshop.* Special Publication Number 2. Ann Arbor, Michigan: The University of Michigan Museum of Zoology.

Schmidt-Nielsen, K. (1984). *Scaling: Why is Animal Size so Important?* Cambridge: Cambridge University Press.

Sneath, P. H. A. (1967). Trend surface analysis of transformation grids. *J. Zool. Lond.* **151,** 65–122.

Sneath, P. H. A. & Sokal, R. R. (1973). *Numerical Taxonomy.* San Francisco: W.H. Freeman and Co.

Simpson, G.G. (1961). *Principles of Animal Taxonomy.* New York: Columbia Univ. Press.

Straney, D. O. (1990). Median axis methods in morphometrics. In *Proceedings of the Michigan Morphometrics Workshop.* eds. F. J. Rohlf & F. L. Bookstein. Special Publication Number 2. Ann Arbor, Michigan: The University of Michigan Museum of Zoology.

Thompson, D. W. (1917). *On Growth and Form.* Cambridge: Cambridge University Press.

Van Valen, L. (1982). Homology and causes. *J. Morphol.* **173,** 305–12.

Wagner, G. P. (1994). Homology and the mechanisms of development. In *Homology: The Hierarchical Basis of Comparative Biology.* ed. B. K. Hall San Diego: Academic Press.

Webber, R. L. & Blum, H. (1979). Angular invariants in developing human mandibles. *Science* **206,** 689–91.

Wood, B. A. (1976). The nature and basis of sexual dimorphism in the primate skeleton. *J. Zool. Lond.* **180,** 15–34.

Yasui, K. (1986). Method for analysing outlines with an application to recent Japanese crania. *Am. J. Phys. Anthrop.* **71,** 39–45.

Young, I. T., Walker, J. E. & Bowie, J. E. (1974). An analysis technique for biological shape. *Medinfo 74,* M.I.T. 843–9.

Zahn, C. T. & Roskies, R. Z. (1972). Fourier descriptors for plane closed curves. *IEEE trans. on Computers.* C-**21,** 269–81.

Zienkiewicz, O. C. (1971). *The Finite Element Method in Engineering Science.* London: McGraw-Hill.

5

Phase Angles, Harmonic Distance, and the Analysis of Form

ROGER L. KAESLER
The University of Kansas

5.1 Introduction

The irascible Robin Whatley is my close friend, and an avid English gardener of some appreciable skill. He once commented disparagingly on my tendency to mercilessly prune shrubs into small spheres, irrespective of their natural shapes or the season in which they are likely to bloom. "The trouble with you, Kaesler," he said, "is that you are not a horticulturist. You're a . . . geometrician." Whatley was absolutely right, and the reason that Fourier descriptors of shape appeal to me, and to so many others who measure shapes, is precisely because we are geometricians and are inclined to favor a quantitative method that has a strong geometrical component. In spite of ourselves, we remain somewhat baffled by matrix algebra, while accepting that it is probably a good thing. But to the geometrically inclined, principal component analysis is not so much about matrix algebra as about football-shaped clouds of points suspended in hyperspace, and discriminant function analysis is about, somehow, algorithmically turning those footballs into basketballs. Moreover, those of us who teach have discovered that most of our students are more comfortable, at least initially, with geometrical models of multivariate morphometrics than with purely algebraic ones. Fourier analysis resolves shapes into a set of geometrically comprehensible, additive, sine and cosine curves that can be readily visualized. It is not surprising, therefore, that many have clung to its use in spite of the criticism that has been leveled at the application of Fourier analysis to morphometrics.

I have three purposes in writing this chapter. First, I want to examine some of the difficulties one faces in trying to use Fourier descriptors to study morphology and to address some of the drawbacks of the way Fourier descriptors of shape have been applied. These drawbacks form the rationale behind much of the recent criticism of the method. Second, I want to briefly mention some of the advantages that the method offers; the characteristics that keep users coming back to Fourier analysis in spite of its shortcomings. Third, I want to demonstrate a way in which the use of Fourier descriptors can be made more effective by employing harmonic distance analysis. This technique was introduced a number of years ago in an obscure publication (Kaesler and Maddocks, 1979; 1987). Harmonic distance analysis evades some of the more serious problems with Fourier descriptors as they are typically used, especially when the intent is to consider distance measures and phenetics.

5.2 Criticisms of the use of Fourier descriptors

5.2.1 Fourier analysis: a life of its own and getting started

Two unfortunate aspects of Fourier analysis deserve brief mention before we focus attention on the more substantive criticisms of the past decade. These aspects hamper application to biology and other fields, and both have to do with the difficulty of getting started. First, Fourier analysis is no longer discussed in the mathematical or engineering literature as simply a means of resolving complex shapes into easily comprehended, trigonometric components. For at least three decades it has had a life of its own in which it is difficult for the non-mathematician to participate (see e.g., Lanczos, 1966). To avoid frustration, the reader is well advised to turn to the older engineering literature. This literature was written when engineers still used slide rules, tables of logarithms of trigonometric functions, and were still concerned with resolving complex shapes into their trigonometric components. Useful references include chapters in old books by Doherty and Ernest (1936), Sohon (1944), Salvadori and Miller (1948), and especially Gaskill (1958), Panofsky and Brier (1965), and the *Handbook of Chemistry and Physics* edited by Hodgman (1958). These old books are almost never checked out of the library when you want to use them!

A more serious obstacle is the fact that authors of much of the geological literature in which Fourier analysis is applied, myself included (Kaesler and Waters, 1972), have not paid sufficient attention to detail. Without dealing with these shortcomings in particular, suffice it to say that notation varies widely among authors. Perhaps as a result, published equations and notation are sometimes not consistent even within the same publication and in some instances are simply incorrect. The a, b, and c terms are commonly turned around, equations for the phase angles are in disarray, minus signs are missing (in my own instance the result of pressed-on symbols having dropped off in transit), and subscripts are likely to be either missing or poorly defined. It is apparent that some authors, having completed their analyses and written their papers, did not work their way through an exercise using the equations they were about to submit for publication. Thus, although the conclusions these authors have reached seem to be perfectly sound, one must exercise caution in attempting to apply the equations from the published, geological literature.

5.2.2 Fourier analysis as a transformation

In his original paper on eigenshape analysis, Lohmann (1983) compared the use of eigenshapes with Fourier analysis, of course emphasizing the advantages that his new method brought to the study of shapes. There are, of course, many similarities between the two methods, especially in their origins. Nevertheless, for purpose of discussion, it is useful to discriminate between Fourier analysis *sensu stricto*, in which an outline is broken down into successively complex, regular, geometric shapes, and eigenshape analysis, where successive shapes explain decreasing amounts of the variance and there is no intervening reliance on the geometrical configurations.

Full and Ehrlich (1986) made a complete and comparatively vitriolic response that stressed the shortcomings of eigenshape analysis without addressing in as much detail the strengths of Fourier analysis (see also Ehrlich and Full, 1986). They considered the differences between biological homology on the one hand, and mathematical or geometrical homology on the other. Homology presents problems for eigenshape analysis, Fourier analysis, and, indeed, any method that does not focus on homologous landmarks. In a further discussion, Rohlf (1986) considered the relationships among eigenshape analysis, Fourier analysis, and the analysis of coordinates. His abstract summed things up rather well: "When *all* eigenvectors and *all* harmonics are retained, both approaches represent orthogonal rotations of the same points. Thus, distances between pairs of shapes (and any multivariate analysis based on distances) must be the same for both analyses" (1986:845).

Thus, if one plans a principal component analysis, as is typically done in conjunction with Fourier descriptors, the Fourier analysis seems to be an unnecessary computational step unless, as Rohlf pointed out (1986:851), "the harmonics are of interest." Rohlf was skeptical of the value of trying to interpret individual harmonics, but this may stem from his being less exclusively geometrically inclined than some of the rest of us. Clearly, however, his evaluation is sound. The onus is on those of us who would like to show the value of interpreting individual harmonics.

5.2.3 The data: Heterocypris incongruens

Figure 5.1 shows the outline of the right valve of an adult female of the parthenogenetic, freshwater, ostracode species *Heterocypris incongruens*. Ostracodes have been used as an example in all aspects of this study because they hold special in-

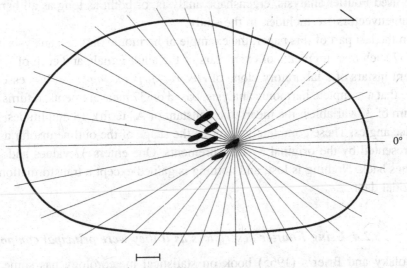

Fig. 5.1 Outline of the rigid, calcareous carapace of an adult female of the parthenogenetic, freshwater, ostracode species *Heterocypris incongruens*. Thirty-six rays were drawn from the center of the anterior, mandibular muscle scar. The outline was digitized where each ray intersected it, and the lengths of the so-called radii were computed. These 36 measurements on this and other specimens form the basis of all the harmonic analysis done herein.

terest for me and because they have a rigid, calcite carapace that lends itself to Fourier analysis; a closed margin; determinate growth through a discrete number of molt stages (instars); a terminal adult stage beyond which there is no additional growth; a long geological history; and, as a group, a very broad range of environmental tolerance.

Thirty-six equally-spaced radii were drawn from the center of the anterior, mandibular muscle scar. The Cartesian coordinates of the intersections of these radii with the outline were digitized, and the length of each computed. One often sees the centroid of the closed shape used as the origin for measurements. Here, however, the mandibular muscle scar, rather than the centroid, was chosen as the origin because a homologous muscle scar occurs on the valves of nearly all ostracodes. Moreover, use of the centroid causes the first harmonic to have an amplitude of zero. Among ostracodes, however, the first harmonic has a special biological appeal because of the progressive elongation of the valves as a result of the posterior, ontogenetic addition of thoracic appendages, and ultimately the development of large and complex, posteriorly located, reproductive organs.

If one counts the radius at both 0° and 360°, one has 37 measurements. Using these 37 measurements, a comparison of the shape of this ostracode and another one, using a measure of taxonomic distance, would give the same results whether one used Fourier analysis, eigenshape analysis, or both, as long as all harmonics or eigenvectors are included in the study.

In the last part of this paper, the example of harmonic distance analysis is based on 37 such measurements of exemplars of the adult female and each of eight different instars of a laboratory clone of *Heterocypris incongruens*. It is essential to note that a complete harmonic analysis, based on 37 measurements, returns a maximum of 37 variables: the mean (the constant or A_0 term), 18 amplitudes, and 18 phase angles. These completely describe the shape of the outline insofar as it was represented by the original 37 measurements. One enters 37 values and gets 37 values back. Nothing is lost, and nothing is gained except a transformation of the original data.

5.2.4 Using Fourier descriptors as if they were principal components

Panofsky and Brier's (1965) book on statistical meteorology has some useful things to say about Fourier analysis. They concluded their brief discussion with an illustrative example in which they computed Fourier analysis of the average hourly temperatures in New York in January, 1951. The first harmonic accounted for 86 percent of the variance. The second accounted for 9 percent, and they concluded (p. 134) that "no additional harmonics need be computed."

This approach to Fourier analysis is equivalent to using the harmonics as if they were principal components except that, unlike principal component analysis, the harmonics do not, in general, explain successively less of the variance. Readers who can recall the deplorably primitive state of computational science in 1965 will understand why Panofsky and Brier (1965), and others of their era, were eager to find ways to avoid further computation. However, while avoiding unnecessary computation, they seem to have set the stage for discarding information. Thus, even Rohlf (1986:848) suggested the use of either Fourier analysis or principal component analysis "to reduce the dimensionality of data and the cost of computer processing," although Ehrlich et al., (1983:204) correctly pointed out that such computations "take essentially no time at all on a modern computer." Thus, we see two opposing forces at work, one wanting to describe shapes more fully and the other wanting to reduce the number of variables in a problem to make it easier to discuss.

Whereas the use of principal component analysis can be said to reduce the dimensionality of morphological space, discarding harmonics as if they were analogous to principal components should be, more properly, regarded as simply throwing away information. Figure 5.2 shows three shapes computed from the measurements in Figure 5.1. Figure 5.2a has been regenerated from the mean and the first three harmonics; Figure 5.2b is from the mean and the first five har-

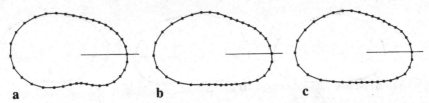

Fig. 5.2 Outlines reconstructed from data of Figure 5.1. Computations involving some or all harmonics. Mean and first three harmonics (Figure 5.2a). Mean and first five harmonics(Figure 5.2b). Mean and all 18 harmonics (Figure 5.2c).

monics; Figure 5.2c uses the mean and all 18 harmonics and thus reproduces the shape in Figure 5.1 insofar as it is captured in the 37 measurements. Both Figure 5.2a and Figure 5.2b look ostracodal or, more precisely, leguminous; and both resemble Figure 5.2c This may be interesting, but it is not sufficiently informative for any further computational analysis. If asked to sketch the outline of a generalized, freshwater ostracode, any specialist on the Ostracoda would draw something like Figure 5.2a or 5.2b. In fact, the outline of the ostracode becomes computationally and biologically interesting only if all the information we have about it is used. Paleontologists and, I suspect, systematic biologists in general, need to avoid throwing away information and basing conclusions on less than we know.

On the other hand, it is conceivable that once the analysis has been completed and all computations finished, some harmonics may have very low power and thus contribute little to the shape. They might safely be discarded without appreciable loss of information, but given the computational power available to most of us, one has to wonder if much is gained by what seems to be a philosophically poor maneuver.

5.2.5 *Perils of ignoring phase angles*

Bookstein et al., (1982) have commented in depth on the use of Fourier analysis in morphometrics. Some of their criticism was aimed at the loss of information that results from the exclusion of phase angles: "half the information of the data base has been intentionally discarded in the course of measurement" (p. 91). In their reply to Bookstein, Ehrlich et al. (1983) described their study of the planktonic foraminifera species *Globorotalia truncatulinoides* and *G. hirsuta*. They reported a consistent orientation of harmonics with respect to the keel of the organism and with respect to each other, a relationship they termed "phase-locking" (p. 205), which they described as "invariant" (p. 204). Although invariant for these species and perhaps for many others, the phase angles, nevertheless, have the potential of conveying a great deal of information about the shape. In the example

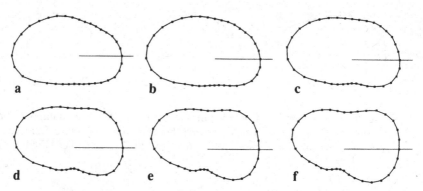

Fig. 5.3 Outlines reconstructed from data of Figure 5.1 using the mean and all 18 harmonics, but with all phase angles altered by varying amounts. No modification of phase angles; identical to Figure 5.2c (Figure 5.3a). Phase angles increased by 10 percent (Figure 5.3b). Phase angles increased by 20 percent (Figure 5.3c). Phase angles increased by 30 percent (Figure 5.3d). Phase angles increased by 40 percent (Figure 5.3e). Phase angles increased by 50 percent (Figure 5.3f).

used here, 18 of the 37 values computed are phase angles, and one ought not to ignore them offhand.

Figure 5.3 shows six shapes computed from the data of Figure 5.1. Figure 5.3a was computed from the mean and all 18 harmonics. It uses both harmonic amplitudes and phase angles and is identical to Figure 5.2c. The remaining five figures were computed by consistently altering the phase angles. In Figure 5.3b the phase angles were increased by 10 percent; in Figure 5.3c the phase angles were increased by 20 percent, and so on up to 50 percent in Figure 5.3f. The outline in Figure 5.3b is vaguely ostracodal, as much so as that of Figure 5.2a. As systematic errors in the phase angles increase, of course, the outlines become increasingly unlike the original (Figure 5.3a) and progressively less ostracodal in appearance. To ignore phase angles is yet another instance of the practice of basing conclusions on less information than one has available. If one discards all harmonics with low amplitudes (often the higher-order terms in the series) and then ignores the phase angles of the few remaining harmonics, the Fourier analysis is likely to result in one's overlooking important and sometimes not so subtle information that is important for characterizing the morphology.

5.2.6 Fourier descriptors, rigid structures, biology, and geology

As with most morphometrical techniques that are used to analyze shapes, Fourier analysis is limited in its application primarily to the study of rigid, skeletal morphology. Thus, in addition to its use in sedimentology (e.g., Ehrlich and Weinberg, 1970; Byerly et al., 1975; Full and Ehrlich, 1982), it has been used widely in osteology (e.g., Lu, 1965; Lestrel, 1980; Lestrel et al., 1993), and invertebrate paleontology (e.g., Kaesler and Waters, 1972; Anstey and Delmet, 1973; Waters, 1977; Prezbindowski and Anstey, 1978; Kaesler and Maddocks, 1979;1987; Healy-Williams, 1983; Healy-Williams et al., 1985). Use in the morphology of unmineralized body parts has been limited primarily to such rigid features as the elytra of beetles (Smirnov, 1927) or to study the distortion of the shapes of such comparatively rigid parts as miospores that may occur as a result of treatment with chemicals in the laboratory (Christopher and Waters, 1974).

Rigid, mineralized, skeletal material from the fossil record is subject to change of shape by various taphonomic agents including necrolysis; abrasion, breakage, and dissolution during biostratinomy; and diagenesis. Although Fourier analysis of shape has promise in the study of these taphonomic effects, I am unaware of specific applications.

5.3 Some advantages of the use of Fourier descriptors

Three aspects of Fourier analysis keep morphometricians interested and coming back. The first of these reasons is simply inertia. Some investigators have programs for Fourier analysis up and running on their computers. The programs seem to be working; the results seem to be interpretable; and, as a consequence, they see no reason to change to another method. Such inertia is all right so long as one gets useful answers to real scientific questions and, especially, as long as one does not invest too much credibility in its defense.

Second, Fourier analysis is easy to understand and simple to program. My own program was written to run on Microsoft Excel 4.0 for the Macintosh computer, a most common and convenient spreadsheet that is also available for PC computers.

The most important reason for the lingering popularity of Fourier analysis, however, goes back to my initial point about geometry. There is something appealing about the geometry of additive, independent, and orthogonal sine and cosine curves. Whether Fourier analysis contributes to the interpretation of biological form or, as Bookstein et al. (1982:92) wrote, makes "systematic differences in form . . . *uninterpretable*," seems to have little bearing on the continued use of the method. Enough morphometricians have enough confidence in the geometry of Fourier analysis to keep the method alive.

Fourier analysis decomposes shapes into a series of additive, orthogonal harmonics. A convenient way to view this is "in reverse", through the successive addition of harmonics. Figure 5.4 shows the mean shape of the data from Figure 5.1 as a circle, and includes five modifications of that circle by the addition of each of the first five harmonics to it. It also shows the effect of successive addition of the harmonics gradually to approximate the original shape. To save space, only

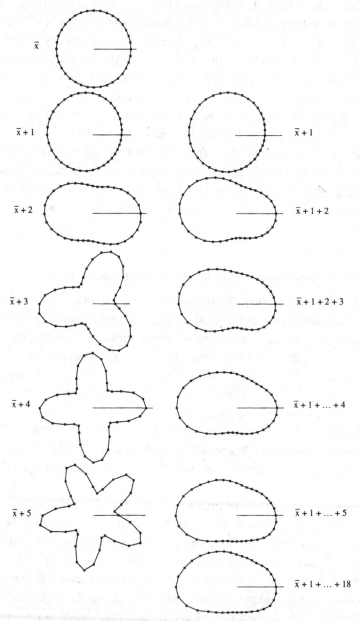

Fig. 5.4 Sequential reconstruction of the outline in Figure 5.1 by successive addition of harmonics to the mean. The circle, top plot on the left, $\bar{x}$, is the A_0 term, or constant, to which the succeeding terms are being separately added to generate the plots shown below. (Editors' note: Technically, the plots on the left are only correct if negative values for the *cos nθ* and *sin nθ* terms are excluded from the plot (and this only applies for even *n*). Otherwise, there would be *2n*, not *n* leaves, as shown here [Selby, 1972:381]).

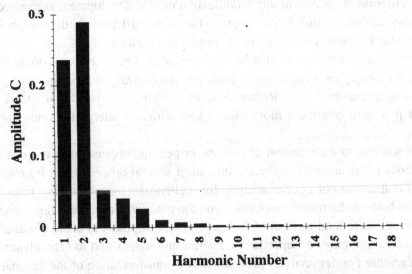

Fig. 5.5 Amplitude (C) versus harmonic number of the 18 harmonics computed from the data in Figure 5.1. The variance explained by each harmonic is proportional to the power, C^2. Note that, unlike principal components, the successive harmonics do not explain decreasing amounts of the variance.

the first five harmonics are shown in the gradual buildup in Figure 5.4, although the sum of the mean and all 18 harmonics is shown for the sake of comparison.

Figure 5.5 shows the amplitude spectrum computed from the data of Figure 5.1. The variance explained by each harmonic is proportional to the square of the amplitude.[1] Note that, unlike principal components, the successive harmonics *do not* explain less and less of the variance. The amplitude spectrum is a useful depiction of the importance of each harmonic in determining the shape of the outline, but it should be used with caution. The danger in relying on it too heavily is that it encourages the discarding of harmonics (those with small magnitude), with a consequent loss of information. Moreover, the amplitude spectrum, by itself, conveys no information about phase angles. Here, for example, depending on one's inclination, one might use only harmonics one and two, one to five, or one to eight.

5.4 Harmonic distance analysis

5.4.1 Avoiding drawbacks; profiting from strengths

How can we avoid the drawbacks of Fourier analysis and profit from its strengths, thus making the method more useful in the description and analysis of shapes?

[1] Editor's note: The variance or power spectrum can also be plotted as power (C^2) versus harmonic number (see Fig. 4.6).

The criterion of success of any modification or new development associated with Fourier analysis is that it must work. That is, it must provide the morphometrician with a means of interpreting the results of Fourier analysis to give useful answers to important biological questions. One must especially avoid doing Fourier analysis simply as a precursor to principal component analysis; in effect, transforming a transformation. Rather, Fourier analysis must be able to stand on its own if it is to become a more useful, geometrically interpreted, morphometric tool.

In addition to the criterion of success, or perhaps because of it, use of Fourier methods in the analysis of shapes must meet several other criteria. It now seems clear that any use of Fourier analysis for multivariate morphometrics must incorporate both the harmonic amplitudes and the phase angles. Moreover, it must use information from all the harmonics that can be computed, to avoid discarding information that may be important in determining shapes and to take advantage of the fact that Fourier analysis provides only a transformation of the original data. This is especially important now that sufficient computer power is readily available, so that one need no longer seek shortcuts to avoid computations. Finally, results of Fourier analysis will be most useful if incorporated into a coefficient of similarity or distance that can be used in conjunction with or in contrast to other phenetic methods.

5.4.2 The harmonic distance coefficient

A great deal of taxonomic information resides in the outlines of ostracodes. Figure 5.6 shows the outlines of three distantly related ostracodes. Their shapes are characteristic of their genera and, to some extent, of the higher taxa to which they belong.

Kaesler and Maddocks (1979; 1987) introduced the harmonic distance coeffi-

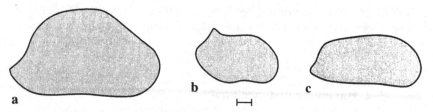

Fig. 5.6 A great deal of taxonomic information resides in the outline of ostracodes. The ostracodes outlined here belong to two superfamilies and three families of the class Podocopida. *Bairdia*, family Bairdiidae, superfamily Bairdiacea (Figure 5.6a). *Loxocorniculum*, family Loxoconchidae, superfamily Cytheracea (Figure 5.6b). *Orionina*, family Hemicytheridae, superfamily Cytheracea (Figure 5.6c). (Modified from Kaesler and Waters, 1972.)

cient in order to assess similarities and differences among ostracodes on the basis of their outlines, and to meet the criteria listed above. They applied the coefficient to a study of 35 species of ostracodes in three genera of the family Macrocyprididae. The method met the criterion of success: using outlines alone, it successfully grouped the ostracodes into six groups of species that Maddocks had recognized independently on the basis of other morphological characters including the morphology of the appendages. The distance coefficient employs amplitudes and phase angles from all computed harmonics. Being a distance coefficient, it lends itself to further computation with such traditional phenetic methods as cluster analysis and nonmetric multidimensional scaling.

The equation for the harmonic distance coefficient is analogous to the average taxonomic distance coefficient (Sokal, 1961) and is shown here as:

$$DH_{jk} = \sqrt{\left(\frac{1}{n}\right) \sum_{i=1}^{n} [(c_{ij} \cos \phi_{ij} - c_{ik} \cos \phi_{ik})^2 + (c_{ij} \sin \phi_{ij} - c_{ik} \sin \phi_{ik})^2]} \quad (5.1)$$

where DH_{jk} is the harmonic distance between species j and k; i is the harmonic number, which varies from 1 to n; c_{ij} is the amplitude of the ith harmonic of species j; and ϕ_{ij} is the phase angle of the ith harmonic of species j; c_{ik} is the amplitude of the ith harmonic of species k; and ϕ_{ik} is the phase angle of the ith harmonic of species k. Figure 5.7 depicts graphically the harmonic distance coefficient showing its similarity to Sokal's (1961) coefficient.

5.4.3 An example

As mentioned above, outlines of exemplar specimens of the adult female and eight instars were digitized according to the plan in Figure 5.1. The ontogenetic series is shown in Figure 5.8, in which specimens are centered on the muscle scars that served as the origin for orientation and measurement. In general, according to Przibram's rule (1931:32), arthropods ought to double their mass with each molt. Many species follow this rule, at least approximately, making grouping of the immature specimens into growth stages rather easy. This is especially so if the specimens have come from a single population or, as in the present instance, from a clone. A convenient linear estimate of this doubling in mass, as expressed in volume, is the factor 1.26, the cube root of two, by which linear dimensions are expected to increase with each stage in the ontogeny.

Here, I use harmonic distance analysis to assess the ontogenetic change of shape of *Heterocypris incongruens*. There are many ways of expressing such change, and some may give better results than are achieved here. This example is intended only to show that the harmonic distance coefficient is a suitable measure of on-

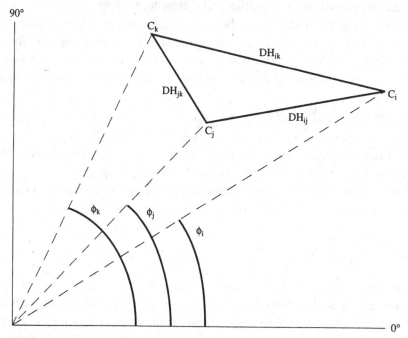

Fig. 5.7 Graphical representation of the taxonomic distance between three species *i*, *j*, and *k* based on amplitudes and phase angles of only one harmonic. The lengths of the dashed lines are proportional to the harmonic amplitudes, *C*, and the sides of the triangle, *DH*, are proportional to the harmonic distances between the three species. In an actual analysis such as the one here, based on 18 harmonics, the harmonic distance would comprise the average of the distance taken over all 18 harmonics. (Modified from Kaesler and Maddocks, 1987).

togenetic changes in the shape of the outline from one instar to the next, and to the adult stage.

In developing his clock model of heterochrony, Gould (1977:246–62) recognized the importance of determining the age of specimens and distinguished between ontogenetic changes of shape and of size (see also McKinney and McNamara, 1991:14). Later, Gould (1988:6) pointed out that "all paleontologists interested in this subject eventually become stymied if they cannot assess absolute ages." Although I cannot judge the absolute ages of these specimens, I know that, in general, ostracodes of this species spend about two days in each stage before they molt. I also know that the outline does not change between molts and that once a specimen reaches adulthood it does not molt again. This is probably as near as a paleontologist can come to determining absolute age, and it is possible in this instance only because the species is extant so that a great deal can be known about its biology, and because the specimens were recently alive, rather than being fossils, and are known to have lived in the same environment.

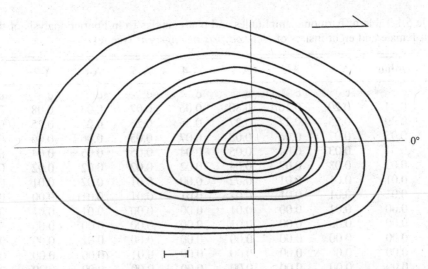

Fig. 5.8 Adult female and eight instars from a clone of the parthenogenetic, fresh-water, ostracode species *Heterocypris incongruens*. The arrow points toward the anterior end, and the scale bar is 0.1 mm. The smallest instar, A-8 (adult minus 8), is on the inside; the largest stage, the adult is on the outside. Note the pronounced difference in shape between the A-3 instar (adult minus three) and the earlier, A-4 instar, as well as the appreciable difference between the A-2 and A-3 instars. These changes in shape coincide with the addition of appendages and the ultimate development of the large and elaborate, posteriorly located reproductive system.

Table 5.1 contains amplitudes and phase angles from the Fourier analysis of the nine growth stages. These values were used to compute the matrix of harmonic distance coefficients in Table 5.2. For the example here, only the distances between successive growth stages are considered, the values that are shown in boldfaced type in Table 5.2, for example, distance of the adult from A-1 (adult minus one), A-1 from A-2, and so on.

Figure 5.9 shows the amplitude spectra of ten harmonics from Fourier analysis of the adult and three instars. The amplitude of the first harmonic is much higher for the adult stage and the A-1 instar than for the two earlier instars considered, A-4 and A-8. This results from the greater elongation of the later growth stages as the thoracic appendages and reproductive organs develop.

Figure 5.10 shows three projections of an ontogenetic trajectory based on: (1) size as measured by the mean radius, or A_0 term; (2) shape, as measured by the harmonic distance coefficient between successive growth stages (see Table 5.2); and (3) age, expressed as stage of growth. For a full discussion of ontogenetic trajectories incorporating these three aspects of heterochrony, see Alberch et al. (1979). Besides the monotonic increase of mean radius during ontogeny,

Table 5.1 Table of harmonic amplitudes and phase angles from Fourier analysis of the adult female and eight instars of *Heterocypris incongruens* (see text).

n	Adult	A-1	A-2	A-3	A-4	A-5	A-6	A-7	A-8
	c	c	c	c	c	c	c	c	c
1	0.24	0.20	0.16	0.30	0.06	0.07	0.03	0.08	0.05
2	0.29	0.25	0.27	0.27	0.25	0.26	0.25	0.25	0.26
3	0.05	0.07	0.06	0.08	0.07	0.05	0.06	0.06	0.05
4	0.04	0.03	0.04	0.05	0.04	0.05	0.05	0.04	0.05
5	0.03	0.03	0.02	0.04	0.02	0.02	0.02	0.02	0.02
6	0.01	0.01	0.01	0.02	0.00	0.01	0.02	0.01	0.02
7	0.01	0.01	0.01	0.01	0.01	0.01	0.01	0.00	0.01
8	0.00	0.01	0.00	0.01	0.00	0.00	0.01	0.01	0.01
9	0.00	0.00	0.00	0.01	0.00	0.00	0.00	0.00	0.00
10	0.00	0.00	0.00	0.00	0.00	0.00	0.01	0.00	0.01
11	0.00	0.00	0.00	0.00	0.00	0.01	0.00	0.00	0.00
12	0.00	0.00	0.00	0.00	0.00	0.00	0.00	0.00	0.00
13	0.00	0.00	0.00	0.00	0.00	0.00	0.00	0.00	0.01
14	0.00	0.00	0.00	0.00	0.00	0.00	0.00	0.00	0.00
15	0.00	0.00	0.00	0.00	0.00	0.00	0.00	0.00	0.00
16	0.00	0.00	0.00	0.00	0.00	0.00	0.00	0.00	0.00
17	0.00	0.00	0.00	0.00	0.00	0.00	0.00	0.00	0.00
18	0.00	0.00	0.00	0.00	0.00	0.00	0.00	0.00	0.00
	Phi	Phi	Phi	Phi	Phi	Phi	Phi	Phi	Phi
1	0.36	3.01	3.70	3.46	2.77	4.63	3.80	5.37	5.10
2	0.24	0.17	0.13	0.17	6.15	6.04	6.03	5.92	6.00
3	2.79	2.21	2.38	2.89	2.00	2.03	1.82	2.04	1.89
4	6.11	6.00	5.94	5.81	5.66	5.51	5.34	5.44	5.56
5	2.48	1.84	2.26	1.98	1.59	1.08	1.66	1.62	1.63
6	5.11	4.24	4.05	4.69	4.03	3.21	2.93	3.86	3.99
7	2.27	1.83	1.63	1.10	1.77	0.51	0.33	0.86	2.54
8	3.84	5.36	4.35	3.96	3.13	2.96	2.06	3.01	2.78
9	0.79	5.50	2.60	6.25	2.15	2.36	4.41	3.71	4.81
10	0.99	2.78	0.31	1.64	3.83	3.94	1.03	3.17	2.92
11	4.97	1.03	4.23	5.48	1.28	2.07	4.00	4.12	4.48
12	3.38	6.28	3.67	1.05	3.87	3.78	4.52	4.71	5.98
13	4.53	4.10	4.60	3.58	4.15	0.95	1.34	6.15	3.56
14	0.32	0.20	5.80	3.73	1.45	3.89	4.56	3.91	0.01
15	3.13	6.11	1.38	1.55	2.22	5.55	0.77	0.26	0.72
16	2.73	3.40	5.73	4.48	4.57	1.44	2.88	4.04	1.29
17	2.88	2.62	2.96	1.09	2.80	5.50	3.00	0.58	5.63
18	3.14	6.28	4.71	6.28	3.14	6.28	6.28	3.14	6.28

Table 5.2 Matrix of harmonic distance coefficients from analysis of the adult female and eight instars of *Heterocypris incongruens* (see text).

	Adult	A-1	A-2	A-3	A-4	A-5	A-6	A-7	A-8
Adult	0.000								
A-1	0.102	0.000							
A-2	0.095	0.032	0.000						
A-3	0.128	0.039	0.037	0.000					
A-4	0.073	0.039	0.037	0.067	0.000				
A-5	0.074	0.059	0.040	0.074	0.028	0.000			
A-6	0.073	0.051	0.041	0.073	0.016	0.016	0.000		
A-7	0.067	0.071	0.054	0.088	0.035	0.016	0.022	0.000	
A-8	0.068	0.062	0.047	0.081	0.027	0.012	0.015	0.011	0.000

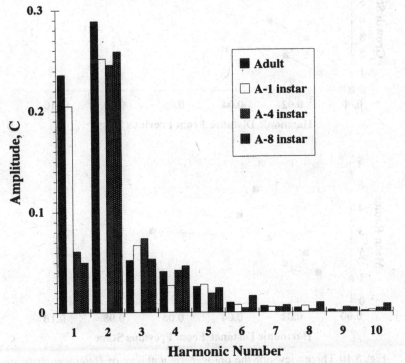

Fig. 5.9 Amplitude spectra of 10 harmonics from Fourier analysis of the adult and three instars, A-1, A-4, and A-8. Amplitudes are similar for all groups except the first. The high amplitude values for the first harmonic of the adult and A-1 instar are the result of elongation of the carapace as thoracic appendages and reproductive organs are added to the posterior portion of the animal.

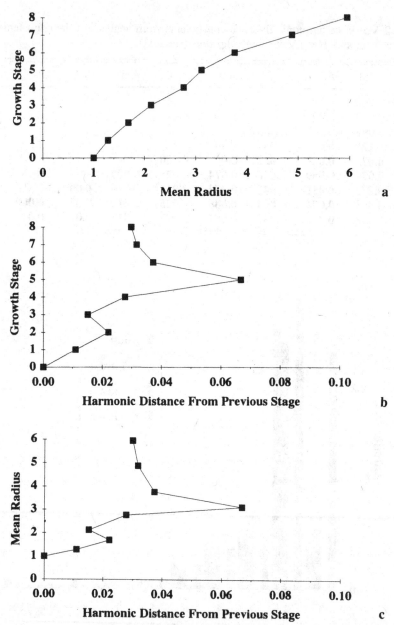

Fig. 5.10 Three views of the ontogenetic pathway of *Heterocypris incongruens* that provide information on the relationships between three important aspects of heterochrony: size (mean radius), shape (harmonic distance from previous stage), and age (growth stage). Growth stage versus mean radius in mm, where growth stage 8 indicates the adult, 7 the A-1 instar, and 0 the A-8 instar (Figure 5.10a). Growth stage vs. harmonic distance coefficient from previous growth stage (Figure 5.10b; see Table 5.2). Mean radius (relative to the size of the A-8 instar) versus harmonic distance coefficient from the previous stage (Figure 5.10c). In both Figures 5.10b and 5.10c, note the comparatively high harmonic distance of stage 5 (A-3) from stages 4 (A-4) and 6 (A-2).

Figure 5.10a has two pronounced features of biological interest. One is the appreciable slowing of the rate of increase of mean radius between stages 4 (A-4) and 5 (A-3). Figure 5.8 shows that these two instars are profoundly different in shape. The A-3 instar has a more adult-like shape than either the A-4 instar that preceded it or, to a lesser extent, the A-2 instar that succeeded it. The other feature of Figure 5.10a is the increased rate of change of the mean radius during the last two molts, presumably the result of development of the final, adult morphology with a complete set of appendages and complex reproductive organs.

Figure 5.10b has harmonic distance coefficients on the abscissa. These show the difference in shape between each growth stage and the previous one. For the A-8 instar, the difference in shape is 0, there being no previous instar. Figure 5.10c shows the same harmonic distances plotted against mean radius relative to the mean radius of the A-8 instar. The pronounced feature of both graphs is the great difference of shape from the A-4 to A-3 to A-2 stages (see also Figure 5.8.). The A-3 instar is the first one in which the posterior portion of the valve has expanded and is thus similar in form to the adult stage. Morphology of the A-2 instar reverses the trend somewhat. The differences in shape of the A-4 and A-2 instars from the A-3 in no way imply that the A-4 and A-2 instars have similar outlines. In this sense, Figures 5.10b and 5.10c may be somewhat misleading unless one continuously bears in mind that the comparisons are sequential. One must refer to Table 5.2 for measures of harmonic distance among other instars.

5.5 Discussion and conclusions

Is anything gained by using Fourier analysis, even in conjunction with harmonic distance analysis, instead of simply using the raw data and computing taxonomic distances? Mathematically, it seems that little is gained—perhaps nothing. The use of Fourier analysis, however, allows the geometrically inclined to keep an intellectual hold on the data. In a sense, the Fourier analysis is a black box that most such people can comprehend. The human mind is very good at pattern recognition. Fourier analysis can be a useful tool if it helps resolve patterns into their component parts.

Harmonic distance analysis allows one to do more with Fourier analysis than would otherwise be possible. It can be used to compute matrices of distance coefficients that can facilitate comparisons of such shapes as outlines that do not have homologous landmarks. It uses all the information present in the original data and discards nothing. Moreover, it does not require the user to interpret individual harmonics, although it does not preclude doing so. Finally, it retains the geometrical underpinning of Fourier analysis with which many investigators are comfortable.

5.6 Acknowledgments

I am pleased to acknowledge Drs. Ross A. Black and Antonis D. Koussis for their early suggestions regarding programming as this project was getting underway. I am especially grateful, however, to Professor Floyd W. Preston, who, on his own time in the early summer of 1963, taught me how to program in FORTRAN II "because it is a good idea for a geologist to know programming," and who, since then, has always been willing to take the time to help this geometrician over the rough spots. He did it again with this project more than 30 years later.

References

Alberch, P., Gould, S. J., Oster, G. F. & Wake, D. B. (1979). Size and shape in ontogeny and phylogeny. *Paleobiol.* **5**, 296–317.

Anstey, R. L. & Delmet, D. A. (1973). Fourier analysis of zooecial chamber shapes in fossil tubular bryozonas. *Geol. Soc. Am. Bull.* **84**, 1753–64.

Bookstein, F. L., Strauss, R. E., Humphries, J. M., Chernoff, B., Elder, R. L. & Smith, G. R. (1982). A comment upon the uses of Fourier methods in systematics. *Sys. Zool.* **31**, 85–92.

Byerly, G. R., Mrakovich, J. V. & Malcuit, R. J. (1975). Use of Fourier shape analysis in zircon petrogenetic studies. *Geol. Soc. Am. Bull.* **86**, 956–8.

Christopher, R. A. & Waters, J. A. (1974). Fourier series as a quantitative descriptor of miospore shape. *J. Paleontol.* **48**, 697–709.

Doherty, R. E. & Ernest, G. K. (1936). *Mathematics of Modern Engineering*, Vol. 1. New York: John Wiley and Sons.

Ehrlich, R. & Full, W. E. (1986). Comments on "Relationships among eigenshape analysis, Fourier analysis, and analysis of coordinates" by F. James Rohlf. *Math. Geol.* **18**, 855–7.

Ehrlich, R., Pharr, R. B., Jr. & Healy-Williams, N. (1983). Comments on the validity of Fourier descriptors in systematics: A reply to Bookstein et al. *Sys. Zool.* **32**, 202–06.

Ehrlich, R. & Weinberg, B. (1970). An exact method for characterization of grain shape. *J. Sed. Petrol.* **40**, 205–12.

Full, W. E. & Ehrlich, R. (1982). Some approaches for location of centroids of quartz grain outlines to increase homology between Fourier amplitude spectra. *Math. Geol.* **14**, 259–70.

Full, W. E. & Ehrlich, R. (1986). Fundamental problems associated with "eigenshape analysis" and similar "factor" analysis procedures. *Math. Geol.* **18**, 451–63.

Gaskill, R. E. (1958). *Engineering Mathematics*. New York: Henry Holt and Company.

Gould, S. J. (1977). *Ontogeny and Phylogeny*. Cambridge, Massachusetts: Belknap Press.

Gould, S. J. (1988). The uses of heterochrony. In *Heterochrony in Evolution*. ed. M. L. McKinney. New York: Plenum Press.

Healy-Williams, N. (1983). Fourier shape analysis of *Globorotalia truncatulinoides* from late Quaternary sediments in the southern Indian Ocean. *Marine Micropaleont.* **8**, 1–15.

Healy-Williams, N., Ehrlich, R. & Williams, D. F. (1985). Morphometric and stable isotope evidence for subpopulations of *Globorotalia truncatulinoides*. *J. Foraminiferal Res.* **15**, 242–53.

Hodgman, C. D. (1958). *Handbook of Chemistry and Physics*. Cleveland, Ohio: Chemical Rubber Company.

Kaesler, R. L. & Maddocks, R. F. (1979). Carapace outline of macrocypridid Ostracoda: Harmonic distance analysis. *Geol. Soc. Am. Abstr.(Prog.)* **11**, 453.

Kaesler, R. L. & Maddocks, R. F. (1987). Preliminary harmonic analysis of outlines of recent macrocypridid Ostracoda. In *VII International Symposium on Ostracodes, Additional Communications and Discussions*. Belgrade 1979. *Comptes Rendus, Des Seances de la Societe Serbe de Geologie*, Pour l'annee 1983 (1984),169 -74.

Kaesler, R. L. & Waters, J. A. (1972). Fourier analysis of the ostracode margin. *Geol. Soc. Am. Bull.* **83**, 1169 -78.

Lanczos, C. (1966). *Discourse on Fourier Series*. New York: Hafner Publishing Company.

Lestrel, P. E. (1980). A quantitative approach to skeletal morphology: Fourier analysis. *Soc. Phot. Inst. Engrs. (SPIE)* **166**, 80–93.

Lestrel, P. E., Bodt, A. & Swindler, D. R. (1993). Longitudinal study of cranial base changes in Macaca nemestrina. *Am. J. Phys. Anthrop.* **91**, 117–29.

Lohmann, G. P. (1983). Eigenshape analysis of microfossils: A general morphometric procedure for describing changes in shape. *Math. Geol.* **15**, 414–21.

Lu, K. H. (1965). Harmonic analysis of the human face. *Biometrics* **21**, 491–505.

McKinney, M. L. & McNamara, K. J. (1991). *Heterochrony. The Evolution of Ontogeny*. New York: Plenum Press.

Panofsky, H. A. & Brier, G. W. (1965). *Some Applications of Statistics to Meteorology*. University Park: Pennsylvania State University Press.

Prezbindowski, D. R. & Anstey, R. L. (1978). A Fourier-numerical study of a bryozoan fauna from the Threeforks Formation (Late Devonian) of Montana. *Journal of Paleontology* **52**, 353–69.

Przibram, H. (1931). *Connecting Laws in Animal Morphology*. London: University of London Press.

Rohlf, F. J. (1986). Relationships among eigenshape analysis, Fourier analysis, and analysis of coordinates. *Math. Geol.* **18**, 845–54.

Salvadori, M. G. & Miller, K. S. (1948). *The Mathematical Solution to Engineering Problems*. New York: McGraw-Hill.

Selby, S. M. (1972). *Standard Mathematical Tables*. (20th ed.) Cleveland, Ohio: The Chemical Rubber Co.

Smirnov, E. (1927). Mathematische Studien über individuelle und Kongregationenvariabilität. *Verhandl. V. Int. Kingr. Verebungswiss.* **2**, 1373–92.

Sohon, H. (1944). *Engineering Mathematics*. New York: Van Nostrand.

Sokal, R. R. (1961). Distance as a measure of taxonomic similarity. *Syst. Zool.* **10**, 70–9.

Waters, J. A. (1977). Quantification of shape by use of Fourier analysis: The Mississippian blastoid genus *Pentremeties*. *Paleobiol.* **3**, 288–99.

Part two

Applications of Fourier Descriptors

6

Closed-form Fourier Analysis: A Procedure for Extracting Ecological Information from Foraminiferal Test Morphology

NANCY HEALY-WILLIAMS, ROBERT EHRLICH AND
WILLIAM E. FULL
South Carolina Commission on Higher Education
University of South Carolina
The Wichita State University

6.1 Introduction

The wide range of morphologies seen among taxa, as well as intraspecific variability, when combined with the diversity of information carried by morphology, ensures that no "one size fits all" method exists for morphologic characterization. In some cases, interest resides in determining how a single species' form has changed in response to environmental parameters; the organism being used solely as a proxy indicator. In other cases, shape components linked to such environmental factors will be excluded in order to determine the nature and rates of *change* of the genome as reflected in morphology. In still other cases, the morphology may be of interest in terms of some sort of functional efficiency with respect to locomotion or musculature.

A task of the investigator is to choose the best procedure for the particular experimental situation. Paramount in any investigation is choosing the optimum method to define statistically significant morphological changes. Forms that depart from classic Euclidean (geometric) regularity commonly can generate a large number of numeric descriptors. Usually, for any given research objective, only a subset of such variables are applicable. Relevant descriptors are termed "features," and feature selection is a critical first step in any morphological analysis. Features may be known *a priori* on the basis of prior research on the taxon or by analogy with related taxa. However, a set of features relevant for one objective (e.g., functionality) may not be relevant for another (e.g., sexual dimorphism). Some features may represent well-defined "landmarks" in that they can be precisely and unambiguously located on a specimen.

In other cases, biomorphs may lack such a pin-point definition — a common problem in paleontology. In paleontology we are restricted to examining the fossilized remains of the once-living organism, and may not have a full grasp of the significance of morphologic characteristics; or, no unambiguous features may be present. In this situation a prudent strategy is to generate a large number of shape descriptors and use statistical methods to identify morphologic characteristics that covary with changes in the system of interest.

Closed-form Fourier analysis is an efficient way to extract morphologic information from certain common classes of shapes. Description of aspects of shape using a Fourier transform is just one of a large set of numerical procedures appropriate for morphologic analysis. For suitable organisms it is a powerful tool for extracting information from morphology. A bonus is that the Fourier series representation completely describes the boundary morphology of a specimen, down to some limiting resolution. That is, the shape can be reconstructed from the series with high precision.

6.1.1 Some shape considerations

Shape can be defined as the spatial configuration of area or volume. In these terms, shape is the basic foundation of geometry, and thereby one may consider the definition of shape to be solely within the provenience of the professional geometer. However, this is not the case because many investigators in the physical and biological sciences are concerned with the idea of shape as an information carrier. Most natural shapes are of little interest to geometers or topologists because they are neither regular and symmetrical, nor so complex as to be interesting topologically. Therefore, the professional mathematician sees the characterization of natural shapes (commonly 2-D outlines) as a simple exercise in empiricism, whereas the scientific researcher views it as an important method for gaining useful information regarding biological or physical form.

The description of shape or form increases in complexity as the object departs from regularity. The shapes of regular polygons and polyhedra can be described exactly using the symmetry operations inherent in group theory. As an object departs from such regularity, the number of shape parameters necessary to define the shape increases rapidly. In theory, the number of such parameters necessary to describe 2-D shapes in general is infinite. This potentially large number of descriptors is in one sense a good thing in that it implies that nonregular shapes can carry a great deal of information. However, in the context of an individual investigation, this potential richness of the "measurement space" may get in the way of discovering the small amount of information of current interest. That is, the objective of most scientists is not necessarily to describe such shapes completely, or even extract all of the information present, but to identify a small set of attributes of the shape that are information-rich with respect to the problem at hand. Other attributes may carry information irrelevant to the problem or may represent noise. So, the task in shape analysis, in the absence of *a priori* information, is to systematically sort through a potentially large number of shape-describing variables in order to identify a subset of relevant variables. Variables that are, in turn, ultimately sensitive to the factors of interest — an exercise in multivariate data analysis.

6.1.2 Approaches to shape characterization

Two approaches have been taken to simplify the process of extracting information from shape measurements. In one approach, certain characteristics of the shape are defined, *a priori,* to contain all or most of the relevant information, and so shapes are compared via sets of "homologous features," often characterized by landmarks. In the other approach, a subset of features is extracted from a larger "complete" set by comparison of shape data within a carefully selected set of objects.

The first approach has some basic limitations but is potentially very powerful. It requires, first of all, an assumption about the importance of the suite of features, and secondly, the object to be measured needs to possess clear geometric cues for their location. This approach requires the *a priori* establishment of the definition of shape and assumes that all of the critical information has been contained within the established metrics.

The second approach avoids the potential pitfalls of landmark analysis, but requires more effort to determine the reasons for the variability of one or another shape descriptor. This approach is particularly useful for taxa that change shape without regard to homologous points, or lack such points. It also requires an analysis of a large number of objects chosen to display a trend or change in geometry with respect to some external frame of reference (e.g., changes in temperature, location, behavior). Of course, once identified, such shape features may be used as an *a priori* set of characters in subsequent analyses, or may assist in defining homologous points. One of the more obvious differences between this and the previous approach is that the latter method does not make assumptions about where the critical information is contained within the set of shape descriptors.

The use of a closed-form Fourier transform to characterize geometry is an example of the second case. After examination of several methodologies, we chose this shape characterization in light of our objectives and of the geometry of the organism of interest — tests of foraminifera. Foraminifera are protozoans that are often enclosed in a calcareous test (shell), which consists of the accumulation of a series of progressively larger subspherical chambers. The shape of the chambers, the degree of size change from earlier to later chambers, and the spatial relationships among chambers are factors intrinsic in controlling the shape of the test. One of our objectives has been to derive paleoceanographic parameters from the morphometry of specimens based on a single species of foraminifera collected at carefully selected locations and/or time intervals.

We measure the profile of the foraminiferan oriented in a standard position. Of the several hundred thousand specimens that we have analyzed in our laboratory, we digitize them in the frontal view which is the most common view for photographic presentation. Some species require additional or alternate views rather than the frontal one in order to examine particular aspects of the test. For example, the planktonic species *Globorotalia truncatulinoides* was examined in side view to determine changes in inflation of the cone-shaped test (Healy-Williams, 1983). Side views have been used to examine the degree of inflation of the test for the benthic foraminifera *Cibicidoides (floridanus) pachyderma* (Healy-Williams et al., 1988). Care needs to be taken to orient the specimen in the same axis so as not to confound the Fourier series. Our research has indicated that a variance of about 10° from the true axial position results in a slight effect on the Fourier series, with this effect increasing with a greater angle of variance (Pharr, 1983).

The chamber arrangement of foraminifera results in relatively smooth outlines, which lack homologous points (in the sense of Bookstein, 1986) for a valid landmark-based analysis (Full and Ehrlich, 1986). Other methodologies based on the concept of geometrical homology represent attempts to circumvent this problem. The application, and perhaps the concept in general, of geometrical homology as discussed by Lohmann (1983) and Bookstein (1986, 1993), is invalid as demonstrated by Full and Ehrlich (1986), and on fundamental mathematical grounds (Lele, 1993).

The Fourier transform requires no knowledge of such homologous features, and so is preferable for use with foraminifera. In addition, Fourier analysis has more than a century of history in the physical sciences, and we can draw on the wisdom of the past in our measurement and analysis. This is not to say that the Fourier approach is the only correct approach under all circumstances. Even within the domain of the Fourier transformation there are choices of the type of basis for the series. For example, our present analytic scheme uses standard trigonometric coefficients (sines and cosines). Other possible schemes might use Haar and Walsh coefficients (triangular and rectangular waves as opposed to the smooth cosine and sine waves). Our experience has shown that the application of the Fourier transform has produced meaningful solutions, and that the exploration of other approaches has not been driven by the failure of the Fourier technique.

6.2 The closed-form Fourier transform[1]

The closed-form Fourier transform produces a trigonometric series that, given enough terms, will approach any observed function as closely as necessary. In general form, the transform consists of both sine and cosine terms and can be defined mathematically in terms of a real and imaginary component (i.e., as a se-

---

[1] A FORTRAN routine that computes the Fourier algorithm is available from the authors.

ries of complex numbers). For some objects whose boundaries are generally convex outwards, the use of polar coordinates with the origin at the center of gravity permits a simplified series consisting only of cosine and phase terms. The main requirements that permit the comparison between individual shape components for two or more outlines are: (1) that the center of the polar coordinate system coincide with the center of gravity; and (2) that any radius vector emanating from this center of gravity intersects the defined boundary only once as it sweeps around the object (Full and Ehrlich, 1982). The first of these main conditions implies that we have "homology" in the sense that similar shapes will yield almost identical amplitude spectra, and the second requirement implies that the boundary can be represented as a single-valued function when viewed from the center. Further discussion of this and other requirements can be found in the next section.

6.2.1 Closed-form Fourier equation

The equation describing the decomposition of a closed shape is:

$$r(\theta) = r_0 + \sum_{n=1}^{N} r_n \cos(n\theta - \phi_n), \tag{6.1}$$

where:

$r(\theta)$ represents the mathematical function of the radius r at an angle θ in the space defined by polar coordinates

r_0 is the average radius of the outline defined in the polar coordinate system

n is the individual shape component number ($n = 1, \ldots, N$) (commonly referred to as a harmonic number)

N is the total number of shape components

r_n is the magnitude of the contribution of the nth shape term (commonly referred to as the harmonic amplitude)

θ represents the polar angle of the individual radius represented by $r(\theta)$, and

ϕ_n represents the phase angle associated with harmonic n.

Equation 6.1 above represents the rendering of the outline defined by a set of discrete points into a mathematical series. The r_i and ϕ_i are determined using a discrete finite Fourier transform. The equations associated with this calculation are:

$$a_j = \frac{2}{m} \sum_{i=1}^{m} F(i)\cos\left(\frac{2\pi j}{m}\right), \tag{6.2}$$

and

$$b_j = \frac{2}{m} \sum_{i=1}^{m} F(i)\sin\left(\frac{2\pi j}{m}\right), \tag{6.3}$$

where:

a_j represents the real jth component of the Fourier transform
b_j represents the complex jth component of the Fourier transform
m is the number of polar vectors defining the shape being analyzed, and
$F(i)$ is the length of the radius at θ_i (the raw data consists of a series of angles (θ) and radii (F)).

The components derived in Equation 6.2 and 6.3 are related to the components of Equation 6.1 by:

$$r_j = \sqrt{a_j^2 + b_j^2}, \tag{6.4}$$

$$\phi_j = \tan^{-1}\left(\frac{b_j}{a_j}\right), \tag{6.5}$$

and

$$r_0 = \frac{\sum\limits_{k=1}^{m} F(k)}{m}. \tag{6.6}$$

6.2.2 Separate contributions of Fourier series terms

Each term of the series represents the contribution (an "amplitude") of a single component to the shape (Figure 6.1). The first term r_0 represents the contribution of a circle centered at the center of gravity (centroid) — that is, it is a size term. We conventionally divide all r_i terms by r_o, and so permit objects with similar shapes but different sizes to have equivalent values in their amplitude spectra. The first amplitude r_1 represents the contribution of a circle tangent to the center and will have a value of zero if the origin is located exactly at the center of gravity. As will be discussed below, computational procedures exist to ensure low values of the amplitude of this harmonic.

The second component of the series, represented by the family of curves defined by the $cos(2\theta)$ function, is a "four-leaf clover." When this function is drawn in polar coordinates (Figure 6.2), one will notice that as θ goes from 0 to 2π, the trajectory of the individual lobes of the $cos(2\theta)$ curve are alternately defined by positive and negative radii. That is, the first lobe is defined by radii that are positive, the second lobe is defined by negative radii, the third lobe is defined by positive radii, and the fourth lobe is defined as negative radii. When the circle of radius r_0 is added to this term (Figure 6.2), the net effect is to "pull in" the circle in the direction of the negative lobes, while at the same time extending the circle in the direction of the positive lobes. The practical outcome of this tug-of-war is that the second harmonic measures the degree of elongation of the object.

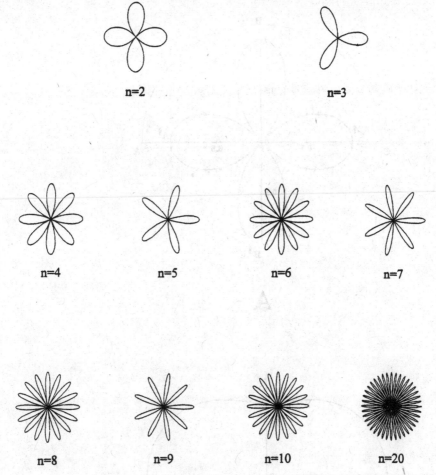

Fig. 6.1 Examples of shapes comprising the individual Fourier components used in closed-form Fourier shape analysis. Each of the terms is of the form $cos(n\theta-\Phi)$ where n is the harmonic number, θ is the polar angle (0–2π), and Φ is the phase angle. Each term in the above fig. has a phase angle radius of 0.0 radians.

The third harmonic represents the contribution of a "trefoil" (a three-leafed clover) to the empirical shape (Figure 6.1); it is a measure of triangularity. Similarly, the fourth harmonic measures quadrateness (even though it is defined by eight lobes), the fifth pentagonality, and so forth. In general, the nth harmonic represents the contribution to the empirical shape of an n-lobed form. As the series progresses, relatively large-scale features, measured by lower harmonics, give way to smaller-scaled "bumpiness" measured by higher harmonics.

The information, then, lies in the amplitudes of the harmonics (the harmonic amplitude spectrum) and the phase angles that mutually relate them. Taken together, the empirical shape can be reconstructed to any degree of precision.

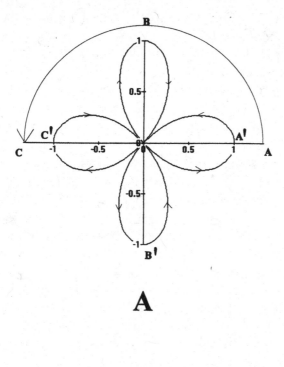

A

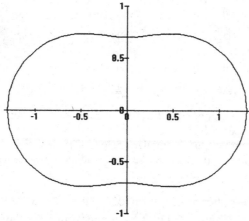

B

Fig. 6.2 Example of the second harmonic leading to elongation. In Figure A, the polar angles vary uniformly between 0 and 2π, producing the trace path shown along arc ABC (defined for only $0-\pi$) producing the curve trace shown. Points A′, B′, and C′ correspond to polar angles represented by points A, B, and C respectively. In B, the curve shown in A has, using a harmonic amplitude of 0.3, been added to a circle of radius 1.0, producing the elongated figure shown.

6.3 Digitization

The raw material for the Fourier transform is the set of coordinates spaced along the periphery (i.e., digitization of the boundary). For our purposes, equal angular divisions between radii are used from the center of gravity. To the extent possible, the spacing of the points (the "frequency") should be at least half the wavelength of the highest harmonic having a significant amplitude (the "Nyquist sampling theorem").

This requires that the number of radii must be twice the highest-numbered harmonic to be calculated. In practice, determining the highest frequency present in a shape (the last harmonic with a statistically significant value, assuming an infinite number of digitized points were calculated) is not so simple a task. Commonly, objects are digitized at a very high sampling rate using one or another electronic digitizing apparatus, and 2,000 or more edge points can be obtained rapidly and precisely. In the case of our foraminifera there is no sense in calculating a series with 1,000 terms, inasmuch as most of the higher-numbered terms will possess very low amplitudes. Additionally, as the Fourier series is bounded (i.e., converges to the actual shape at some point), there is a practical point where the measured information will have little, if any, significance. In all but extreme cases (never encountered in foraminifera), the significant information is captured by a subset of 96 points, and this subset is subsequently used to calculate our series.

These points are initially selected to subtend equal angles from the center (not necessarily the center of gravity) defined by the original set of edge points. However, when measured from the center of gravity of the subset, the original digitized points seldom subtend equal angles. Therefore, new equal-angle points must be interpolated from the original set of digitized data. The accuracy of this interpolation is a function of the number of digitized vectors defining the perimeter. We have found that at least 400 points are necessary to ensure that the error due to this interpolation is minimal. We generally have from 600–1200 points defining the edge of the foraminifera. As mentioned earlier, the center of gravity of this set of interpolated points should also represent the origin of the polar coordinate system being used. Deviation of the centroid from the origin of the polar coordinate system will result in increased energy (higher amplitude) of the first harmonic (the offset circle). As the first harmonic amplitude increases, the ability to compare from one shape to another is diminished. Relatively small changes in this amplitude can cause relatively large changes in statistical significance among populations of shapes, even at the lower harmonic level (see Full and Ehrlich, 1982).

6.3.1 An iterative method for centroid convergence

The centering error (as it is called) is a source of noise that we wish to minimize if we are to extract any but the most obvious information from our system. Hence, an iterative process is required in order to locate a subset of points subtending equal angles from the center of gravity. Specific details of how this is done can be found in Full and Ehrlich (1982). The generalities of the procedure are outlined as follows:

1. Estimate an initial guess for a center of gravity. This initial guess is determined using the entire set of original digitized points defining the outline of the shape. The equations for this are:

$$\bar{x} = \frac{\left(\sum_{i=1}^{n^*} x_i\right)}{n^*} \tag{6.7}$$

and

$$\bar{y} = \frac{\left(\sum_{i=1}^{n^*} y_i\right)}{n^*}, \tag{6.8}$$

where $\bar{x}$ and $\bar{y}$ represent the coordinates of the centroid; x_i and y_i are the digitized coordinates of the outline; and n^* is the total number of these digitized pairs.

2. Transform the digitized coordinates to polar coordinates with the putative centroid as center.
3. Using the first digitized point as the starting position, interpolate m edge points subtending equal angles about the centroid. Linear interpolation is acceptable when the number of edge points exceeds 400–600.
4. Calculate a new centroid (center of gravity) of the m interpolated points.
5. Compare this centroid to the previously calculated centroid. If the centroids are within an arbitrary distance (currently 0.0007 pixel distance is used), then use these points to calculate the Fourier series. Otherwise, continue with this procedure.
6. Calculate a new guess for the centroid by locating the point at the end of a line segment drawn from the center of gravity of the subset of m points through the center of gravity (the Evan's method). This point is the new candidate centroid.
7. Move back to the fifth step, above, and compare with the previously calculated center of gravity.

Under most conditions encountered with foraminifera, convergence occurs within ten iterations. Additional procedures are built into this iterative scheme that ensure convergence of very irregular shapes (discussed in Full and Ehrlich, 1982), although these irregularities are not encountered with the relatively smooth edges of foraminifera.

Given the 96 edge points, we only calculate values for only the first 24 harmonics. The extra edge points are used to reduce the chances of aliasing, hence minimizing the chance of violating the Nyquist sampling theorem. The Fourier transform produces a bounded orthogonal series, and much of the "energy" that would be involved in aliasing is implicitly present in the higher terms. Additionally, the chance of noise being introduced by the digitizing process (the usual digitizing grid is rectangular, thereby introducing a calculable level of aliasing) is minimal as well. This latter type of noise is further minimized by the choice of the number of points used to define the outline (600–1200 points normally).

6.3.2 *Analysis of Fourier data*

Once the digitizing and computation of the Fourier transform procedure are complete, the resulting harmonic amplitudes and phase angles are stored as output, and represent base data that will be subsequently analyzed.

The next step of the Fourier procedure is information extraction. This can be performed in many different ways depending on the parameters of the investigation. In our investigations of foraminifera shape, the amplitude spectrum is easier to deal with, in contrast to the phase angles. Analysis of the angular data can be facilitated by simple inspection, if the polar coordinate system can be "anchored" by placing the origin at the same homologous point on every specimen (i.e., if the same starting position is used for each shape being digitized). In the case of foraminifera this is not possible, given their relatively smooth outlines. However, relative phase can be calculated by relating higher-ordered phase angles to the position of the first harmonic. Even this calculation is tedious in practice. Furthermore, because the tangent function (Eq. 6.5) is periodic, there is a reduction in the amount of information that can be carried by the phase angle, especially with higher harmonics. That is, for the fifth harmonic, phase angles of $\pi/5$, $2\pi/5$, $3\pi/5$, $4\pi/5$, and π will produce exactly the same external shapes. In other words, rotations about the centroid through multiples of the phase angle cannot be distinguished from each other.

The real information in phase, as previously mentioned, is not in the value of an individual phase component, but lies in the relative relationships of the components amplitude and phase to each other. Thus, shape information may exist in the amplitude spectrum, the phase spectrum, or in their combinations. Because the amplitude spectrum can be analyzed by any of an assortment of multivariate procedures, it is prudent to begin the analysis by utilizing the amplitudes, and saving the phase spectrum only if issues are not resolved using amplitudes. This is the approach we have adopted. Although we have examined the phase data (Pharr, 1983), we have determined that a greater degree of statistically significant information resides in the amplitude spectrum in comparison to the phase data.

6.3.3 Harmonic amplitude spectra

The simplest method of analysis involves visual inspection of the plot of the logarithm of the amplitude versus harmonic number. Figure 6.3 depicts the harmonic amplitude spectra for two species of planktonic foraminifera. Inspection of the figure shows that, in terms of the amplitude spectrum, the shapes differ more at some harmonic numbers than others, and that generic information may be carried by harmonics different from those carrying intraspecific information.

In terms of the specimens compared in Figure 6.3a, the two specimens of *Globorotalia truncatulinoides* differ at harmonic two, which is an indicator of length. The primary difference between these two forms of *G. truncatulinoides* is coiling direction, with the longer one coiled sinistrally (i.e., to the left). This distinction was easily detected from visual inspection of the amplitude spectra but earlier biometric studies of this species, using length/width ratios (Kennett, 1968; Takayanagi et al., 1968), did not discern that left-coiled specimens were longer than right-coiled specimens.

Our result is significant for two reasons. First, in this species, and perhaps others which have coiling direction changes, the two variants are not mirror images of one another and may not behave the same biologically. Therefore, morphologic studies that assume that coiling variants are identical, except for coiling direction (Lohmann and Malmgren, 1983), carry unwarranted assumptions concerning the "plasticity" of this species. Second, it emphasizes the ease of "feature extraction" using the Fourier amplitude spectrum, as well as the use of this tool for exploratory data analysis.

Figure 6.3b shows the amplitude spectra for *Globorotalia inflata* collected from plankton tows from the western North Atlantic. The specimen from MOC38 (0–25 m water depth) differs from the one taken from MOC62 (850–1000 m) in the degree of elongation, triangularity, and quadrateness (harmonics 2, 3, and 4 respectively), as well as in other harmonics. These results were used to infer that this species continues calcification as it occupies progressively greater depths, a point of previous disagreement (Hemleben et al., 1985; Fairbanks, personal communication). The results based on the Fourier analysis indicate that calcification in this species is controlled by physical factors associated with depth and strength of the thermocline (Healy-Williams, 1990).

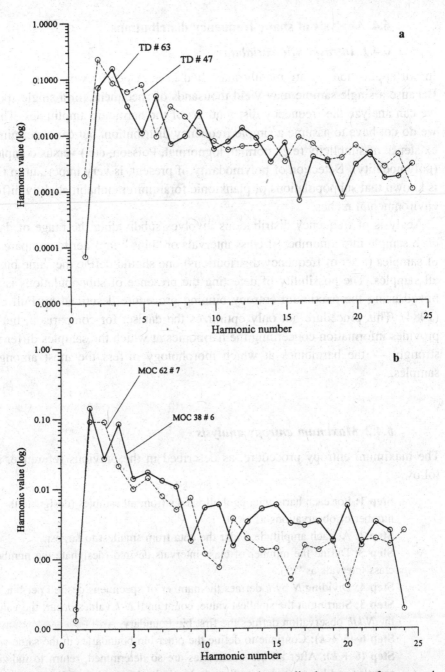

Fig. 6.3 The logarithmic plot of the harmonic amplitude spectra of two specimens each of the planktonic foraminifera *Globorotalia truncatulinoides* (6.3a) and *Globorotalia inflata* (6.3b).

6.4 Analysis of shape frequency distributions

6.4.1 Intraspecific variability

In our application we are mainly interested in intraspecific variability in shape. Because a single sample may yield thousands of specimens for a single species, we can analyze the frequency distributions of the harmonic amplitudes. That is, we do not have to assume a simple frequency distribution, but can determine the existence of simplicity (e.g., normal, lognormal, Poisson, etc.) versus complexity (polymodality). Detection of polymodality, if present, is very important in that it is known that subpopulations of planktonic foraminifera inhabit slightly different environmental niches.

Analysis of frequency distributions involves subdividing the range of data in each sample into a number of class intervals or "bins." In order to compare a set of samples (a set of frequency distributions) one should define the same bins for all samples. The possibility of detecting the presence of subpopulations is optimized using the maximum entropy binning procedure described by Full et al., (1984). This procedure not only optimizes the dataset for comparison, but also provides information concerning the harmonics at which the samples differ most strongly — the harmonics at which morphology differs the most among the samples.

6.4.2 Maximum entropy analysis

The maximum entropy procedure, as described in the previous reference, is as follows:

> **Step 1:** For each harmonic, pool all values from all samples (designate the total number of observations as "N").
> **Step 2:** At each amplitude, order the data from smallest to largest.
> **Step 3:** Define the number of class intervals desired (designate the number of class intervals as "k").
> **Step 4:** Dividing N by k defines the number of specimens desired per bin.
> **Step 5:** Starting at the smallest value, count until N/k values occur; the value of the N/kth observation defines the first bin boundary.
> **Step 6–(5 + k):** Continue to define the other bin boundaries in the same way.
> **Step (6 + k):** After all bin boundaries are so determined, return to individual samples and bin according to boundaries derived from the pooled data.

Consistent algorithms have to be defined for binning, taking into account what happens if the N/kth and the (N/kth + 1) observations are the same (i.e., in which bin should the [N/kth + 1] observation be located). With the variability expressed in foraminifera, problems arising out of these situations are rarely, if ever, encountered.

The product of this algorithm is a set of frequency distributions for each harmonic, each binned according to the maximum entropy algorithm. This means that at each harmonic the binned data are in a state of maximum unbiased contrast (Full et al., 1984). Information entropy of a frequency distribution refers to the degree to which the data are spread evenly across all bins. A distribution is in a state of maximum entropy if equal amounts of data occur in all bins (an individual harmonic amplitude has an equal probability of being contained within any interval). Entropy itself is the sum of the products of the number of intervals times the negative of the logarithm (commonly the natural log) of the proportion of observations in each interval. Knowing the number of class intervals, the maximum possible entropy can be calculated, and the ratio of the observed and maximum values, the "relative entropy," can be calculated. A relative entropy of unity is maximum.

The mean relative entropy can be calculated across all samples for each harmonic, and the squared relative entropy values are commonly plotted against harmonic number (Figure 6.4). Harmonic(s) displaying *minimal* entropy-squared val-

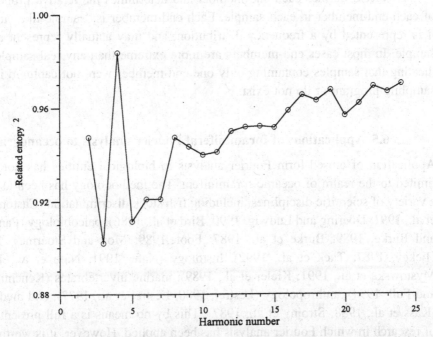

Fig. 6.4 Plot of relative grand entropy versus harmonic number for a digitized foraminiferal sample. Harmonics displaying minimal values represent harmonics that display maximum contrast (e.g., contain the maximum amount of shape information among samples).

ues represent harmonics displaying maximum contrast among samples; that is, frequency distributions that display great differences between intervals. Entropy, in this case, is a measure of contrast between intervals of a frequency plot for a specific harmonic number. Because the samples are selected according to a set of external criteria (such as along a temperature gradient), harmonics displaying minimal entropy commonly define shape components that most efficiently carry information relevant to the problem at hand. For statistical purposes, plots of harmonic number versus the chi-square statistic have also been useful.

In maximum entropy format, the set of frequency distributions for any harmonic of interest is analyzed to determine the nature and extent of polymodality. Space prevents discussion of the principles behind these procedures involving the programs SAWVECA and SAWVECB (but see Evans et al., 1992). General principles involved with the procedure can be found in Ehrlich and Full (1987). The program SAWVECA determines the number of subpopulations (or "end-members"). If samples have identical frequency distributions that only differ with respect to the position of the mean, SAWVECA will report the fact; that is, end-members (subdistributions) will not and cannot be forced onto the dataset by the algorithm.

SAWVECB defines each end-member and determines the relative proportion of each end-member in each sample. Each end-member is "sample-like" in that it is represented by a frequency distribution, and may actually represent a real sample. In most cases end-members are more extreme than any real sample, indicating that samples containing only one end-member were not captured in the sampling program or do not exist.

6.5 Applications of foraminiferal Fourier analysis to oceanography

Application of closed-form Fourier analysis to biological entities has not been limited to the realm of oceanic foraminifera. The methodology has been used in a variety of scientific disciplines, including fish stock discrimination (Castonguay et al., 1991; Doering and Ludwig, 1990; Bird et al., 1986), paleobiology (Pandolfi and Burke, 1989; Burke et al., 1987; Foote,1989; Mou and Stoermer, 1992; Hickey, 1987; Tack et al., 1992), histology (Nafe, 1991; Nafe et al., 1991; Mystowska et al., 1991; Kieler et al., 1989), marine invertebrates (Kenchington and Full, 1994), anthropology (Lestrel, 1980; Lestrel et al., 1977), and medicine (Kass et al., 1988; Strojny et al., 1987). This by no means is a full presentation of research in which Fourier analysis has been applied. However, it is worth noting that most of the shapes quantified in these investigations represent fairly smooth outlines, which lack distinct homologous points with which to perform landmark analysis.

We have used closed-form Fourier analysis to examine the morphologic variation exhibited by several hundred thousand foraminiferal tests. These studies have been wide ranging and have analyzed both fossil and living material in order to improve the application of these organisms in paleo-environmental reconstructions and biostratigraphy. Due to their abundance and widespread distribution, both benthic- and planktonic- dwelling foraminifera have been used in a variety of geologic investigations. Benthic foraminifera evolved in the Cambrian and planktonic foraminifera in the Mesozoic. This long-ranging evolutionary development has provided geologists with a wealth of stratigraphic and environmental information with which to decipher marine sedimentary sequences.

Planktonic foraminifera dwell in the upper portions of the water column within the mixed layer, and often reach maximum abundance within the chlorophyll maximum zone. Although these organisms reside in the upper water column, they have developed species-specific depth ranges. For example, *Globorotalia truncatulinoides* is considered to be a deep dweller, whereas the symbiont-bearing species *Globigerenoides ruber* is a surface dweller. As surface dwellers of upper waters, their distribution patterns have been linked to water masses and ecosystem/ecotone boundaries. These distribution patterns have therefore allowed foraminifera to not only provide stratigraphic information, but also information on the paleoecology of the global ocean.

6.5.1 Morphologic information from foraminifera

Morphologic variation of foraminiferal tests has been noted in nearly all fossil and extant species, and represents the product of both heredity and environment. The phenotype has been considered by numerous investigators to represent another source of biostratigraphic and paleoecological information (Bandy, 1960; Kennett, 1976; Scott, 1974; Parker, 1962; Reyment, 1980). Early studies examining this variation relied on exacting linear measurements, which often did not provide significant information on test morphology. The selection of characters to be measured (length/width, distance between chambers, chamber suture angle, aperture size/position) requires careful consideration, which is often not undertaken in such investigations (Reyment, 1980). The application of quantitative methods to the field of foraminiferal morphologic investigations has allowed the analysis of large numbers of specimens, but, more importantly, by measuring the shapes completely it ensured that the information-carrying aspects of the test were captured. We provide here an example from the planktonic realm, which demonstrates how we can extract ecological information from foraminiferal test morphology.

G. truncatulinoides is a planktonic species that we have examined in detail, and it affords an excellent example of how closed-form Fourier analysis can provide us with new information regarding the habitat within which these organisms reside. *G. truncatulinoides* were examined from sediments underlying subtropical to sub-antarctic (22°–47°S) water masses in the southern Indian Ocean (Healy-Williams, 1983; Healy-Williams and Williams, 1981; Healy-Williams et al., 1985). Samples were retrieved from a region that features several distinct water mass boundaries including the Subtropical Convergence Zone (STC), which separates subtropical and subantarctic water masses (Figure 6.5). The STC is defined by strong temperature (10°–14°C) and salinity (35.5°/oo–34.7°/oo) gradients. Specimens to be analyzed were subdivided according to specific size ranges to monitor possible allometric effects, along with coiling direction (sinistral or dextral).

Maximum entropy analysis of the 23 harmonic amplitude spectra for 3400 specimens were analyzed to determine which geometric shape components (harmonics) contained significant morphologic information. Harmonics 2 and 3 were deemed to be the main information carriers, representing test length and inflation,

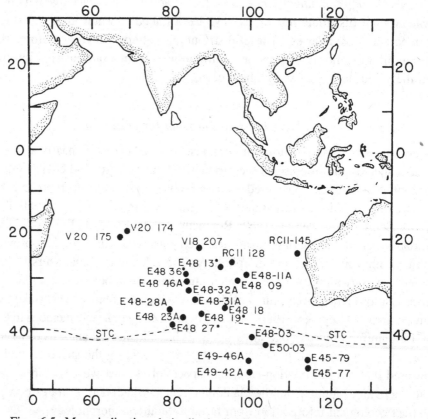

Fig. 6.5 Map indicating latitudinal distribution of samples used in the *Globorotalia truncatulinoides* investigation. Note the location of the Subtropical Convergence (STC). (From Healy-Williams, 1983, 1984; Reprinted by permission from Elsevier Science Publishers.)

respectively. *G. truncatulinoides* is a cone-shaped foraminiferan which we examined in side view. Previous length/width measurements of *G. truncatulinoides* (Takayanagi et al., 1968; Kennett, 1968) had shown a gradational relationship of test length and height with latitude. Application of quantitative techniques to the varied morphology of this species allowed a more precise analysis of the modes of morphologic variation and the relationship of morphologic variants to water mass conditions.

The results presented here for *G. truncatulinoides* represent typical distribution patterns for morphologic variants for the several hundred thousand planktonic and benthic foraminifera examined by us to date. Judging from these results the use of means could mask important paleoenvironmental information. Examination of the frequency distributions for harmonics 2 and 3 indicated bimodal to polymodal structure of the populations with latitude. Such complexity of structure had not been expected, although nonnormal distributions had been observed in foraminiferal populations (Malmgren, 1979). For most investigations of foraminifera, however, unimodal distribution patterns are assumed, allowing the use of arithmetic means to describe populations. Existence of multi-modal population distributions indicates that foraminiferal populations must be viewed as consisting of numerous subgroups that have potential paleobiologic significance.

6.5.2 Morphology and upper water column dynamics

The question to be asked is, how can such complexity arise? As discussed above, little is known about foraminiferal biology. What is known is limited due to the inability to maintain reproducing cultures of planktonic foraminifera under laboratory conditions (Hemleben et al., 1989). Plankton tows provide important information on abundance and depth distribution, but represent only a small time-slice. Previous studies have shown that a seasonal component exists for the presence of morphologic variants (Reynolds and Thunell, 1986; Keller, 1978). When examining foraminifera from a typical 1-cm^3 sediment sample we are looking at contributions from the entire water column over some time period, determined largely by the accumulation rate and extent of dissolution at that site. The resultant sediment sample is the net end product of the overlying water conditions and postdepositional processes occurring on the seafloor. The upper water column itself is a dynamic region influenced by solar and atmospheric conditions. Such alteration in surface water conditions affects the flora and fauna growing within these waters in terms of seasonality, annual cycles, an so on. Thus it is not surprising that foraminiferal populations consist of complex distributions resulting from changing hydrographic conditions. "Decoding" this complexity affords the opportunity of obtaining a richer database from which to extract paleoecological information.

Test elongation was shown (Healy-Williams, 1983) to be highly correlated with coiling direction (r = 0.91), with the sinistral forms being longer. Attempts have been made to relate foraminiferal coiling changes with surface water temperature. It has been shown for this species that coiling changes do not occur simply as a response to water temperature (Ericson et al., 1954; Cifelli, 1971; Thiede, 1971) and may be related to salinity (Thiede, 1971). The degree of inflation of the conical test, however, was highly correlated with the environmental parameters of temperature (r = 0.85) and salinity (r = 0.88), and provides a useful marker for changes in the physical-chemical properties of the water mass.

SAWVECA, a vector analysis program (discussed in Section 6.4.2), determined that the frequency distributions were comprised of mixtures of three morphologic groups, termed subpopulations. The normalized varimax loadings accounted for 90% of the variance displayed by the distributions (Healy-Williams et al., 1985). The proportion of each of these three subpopulations are shown in ternary diagrams (Figures 6.6 and 6.7) for each harmonic. Figure 6.6 indicates that for length,

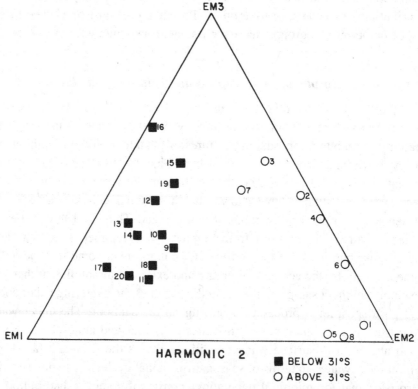

Fig. 6.6 Ternary diagram of the relative proportion of harmonic 2 for each end-member in each sample analyzed for *Globorotalia truncatulinoides*. (From Healy-Williams et al., 1985; Reprinted by permission of the Cushman Foundation for Foraminiferal Research.)

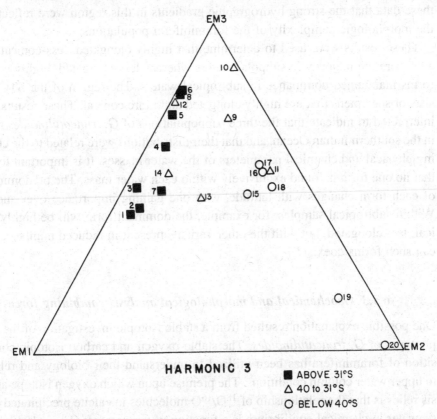

Fig. 6.7 Ternary diagram of the relative proportion of harmonic 3 for each end-member in each sample analyzed for *Globorotalia truncatulinoides*. (From Healy-Williams et al., 1985; Reprinted by permission of the Cushman Foundation for Foraminiferal Research.)

the samples can be described as binary mixtures: (1) below 31°S a mixture of subpopulations 1 and 3; and (2) above 31°S a mixture of subpopulations 2 and 3. Although in a region of ecological complexity, 31°S is not situated at a major water mass boundary, yet foraminiferal morphology indicates a demarcation at this latitude (Bé, 1969; Bé and Tolderlund, 1971; Bé and Hutson, 1977; Prell et al., 1979; Williams et al., 1985). It is at 31°S that there is an abrupt change in coiling direction, which is highly correlated with test length.

Changes in the degree of test conicalness reflect the ecological diversity of the waters of this sector of the southern Indian Ocean (Figure 6.7). *G. truncatulinoides* in waters above 31°S are mixtures of subpopulations 1 and 3, and those below 40°S are mixtures of subpopulations 2 and 3. Specimens from the dynamic region between 31°S and 40°S, which includes the STC, are bimodal mixtures of the three subpopulations but do not display a latitudinal trend. We concluded from these data that the strong hydrographic gradients in this region were reflected in the morphologic complexity of the foraminiferal populations.

These results were used to determine that highly elongated, less-conical variants were the majority in subpolar waters whereas less-elongated, highly conical forms maintained dominance in subtropical waters. The region of the STC consists of specimens that are highly elongated and more conical. These results were interpreted to indicate that the three subpopulations of *G. truncatulinoides* reside in the southern Indian Ocean, and that their distributions were related to the changing physical and chemical parameters of the water masses. It is important to note that no one form is found exclusively within each water mass. The predominance of each form changes with latitude, with one gaining importance over another. Within subtropical samples, for example, the dominant form will be highly conical, less elongated, but with the other variants present in reduced numbers. How can such forms coexist?

6.5.3 Geochemical and morphological analysis:combining forces

One possible explanation resulted from a stable isotopic investigation of the morphotypes of *G. truncatulinoides*. The stable oxygen and carbon isotopic composition of foraminifera has been utilized to understand their biology and relation to upper water column conditions. The premise upon which oxygen isotope analysis relies is that the relationship of $^{18}O/^{16}O$ molecules in calcite precipitated from seawater in chemical equilibrium is a function of temperature (Urey, 1947). This relationship allows the assessment of $^{18}O/^{16}O$ ratios in terms of water temperature at the time of test calcification. This technique has been employed to estimate the vertical depth stratification of planktonic species in the water column, as well as to make paleoclimatic reconstructions of the global ocean.

Carbon isotopic composition of planktonic foraminifera is considered to reflect a "vital effect," essentially a divergence from equilibrium caused by physiological components. It has been shown, in symbiont-bearing foraminifera, that the $\delta^{13}C$ value is a function of symbiont-photosynthetic activity and irradiance level in the water column. Thus, deviation from thermodynamic equilibrium will be dependent upon the depth in the water column in which the species resides.

An even stronger methodology for foraminiferal morphologic investigations can be obtained by incorporating the use of two quantitative techniques (shape analysis plus stable carbon and oxygen isotopes) to delineate the meaning of shape variants in terms of ecology. The closed-form Fourier analysis results were utilized to delineate morphologic variants typical of the subpopulations to be analyzed isotopically. It was determined (Healy-Williams et al., 1985) that the depth stratification (depth habitat) of *G. truncatulinoides* within the water column shallowed (decreased depth range) as specimens were compared to the hydrography from

22°S to 47°S. This species' depth habitat followed the shoaling of the $26.6\sigma_t$ isopycnal, oxygen minimum and nitracline. These major core layers are found at an average water depth of 400 m at 20°S, and shallow to between 0 to 100 m at 50°S.

Healy-Williams et al.'s (1985) interpretation of these data was that this species is capable of expanding one morphologic group over another as the niche varies with latitude. To confirm this interpretation, they analyzed the stable oxygen and carbon isotopic composition of the two end points of the subgroups from the same sediment sample. For each sample analyzed over the latitudinal range, the highly conical morphotype was depleted in ^{18}O and ^{13}C in relation to the more-compressed forms from the identical sediment sample (Figure 6.8). The isotopic separation of specimens analyzed from the same sediment sample validates the interpretation that the morphotypes of *G. truncatulinoides* reside within different subniches in the water column. If the morphotypes displayed similar or widely varied isotopic values, one could have interpreted this to demonstrate that *G. truncatulinoides* resided in a wide depth range. The fact that this species' morphotypes secrete their calcite tests with a consistently different isotopic value clearly demonstrates that the morphotypes reside in different portions of the water column.

6.5.4 Morphology and paleoceanography

It was determined in the above-described investigation of *G. truncatulinoides* that this species spans the southern Indian Ocean in a series of stepwise shifts in morphology and isotopic composition. These shifts are related to the changing hydrographic conditions of the upper water column. In the colder waters of the subantarctic, this species shallows its niche to the upper 100 m, where nutrient levels are five times that found in subtropical waters. We conclude that such information becomes a powerful tool in tracing past hydrographic conditions of the upper water. If one subpopulation is dominant within a specific hydrographic regime, what occurs in the population structure as one water mass expands at the expense of another?

During the Last Glacial Maximum (LGM) the STC shifted northward between 2° and 5°, caused by an expansion of polar waters at the expense of the subpolar water mass. Examination of specimens of *G. truncatulinoides* at selected isotope stages from a core located beneath the present day STC demonstrates distinct shifts in test conicalness (Harmonic 3) through time (Figure 6.9). The large shift during Stage 2 represents an influx of less-conical/more-elongated forms as colder waters expand into the region of 38°S. Changes in elongation are not as strong due to the consistently high percentages of sinistral forms in the sample. Similar analyses were performed on specimens located within the Central Gyre, which experienced little change during LGM. Specimens analyzed displayed little morphologic variability during the last 60,000 years, indicating fairly stable hydro-

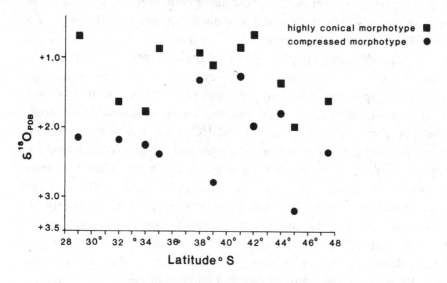

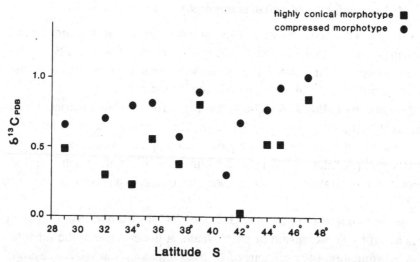

Fig. 6.8 Oxygen and carbon isotopic values for the highly conical and compressed morphologic variants of *Globorotalia truncatulinoides*. Each variant was analyzed separately from within the sediment sample and the same size fraction. In all cases, the highly conical forms are more depleted in [18]O and ^{13}C in comparison to the compressed forms. (From Healy-Williams et al., 1985; Reprinted by permission of the Cushman Foundation for Foraminiferal Research.)

CORE   48-27

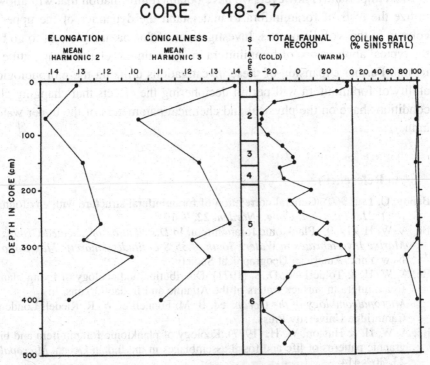

Fig. 6.9 Correlation of harmonics 2 and 3 amplitude values at selected intervals down core. Note the strong shift in shape during the glacial maximum in Stage 2. (From Healy-Williams, 1983, 1984; Reprinted by permission from Elsevier Science Publishers.)

graphic conditions in the region. Such results clearly demonstrate that foraminiferal morphology can be used to trace the hydrographic dynamics of the upper water column and that Fourier analysis is a suitable tool for extracting these data.

6.6 Conclusions

The Fourier method of analyzing shape is an extremely efficient methodology for extracting morphologic data from foraminifera. The relatively smooth outlines of these organisms do not lend themselves to other morphologic procedures that rely on consistently definable landmarks. Fourier analysis has also been advantageous for our investigations in that it permits analysis of a large number of specimens, thus allowing us to potentially capture the wide range of variability exhibited by these microorganisms. That is, several hundred to a thousand specimens can be digitized per day, with subsequent analysis performed "off line" (Healy-Williams, 1983).

Most importantly, however, we have obtained information that will allow us to utilize the tests of foraminifera to understand the dynamics of the upper water column of the world's oceans. Investigations of this nature allow us to go beyond the typical applications of foraminifera in determining relative temperature of the region. Further expansion of these investigations utilizing the morphologic variability of foraminifera will permit deciphering the effects that changing climatic conditions have on the physical and chemical parameters of the upper water column.

References

Bandy, O. L. (1960). General correlation of foraminiferal structure with environment. *Rept. 21st Int. Geol. Cong., Norden.* **22**, 7–19.

Bé, A. W. H. (1969). Planktonic foraminifera. In *Distribution of Selected Groups of Marine Invertebrates in Waters South of 35°S Latitude, Antarctic Map Folio 11.* New York: American Geographical Society.

Bé, A. W. H. & Tolderlund, D. S. (1971). Distribution and ecology of living planktonic foraminifera in surface waters of the Atlantic and Indian Oceans. In *Micropaleontology of the Oceans* ed. B. M. Funnell & W. R. Riedel. London: Cambridge University Press.

Bé, A. W. H. & Hutson, W. H. (1977). Ecology of planktonic foraminifera and biogeographic patterns of life and fossil assemblages in the Indian Ocean. *Micropaleont.* **23**, 369–414.

Bird, J. L., Eppler, D. T. & Checkley, D. M. (1986). Comparison of herring otoliths using Fourier-series shape analysis. *Canad. J. Fisheries* **43**, 1228–34.

Bookstein, F. (1986). Size and shape spaces for landmark data in two dimensions. *Stat. Sci.* **1**, 181–242.

Bookstein, F. (1993). Higher order features of shape change for landmark data. In *Proc. Mich. Morphometrics Workshop* ed. F. J. Rohlf & F. L. Bookstein. Ann Arbor: University of Michigan Museum of Zoology.

Burke, C. D., Full, W. E. & Gernant, R. E. (1987). Recognition of fossil freshwater ostracodes: Fourier shape analysis. *Lethaia* **20**, 307–14.

Castonguay, M., Simard, P. & Gagnon, P. (1991). Usefulness of Fourier analysis of otolith shape for Atlantic mackerel (*Scomber scombrus*) stock discrimination. *Canad. J. Fisheries* **48**, 296–302.

Cifelli, R. (1971). On the temperature relationships of planktonic foraminifera. *J. Foraminiferal Res.* **1**, 170–77.

Doering, D. & Ludwig, J. (1990). Shape analysis of otoliths: a tool for indirect aging of eels, *Anguila anguilla. International Revue der gesmaten m Hydrobiologie* **75**, 736–43.

Ehrlich, R. & Full, W. E. (1987). Sorting out geology - Unmixing mixtures, *IAMG Special Publication*, ed. W. Size. New York: Oxford University Press.

Ericson, D. B., Wollin, G. & Wollin, J. (1954). Coiling direction of *Globorotalia truncatulinoides* in deep-sea cores. *Deep-Sea Res.* **2**, 152–58.

Evans, J. C., Ehrlich, R., Krantz, D. & Full, W. E. (1992). A comparison between polytopic vector analysis and empirical orthogonal function analysis for analyzing quasi-geostrophic potential vorticity. *J. Geophys. Res.-Oceanography* **97**, 2365–78.

Foote, M. (1989). Perimeter-based Fourier analysis: A new morphometric method applied to trilobite cranidium. *J. Paleont.* **63**, 880–5.

Full, W. E. & Ehrlich, R. (1982). Some approaches for location of centroids of quartz grain outlines to increase homology between Fourier amplitude spectra *Int. J. Math. Geol.* **14,** 34–6.

Full, W. E., Ehrlich, R. & Kennedy, S. K. (1984). Optimal configuration and information content of sets of frequency distributions. *J. Sediment. Petrol.* **54,** 117–26.

Full, W.E. & Ehrlich, R. (1986). Fundamental problems associated with "eigenshape analysis" and similar "Factor" analysis procedures. *J. Math. Geol.* **18,** 451–63.

Healy-Williams, N. (1983). Fourier shape analysis of *Globorotalia truncatulinoides* from late Quaternary sediments in the southern Indian Ocean. *Marine Micropaleont.* **8,** 1–15.

Healy-Williams, N. (1990). Morphological changes in living foraminifera and the thermal structure of the water column, Western North Pacific. *Palaios* **4,** 590–7.

Healy-Williams, N. & Williams, D. F. (1981). Fourier analysis of test shape of planktonic foraminifera. *Nature* **289,** 485–7.

Healy-Williams, N., Ehrlich, R. & Williams, D. (1985). Morphometric and stable isotopic evidence for subpopulations of *Globorotalia truncatulinoides. J. Foraminiferal Res.* **15,** 242–53.

Healy-Williams, N., Gary, A. C., Williams, D. F., Monts, L. A., Guilderson, T., Trainor, D. M. & Roosen, E. (1988). *Morphometric Studies Group Industrial Associates Year 2 Report.* Unpublished Contract Report, Univ. of South Carolina.

Hemleben, C., Spindler, M. & Anderson, O. R. (1989). *Modern Planktonic Foraminifera.* New York: Springer - Verlag.

Hemleben, C., Spindler, M., Breitinger, I. & Deuser, W. G. (1985). Field and laboratory studies on the ontogeny and ecology of some globorotalid species from the Sargasso Sea off Bermuda. *J. Foraminiferal Res.* **15,** 254–72.

Hickey, D. R. (1987). Shell shape in Late Pennsylvanian myalinids (bivalvia). *J. Paleontol.* **61,** 290–311.

Kass, D. A., Traill, T. A., Keating, M., Altieri, P. I. & Maughan, W. L. (1988). Abnormalities of dynamic ventricular shape change in patients with aortic and mitral valvular regurgitation - assessment by Fourier shape analysis and global geometric indexes. *Circulat. Res.* **62,** 127–38.

Keller, G. (1978). Morphological variation of *Neogloboquadrina pachyderma* (Ehrenberg) in sediments of the marginal and central north-east Pacific Ocean and paleoclimatic interpretation. *J. Foraminiferal Res.* **8,** 208–24.

Kenchington, E. L. & Full, W. E. (1994). Fourier analysis of scallop shells *(Placopecten magellanicus)* in determining population structure. *Canad. J. Fisheries* **51,** 348–56.

Kennett, J. P. (1968). *Globorotalia truncatulinoides* as a paleoceanographic indicator. *Science* **221,** 1153–6.

Kennett, J. P. (1976). Phenotypic variation in some Recent and late Cenozoic planktonic foraminifera. In *Foraminifera* vol. 2, ed. R. H. Hedley & C. G. Adams. New York: Academic Press.

Kieler, J., Shubis, K., Grzesik, W., Strojny, P., Wisniewski, J. & Dziedzic, A. (1989). Spreading of cells on various substrates evaluated by Fourier analysis of shape. *Histochem.* **92,** 141–8.

Lele, S. (1993). Euclidean distance matrix analysis (EDMA): Estimation of mean form and mean form distance. *J. Math. Geol.* **25,** 573–602.

Lestrel, P. E., Kimbel, W. H., Prior, F. W. & Fleischmann, M. L. (1977). Size and shape of the hominid distal femur: Fourier analysis. *Am. J. Phys. Anthrop.* **46,** 281–90.

Lestrel, P. E. (1980). A quantitative approach to skeletal morphology: Fourier analysis. *Proc. Soc. of Photo-optical Inst. Eng.(SPIE)* **186,** 80–93.

Lohmann, G. P. (1983). Eigenshape analysis of microfossils: A general morphometric procedure for describing changes in shape. *J. Math. Geol.* **15,** 659–72.

Lohmann, G. P. & Malmgren, M. A. (1983). Equatorward migration of *Globorotalia truncatulinoides* ecophenotypes through the Late Pleistocene: gradual evolution or ocean change? *Paleobiol.* **9,** 414–21.

Malmgren, B. A. (1979). Multivariate normality tests of planktonic foraminiferal data. *J. Math. Geol.* **11,** 285–97.

Mou, D. & Stoermer, E. (1992). Separating *Tabellaria* (*Bacillariophyceae*) shape groups based on Fourier descriptors. *J. Phycol.* **28,** 386–95.

Mystowska, E. J., Komar, A., Strojny, P., Rozycka, M. & Sawickli, W. (1991). Fourier analysis of the nuclear and cytoplasmic shapes of living 2-cell murine embryos. *Anal. Quant. Cytol. Histol.* **13,** 209–14.

Nafe, R. (1991). Planimetry in pathology - a method in its own right besides stereology and automatic image analysis. *Exp. Pathol.* **43,** 239–46.

Nafe, R., Roth, S. & Rathert, P. (1991). Fourier analysis as a planimetric procedure—application to malignant and normal urothelial cells with reactive changes. *Exp. Pathol.* **43,** 155–61.

Pandolfi, J. M. & Burke, C. D. (1989). Environmental distribution of colony growth form in the favositid *Pleurodictyum americanum. Lethaia* **22,** 69–84.

Parker, F. L. (1962). Planktonic foraminiferal species in Pacific sediments. *Micropaleontol.* **8,** 219–54.

Pharr, R. B. (1983). Examination of the late Quaternary paleobiology of *Globorotalia truncatulinoides* using Fourier shape analysis. Masters thesis, University of South Carolina.

Prell, W. L., Hutson, W. H. & Williams, D. F. (1979). The Subtropical Convergence and late Quaternary circulation of the southern Indian Ocean. *Marine Micropaleontol.* **4,** 225–34.

Reyment, R. A. (1980). *Morphometic Methods in Biostratigraphy.* New York: Academic Press.

Reynolds, L. & Thunell, R. C. (1986). Seasonal production and morphological variation of *Neogloboquadrina pachyderma* (Ehrenberg) in the northeast Pacific. *Micropaleontol.* **32,** 1–18.

Scott, G. H. (1974). Biometry of the foraminiferal shell. In *Foraminifera,* vol. 1, ed. R. H. Hedley & C. G. Adams. New York: Academic Press.

Strojny, P., Traczyk, Z., Rozycka, M., Bern, W. & Sawicki, W. (1987). Fourier analysis of nuclear and cytoplasmic shape of blood lymphoid cells from healthy donors and chronic lymphocytic leukemia patients. *Anal. Quant. Cytol. Histol.* **9,** 475–9.

Tack, J. F., Vanden, E., Bergh, E. & Polk, P. (1992). Ecomorphology of *Crassostrea cucullata* (Born, 1778) (Ostreidal) in a mangrove creek (Gazi, Kenya). *Hydrobiologia* **247,** 109–17.

Takayanagi, Y., Niitsuma, N. & Sakai, T. (1968). Wall microstructure of *Globorotalia truncatulinoides* d'Orbigny. *Science Report of Tohoku University, 2nd Series (Geology)* **40,** 141–70.

Theide, J. (1971). Variations in coiling ratios of Holocene planktonic foraminifera. *Deep-Sea Res.* **18,** 823–31.

Urey, H.C. (1947). The thermodynamic properties of isotopic substances. *J. Chem. Soc.* 562–81.

Williams, D. F., Healy-Williams, N. & Leschak, P. (1985). Dissolution and water-mass patterns in the S.E. Indian Ocean, Part 1: Evidence from Recent and late Holocene foraminiferal assemblages. *Geol. Soc. Am. Bull.* **96,** 176–89.

7

Fourier Descriptors and Shape Differences: Studies on the Upper Vertebral Column of the Mouse

D.R. JOHNSON

Centre for Human Biology, University of Leeds

7.1 The measurement of biological shape change

Biologists are interested in living things. Morphologists are interested in the shape of living things, the differences in shape between living things, and the transformation of one shape into another.

In recent years there have been a number of developments in techniques for shape description in areas outside the biological sciences, stimulated, in part, by developments in image processing technology. It is now a relatively simple and inexpensive task to extract and store a closely placed series of Cartesian coordinates taken from the outline of a structure using an inexpensive video-digitizer. This newly available technology has led biologists to try various new means for form description which will allow more information derived from the outline data to be compared or studied than permitted by previous methodology (O'Higgins and Johnson, 1988).

Before one can compare forms one must describe them. The descriptions may be purely subjective, in which case the comparisons between forms are also subjective. The trend, however, is toward more objective studies, which implies a quantitative method of form description. There are many ways in which form may be described. Some of these are particularly relevant to biological material, for example, distances between "homologous landmarks": others, such as measures of perimeter, curvature in outline, and so on, are equally applicable to biological or nonbiological material.

The comparison of forms resolves, therefore, into two processes; the description of some aspects of shape, and the comparison of these descriptions between individuals or populations of individuals.

By far the commonest approach to quantitative definition of biological form has been to use measurements or angles taken between defined morphological landmarks. A detailed critique of landmarks is out of place here; suffice it to say that there are problems with "homologous landmarks". For example, many shapes

157

may be envisaged as gentle curves of varying radius, such as the sagittal section of the cranial vault, which may have no landmarks, although it certainly has shape. For such situations an alternative landmark-free method might be preferred.

One such case is the study of the vertebrae of the mouse. We chose to study the upper part of the mouse vertebral column because the mouse is a well-used laboratory animal, and because large numbers of cleaned skeletal preparations were available. The upper vertebral column offered us a series of around eight shapes per mouse (C1–T2 inclusive) forming a metameric series.[1] A metameric series is also often a homologous series: see Goodwin (1993) for an interesting biological and mathematical description of homology. The cervical vertebrae and T1 and T2 of mouse are narrow in an anterio-posterior direction compared to their height and width. Most of the shape information they contain is, thus, preserved in an anterior or posterior two-dimensional projection, and the bones lie naturally flat on a microscope stage. On the other hand, the outline of vertebral shapes are fairly simple; that is, they have few "corners" — sharp changes of curvature that define landmarks accurately, and are thus more suitable for a landmark-free description which would allow multivariate statistical analysis.

We first applied Fourier analysis to the measurement of biological shape in 1985. The first paragraph of the introduction of our first paper on vertebral analysis (Johnson et al., 1985) is worth repeating:

> A common problem facing the morphologist is the comparison of the shapes of complex biological structures in a manner that allows full account to be taken of natural variation. The traditional answer to this problem has been to derive simple quantitative data which are suitable for univariate or multivariate statistical analysis. Traditionally these data have been in the form of linear or angular measurements taken between homologous points, or ratios of such measurements. In addition to the problems inherent in defining homologous points, this approach suffers from the additional disadvantages that it ignores the frequently large intervening regions and that it produces measurements which may be disconnected from each other. In consequence so much information is lost that the original shape, or even an approximation to it, cannot be reconstructed from the data.

This propaganda statement naturally emphasizes the positive points of using Fourier analysis and omits the negatives — that Fourier coefficients tell us nothing of the way in which shape changes or where within the shape these changes are to be found. Indeed, they are inherently unfriendly, so much so that Bookstein et al. (1982) suggested that Fourier data are inherently uninterpretable because a shape change will alter every Fourier component (but see the refutations by Erlich et al., 1983; and Read and Lestrel, 1986).

---

[1] Editor's note: From metamere — one of several similar segments of the animal body.

7.2 Image capture, processing, and storage

Choosing an essentially 2-D system allowed us to side-step the first potential problem, that of recording a shape as 2-D outline data, though 3-D shapes may be reduced to 2-D by methodologies such as those of Yasui (1986) or O'Higgins and Williams (1987). A 2-D outline can be converted to a coordinate stream by projecting it onto a digitizing pad and carefully tracing the outline, or directly by one of the now commonly available image analysis packages. Either will produce a stream of Cartesian coordinates: the former method is slow and laborious and prone to errors. Initially we recorded both inner and outer outlines of the vertebrae. The former, that is, the boundary of the neural canal, was less variable and so less interesting, and was accordingly dispensed with.

7.3 Outline superimposition and problems of fit

Outlines recorded as Cartesian coordinates cannot be directly compared with each other. Coordinates differ in value because of differences in position and orientation, as well as differences in size and shape. If measurements are to be taken from outlines in a way that will allow comparison, they must be invariant in translation and rotation. This requires two landmarks (though not necessarily in the biological sense). Polar coordinates may not be a valid method of representing shapes with large re-entrants[2] since some polar radii will have two potential values. If this appears to be a problem, care should be taken that polar coordinates are always recorded clockwise (or anticlockwise), so that any deformation is at least consistent. It may be worth recording clockwise and anticlockwise polars and averaging the two sets. In practice superimposition and fitting of the outlines is done as follows.

7.3.1 Center of area

From the digitized coordinates, the first landmark calculated for each shape is the center of area, which is found by integration. The shape is then redefined with 128 equiangular polar coordinates centered on the center of area. Each shape is scaled to a standard area by summing all the areas for a shape, dividing by the number of individuals, and multiplying each area by this factor. A new center of area (centroid) of the shape is then found by integration and the shape reexpressed as a number (n) of polar coordinates centered on this point. Some transform programs require that n is a perfect square, so it is sensible to define n in this way.

---

[2] Editor's note: The problem of re-entrants, generally, does not occur with the use of elliptical Fourier functions (see Chapters 2 and 16).

Other programs can accept values of n that are not a perfect square, or varying values of n. We routinely use 128 polars, although 256 or 512 would give a better resolution at the expense of increased computing time.

7.3.2 Rotational fitting

The bones have been laid on the stage of the microscope in no special orientation, so it is necessary to rotate them relative to each other. It would be possible to use the outline of a specific vertebra of the correct type (i.e., a C3 for C3s, a C4 for C4s, etc.) as a template. In practice we achieved better results and fewer misfits (usually rotations through 90 or 180 degrees) by using a simple polygon appropriate to the vertebra rotated (Figure 7.1). The reference axis was taken as the midline of the standard shape.

Fitting is by a modification of the least squares fit method of Sneath (1967). A least squares fit comparison is made between a shape and a standard, then the shape rotated by one polar coordinate and the process repeated until a minimum is found after one complete rotation. Sneath took as his measurement of fit the sum of the squared distances between a number of homologous landmarks expressed in Cartesian coordinates. We use residual area when shapes are superimposed upon their centroids. Residual area is a function of the sum of the differences on each individual radius squared:

$$\alpha = \sum_{k=n-1}^{N} \sqrt{(r_k^2 - R_k^2)^2}, \tag{7.1}$$

where r_k and R_k are corresponding polar coordinates on two shapes, n = number of polar coordinates and n = kth polar.

As a further refinement another fitting stage was introduced. Once a group of shapes has been aligned upon their center of area and rotationally fitted, a mean shape can be generated from the mean of each corresponding polar coordinate. Each shape can then be refitted to the mean for its group.

Group mean outlines may also be fitted on top of each other. The significance of a difference at any polar radius has been traditionally ascertained by a t-test. Alternatively, an estimate of total shape similarity can be made. The use of a t-test in such a context may be objected to on the grounds that polar radii are highly correlated in length. A MANOVA test that deals with such correlations would be a better alternative, but to my knowledge this does not exist.

A single individual of unknown provenance may be assessed for fit to any number of group mean shapes. A bone having a good fit (low sum of differences of squares) with one group and a high sum (of differences of squares) for another is likely to be a member of the first group.

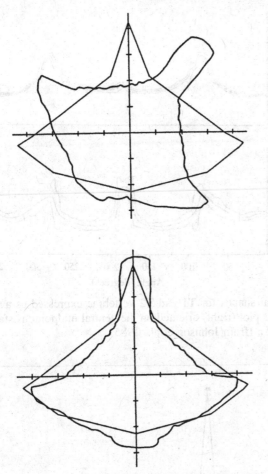

Fig. 7.1 Finding the best fit. The image of a vertebra (top), reconstituted from the Cartesian coordinate stream and reexpressed as polar coordinates about its center of area lies in no particular orientation . Upon it is superimposed a standard shape of equal area orientated about x and y axes. The image of the vertebra (bottom) after a least squares fit has been performed to minimize the residual area. The image of the vertebra is rotated to the position of best fit (From Johnson et al., 1985).

7.3.3 Breaking the continuous shape into a discontinuous wave form

At this point the outline of a scaled, rotated vertebra can be broken at a fixed point (we use the ventral midline, i.e., the point where the ventral outline of the vertebra crosses the central axis of the reference shape) to convert the circular data to a discontinuous wave form (Figure 7.2a). Because the first and last points of the wave form were originally adjacent points on the vertebral outline, they will have equal or nearly equal numerical values.

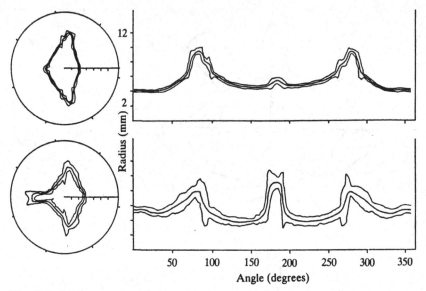

Fig. 7.2a Mean vertebral shapes for T1 and T2 vertebrae expressed as a polar plot (left) and as a linear plot (right) oriented on the ventral midpoint as starting point at 0° to 360°, or 2π (from Johnson et al., 1985).

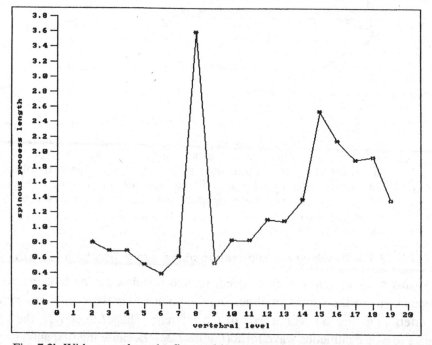

Fig. 7.2b With some data, the first and last data points in a series are not equal in numerical value. Shown here is a plot of spinous process length against vertebral level for vertebrae 2 to 19 in the mouse. Circularizing these data (in effect, joining the value at 19 to value 2) would produce an artifactual step.

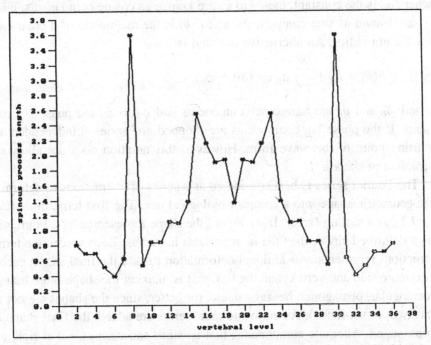

Fig. 7.2c In this case the data in Figure 7.2*b* can be better circularized by reflecting the data. The same data have been plotted on a horizontal axis which now runs from 2 to 36. The values from 19 to 36 merely repeat 2–19 in reverse order. Points 2 and 36 now, of course, have identical numerical values.

Series of observations derived by other means, and where the first and last observations are numerically dissimilar, will introduce a step into the circular data. This can legitimately be removed by circularizing the data by reflecting it. Thus, the values of a polar radius R ($R_1 \ldots R_n$) are represented by a series $R_1 \ldots R_n \ldots R_1$. The reflected data is then modeled by a portion of the Fourier series (Figures 7.2b and 7.2c).

7.4 Form similarity and discrimination

A set of wave forms derived as explained above may be represented by a transformation, which essentially describes the wave form by means of a series of Fourier components.

7.4.1 *The Fourier sequence*

The Fourier series can be expressed as:

$$F(\theta) = a_0 + a_1 \cos\theta + b_1 \sin\theta + a_2 \cos2\theta +$$
$$b_2 \sin2\theta + \ldots + a_n \cos n\theta + b_n \sin n\theta \quad (7.2)$$

where a_0 is the constant, the a_1 to a_n are known as cosine components, the b_1 to b_n are known as sine components, and $F(\theta)$ is the magnitude of a polar radius r for the nth radius. An alternative notation is:

$$F(\theta) = a_0 + \sum_{n=1}^{N} a_n \cos(n\theta + \phi_n), \tag{7.3}$$

where a_0 and a_n are harmonic components and ϕ_1 to ϕ_n are phase lag components. If the phase lag components are dropped, the series is independent of the starting point of the wave form. However, this notation does not allow reconstruction of shapes.

The Fourier series is best considered as a series of instructions to deform a basic geometric shape into a complex biological one. The first term, or a_0 (Eq. 7.2 or 7.3), is a scaling factor. If we regard the shape represented by a Fourier series as a deformed circle, then the a_0 represents its radius. Each successive term (instruction) represents an additional deformation toward the final shape, each with one more side and vertex than the last, that is, making the shape more triangular, rectangular, pentagonal, hexagonal, etc. (or better, since the shapes are not angular, two- to six-lobed). As more and more terms are added, the final shape is approximated. Alternate components, that is, sines and cosines, act at right angles to each other. The arbitrary starting point that we use ensures that the sine components representing left and right asymmetries are thus small for our inherently symmetrical vertebrae, whereas the cosine components are larger and represent symmetrical shape changes. The higher-order members of the Fourier series represent finer and finer perturbations of the shape. The magnitude of these later "fine tuners" are smaller than the earlier components, which have already described the shape well.

7.4.2 Practical considerations — using Fourier series as discriminators

We have now produced a group of Fourier series, one for each shape in our sample and each potentially infinitely long. Two trends, already mentioned, are that early components are larger than later ones and that (in the instance of mouse vertebra) cosine coefficients are larger than sine coefficients. For a listing of the Fourier coefficients obtained from mouse T1 vertebra, contact the author.

First term, the scaling component. Since we have already scaled our vertebral shapes to unit area (i.e., made them all the same size) before expressing them as polar coordinates, this should be constant throughout. In practice any rounding errors that build up are seen here, and the series may need to be renormalized with respect to a_0.

Subsequent sine coefficients. If we are not interested in left and right asymmetries, these may be discarded and the shapes assumed to be bilaterally symmetrical.

Subsequent cosine coefficients. We have seen that the representation of shape reconstructed from a Fourier series depends on the number of coefficients used. A mouse T2 vertebra was reconstructed with 5–60 pairs of sine/cosine coefficients or harmonics. Increasing the numbers of sine/cosine pairs (harmonics) from 5 to 15 refine a rather crude sketch of a vertebral outline into something much more closely resembling the real thing. Beyond 15 pairs, however, little additional realism is gained as the outline just shows high-frequency noise (see Figure 7 in Johnson et ál., 1985).

This suggests that there may be an optimum length for a Fourier series, at which point it best describes a particular shape. We investigated this possibility by using samples of vertebrae from two distinct strains of mice, which Fourier analysis showed to be different in shape. For each strain, 10% of the vertebral sample was removed at random. An attempt was made to classify the vertebrae on the basis of differing numbers of Fourier components (sine and cosine pairs in this experiment). The procedure was repeated ten times (with a different random subsample removed each time) for 5–25 Fourier components, and the total number of bones misclassified was recorded. The results (Figure 7.3) show a curve with a distinct minimum, in this case at 15 pairs, where 92% of the bones were classified correctly. Using a cruder outline (with fewer components) increased misclassification, but so did gilding the lily by using more, presumably because high-frequency noise (produced from large numbers of Fourier coefficients) was added.

This result shows that the ability to discriminate between shapes is greatest for a particular number of Fourier coefficients. Repeating the experiment using dif-

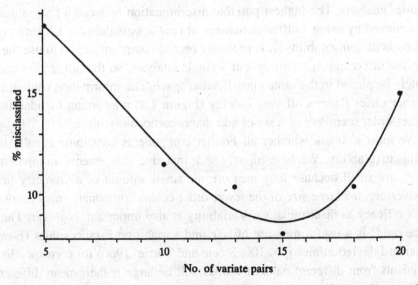

Fig. 7.3 Regression curve of percentage of bones misclassified against number of coefficient pairs used (see text). (From Johnson et al., 1985).

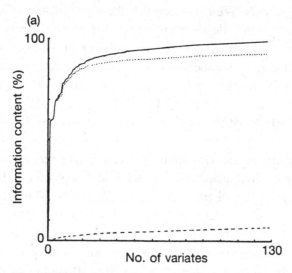

Fig. 7.4 The information content of increasing numbers of sine and cosine components from a Fourier series. Most of the information is carried in the lower-order components, and in the vertebral outlines that are largely symmetric; the sine components (representing asymmetry) are unimportant. Sum of sine and cosine components (solid line), cosine components (dotted line), and sine components (hatched or dashed line). Horizontal axis refers to increasing number of sine and cosine terms (harmonics).

ferent vertebrae (and hence different shapes) gave a slightly different optimum, and using other shapes would no doubt give much larger differences. Discriminant ability varies with shape: this point should always be born in mind when using Fourier analysis. The highest possible discrimination between T1 vertebrae may be achieved by using a different number of coefficients than for T2. However, on the basis of comparability, it is probably best to compromise and to use the same number of components throughout a single analysis, so that all shapes can ultimately be placed in the same classification space. The information content of any Fourier series flattens off very quickly (Figure 7.4) reinforcing the idea that the higher-order members of a series add disproportionately little.

We must also ask whether all Fourier components contribute equally to discriminative ability. We have already seen that sine components, in our test situation, are small because they measure the small amount of asymmetry present. Conversely, the large size of the lower-order cosine components may account for their efficacy as discriminators. Variability is also important, however. The variance ratio[3] is a useful measure of size and variability. F-ratio values (F-ratio = [standard deviation/mean] $\times$ 100; Steele and Torrie, 1980) for corresponding coefficients from different pairs of shapes will be large if their mean difference is

---

[3] Editor's note: Also termed the coefficient of variation or CV (Sokal and Rohlf, 1969).

large and/or their standard deviations small, and small if their means are similar and/or their variabilities high (Figures 7.5a, b). Thus, one may choose to drop certain coefficients from an analysis on the basis of low discriminative ability. However, this may decrease the ability to reconstruct the shape from its Fourier components.

This can be carried out by simply calculating F-ratio values for all coefficients, then sorting them by the F-ratio and choosing the best discriminative number by iteration. Alternatively the SAS procedure Discrim (SAS, 1982) does this more elegantly, choosing the best *n* discriminants.

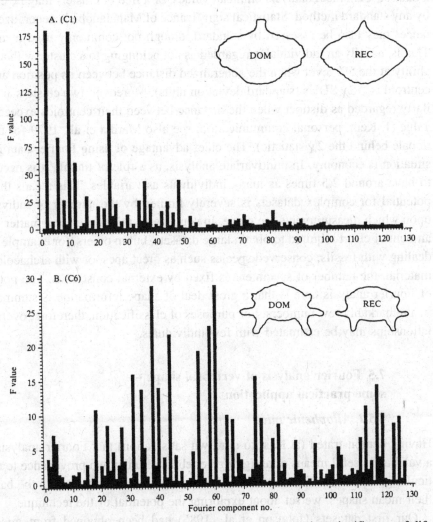

Fig. 7.5 Plots of Fourier component number (*x* axis) against significance of difference between DOM and REC strains (F-ratio, *y* axis) for vertebrae C1 (Figure 7.5a) and C6 (Figure 7.5b). (From O'Higgins et al., 1988).

7.4.3 Using Fourier analyses in multivariate statistics

We have already seen how the Fourier series can give an optimum discrimination between sets of shapes. It can be used to classify individuals as belonging to one group (arbitrary or real) or another, on the basis of similarity. It can also be used to find groupings within datasets. Indeed, providing that all the data have been placed in the same classification space (i.e., that we are comparing like with like, and all individuals have been treated in an identical manner in terms of how many and which coefficients have been selected), it can be treated exactly like any other data. Principal component analysis may be used on the individuals in a dataset, Canonical analysis on mean values, or a tree of clusters may be created by any standard method. Statistical significance of Mahalanobis's generalized distances may also be assessed by standard, though not commonly used measures. That is, usually an individual is regarded as not belonging to a cluster with a probability at the 5% level when the generalized distance between its position and the centroid is $>2\chi$ SDUs (standard deviation units). Moreover, two clusters are similarly regarded as distinct when the distance between their centroids exceeds this value (J. Kent, personal communication; see also Mardia et al., 1979 for the rationale behind the 2χ statistic). The chief advantage of using Fourier data in this situation is economy. In multivariate analysis, as a rule of thumb, it is necessary to have around 2.5 times as many individuals as variables. This means that the potential for complex datasets is severely limited by the number of individuals upon which measurements are made. In some cases this is merely a matter of the amount of effort required to obtain large datasets, but in others, for example when dealing with fossils, conserved species such as great apes, or with archaeological material, the number of specimens is fixed by external constraints. The potential of Fourier analysis here is that a great deal of shape information is summarized in a remarkably few numbers. For purposes of classification, therefore, overall relationships may be estimated with few individuals.

7.5 Fourier analysis of vertebral shape — some practical applications

7.5.1 Allophenic mice

Having demonstrated (at least to our own satisfaction) that Fourier analysis was a valid and useful means of assigning vertebrae of unknown provenance to a particular group, and for discriminating between groups of vertebrae on the basis of their mean shapes, we set about exploring the potential of the technique.

Our first datasets (Johnson et al., 1985) had been obtained from mice homozygous for seven major mutations, any one of which might be classified as a pathology (but none of which affected the skeleton drastically), and their normal litter mates. These mice were chosen because of the rather large differences expected in vertebral shape as a result of the seven known segregating genes and the effects of an unknown number of other segregations. Interestingly, these mice had been initially bred for the production of allophenic chimeras. These are four-parent mice produced by the fusion of two eight-cell embryos of different genotypes. The resultant cell mass, reimplanted into a suitable female, then behaves as a normal embryo, but is made up of a mixture of cells derived from both embryos. These are arranged in groups, or clones of similar cells; the products, ultimately, of one of the original embryonic cells. The question of interest at the time was the fineness of the mix (Grüneberg and McLaren, 1972). Moore and Mintz (1972) had suggested, based on outline drawings of anterior views of vertebrae assessed by eye, that vertebrae might be derived from as few as four clones of sclerotomal cells, of which two, one left one and one right, might be visible in anterior view. In a chimera these clones would all be homozygous, but might both be of one or the other genotype, or one of each. This should produce individual vertebrae, in any one chimeric mouse, which closely resemble one or the other genotype; or a vertebra in which, say, the left half resembles one genotype and the right half resembles the other genotype. We reasoned that such vertebrae should be markedly more asymmetrical than normal.

We looked at the C1 to T2 series using a different set of variates (those with the highest F-ratios) for each bone (since we did not wish to make comparisons at more than one vertebral level) and different numbers of variates (the number that gave maximum discrimination between genotypes at that vertebral level). Sine components were not omitted, since we were interested in bilateral symmetry.

Discrimination between vertebrae from individual mice classified as skeletally chimeric by Grüneberg and McLaren (1972) were tested for fit against each parental group, using the best discriminatory variate set for that vertebra. Multivariate analysis (with the relevant variates as variables) showed that some chimeric vertebrae mapped within 2χ standard deviations of one or the other parental group, some were within 2χ standard deviations of both parental groups, and some were far away from both (Table 7.1). Dendrograms (Figure 7.6) showed three groups of vertebral shapes, one including each parental type and a third (chimeric) which was very different from either parent. Asymmetry was calculated by summing the absolute value of the sine components divided by the number of sine components used. Asymmetry was never significantly greater in the chimeric vertebrae.

The message clearly indicated that the data did not support the hypothesis of Moore and Mintz (1972), because the chimeric vertebrae were clearly not systematically asymmetrical. The fairly tight grouping of the chimeric vertebrae (i.e.,

Table 7.1 Generalized distance matrix (square root of Mahalanobis's distance) for C3 vertebrae from dominant (DOM), recessive (REC), and an illustrative selection of chimeric mice (specimens X1–X21) taken from those shown in Figure 7.5. Values in **bold** type are considered to be members of the DOM or REC parental group since they lie within 2.78 standard deviations (i.e., 2χ SDUs, Mardia et al., 1979) of the parental group mean. X1 and X18 are within this distance of both DOM and REC, and, thus, could be members of either group.

	DOM	REC	X1	X8	X9	X10	X13	X18
DOM	-							
REC	3.23							
X1	**1.97**	**2.65**						
X8	5.01	5.18	4.60					
X9	5.48	4.30	5.23	4.60				
X10	19.18	18.83	20.09	17.41	15.92			
X13	9.91	9.50	13.04	9.49	10.29	9.15		
X18	**1.67**	**2.62**	2.66	4.77	5.31	18.42	10.83	
X21	**2.26**	4.77	2.88	5.18	7.20	20.40	13.09	2.88

their similarity to each other) and the very different vertebral shape achieved by chimeras suggest that:

(1) something is going on in chimeric vertebrae which we do not understand. The initial intuitive guess, that the chimeric phenotype represents the F_1 shape, that is, the shape of vertebrae from a cross between the strains is, of course, wrong since all genes are homozygous in any one cell. The chimera is clearly not an F_1 but is made up of some admixture or other of parental cell types.
(2) whatever is going on is fairly similar in all corresponding vertebrae. Perhaps the size of chimeric patches is small.

From a methodological point of view the misclassification rate between parental types (Table 7.2) was disappointing, reaching 13.4% for C6 (a rather complex shape) even when we maximized discrimination by careful choice of coefficient and coefficient number. This was no doubt due to the constraints imposed by the sample sizes, only 16 of one parental type and 25 of the other being available. Better resolution of parental type would no doubt have helped to remove the "indeterminate" category; that is, those chimeric bones that could, from our evidence, have belonged to either parental type.

7.5.2 The question of intermediacy, and the shape of F_1 vertebrae

The strains studied so far have been specially bred, and carry a number of recessive genes (albeit not frank skeletal shape genes) in one parent. We wondered how shape was inherited in a less-artificial situation. We reasoned that shape might be

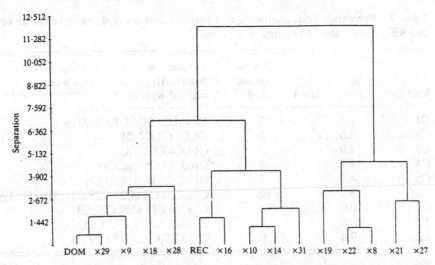

Fig. 7.6 Dendrogram (Ward's method — error sum of squares) of the distributions of the shape of the C4 vertebrae from the DOM and REC strains and 13 allophenic chimeras. The latter specimens (X numbers) form three groups, one clustering around DOM, one around REC, and a third, dissimilar group (from O'Higgins et al., 1986).

inherited similarly to height, although height is one-dimensional. It seemed logical to assume first that shape would be inherited by the transmission of a large number, rather than a small number of genes, and second, that there would either be no dominance, or that dominance, given the large number of genes involved, would be randomly shared between parents. In order to investigate this, we looked also at two inbred strains of mice and an F_1 between them. Each inbred strain would be homozygous for a different set of shape genes and the cross between them heterozygous where a gene differed between parents. In fact we did the investigation twice (O'Higgins et al., 1988; Johnson et al., 1992) using different strains.

Because the height achieved by offspring is related to mean parent height we decided to use mean parental shape as a starting point. Two populations of individuals (parental strains) of different shape can be mapped into multivariate space so that their means are separated. With three populations their relative positions in space will depend upon their morphological relationships. Sneath and Sokal (1973) have suggested that an intermediate between two shapes P_1 and P_2, identified as I, could lie on what may, as a manner of speaking, be envisaged as the direct line in some appropriate taxonomic space between taxa P_1 and P_2. If so, then $(P_1 - I) + (I - P_2)$ must equal P_1P_2 and I is a mathematical intermediate. Alternatively, the sum of the distances $P_1I + P_2I$ could be greater than P_1P_2 and an intermediate, I, could lie *off* the line P_1P_2 (Figure 7.7a). Sneath and Sokal (1973) further suggested that this could be due to overdominance or epistasis, im-

Table 7.2 Percentage misclassification of cervical and upper thoracic vertebrae of DOM and REC mice (after O'Higgins et al., 1986).

Vertebra	% misclassified	No. of variates used	Variates used (in decreasing F-ratio order). All cosines except where marked with an 'S'
C1	1.7	7	C2,C6,C13,C4,C7,C25,C19
C2	3.6	5	C4,C3,C2,C7,C1
C3	4.9	5	C14,C4,C6,C7,C2
C4	12.5	5	C16,C57,C55,S6,C48
C5	4.9	5	C13,C8,C18,S44,S4,C6
C6	13.4	12	C31,C20,C30,C23,C28,C16,S16,C38,C59,C21
C7	0.0	6	C3,C13,C18,C29,C26,C31
T1	2.8	7	C3,C4,C12C,8,C33,C15,C35
T2	9.8	6	C21,C17,C25,C13,C29,C5.

plying that mathematical intermediacy is perhaps the norm. To our knowledge this concept had never been tested.

We had at our disposal the means to test this hypothesis. First, the vertebral shapes of two populations of inbred mice could be represented with polar coordinates (radii) or with Fourier components describing the shapes. Second, a set of intermediate Polar coordinates or Fourier components can then be generated from the equation:

$$C_n = A_n + x\%(B_n - A_n), \tag{7.4}$$

where A_n is the nth polar radius of shape A, B_n is the nth polar radius of shape B, C_n is the nth polar radius of shape C, an intermediate. When $x\% = 0\%$, shape C = shape A; when $x\% = 100\%$, shape C = shape B; and when $x\%$ is intermediate, so is C. Fourier components may be substituted for polar radii. The 50% intermediate generated from the mean of means for each parent will then have all Fourier components exactly intermediate between those of parental strains.

These mathematical intermediates behaved exactly as one might have hoped, mapping at the midpoint of the interparental axis (**I**, Figures 7.7a and 7.7b). F_1s and their parents always formed a triangle. This triangle can be further analyzed by computing an **a** distance, the distance between F_1 and the 50% mathematical intermediate at **I** and resolving this trigonometrically into the **b** and **c** distances. The **c** distance represents the extent to which the F_1 resembles one parent (the nearer one) rather than the other. The **b** distance, which we have termed "new shape," is a measure of how far the F_1 lies from the line P_1P_2, which represents all possible mathematical mixtures (e.g., 10%P_1, 90%P_2, 20%P_1, 80%P_2, etc.) between P_1 and P_2.

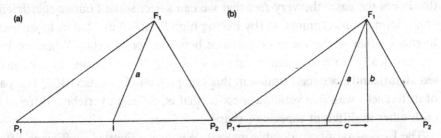

Fig. 7.7a Mean parental vertebral shapes (**P1, P2**) and their mathematical inter-mediate (**I**) will lie in a straight line in classification space. The mean shape of the vertebrae of their offspring (**F₁**) invariably lies off this line so that the par-ents and offspring form the vertices of a triangle. The distance **a** is between the position of the **F₁** and the mathematical intermediate, **I**, which lies at the mid-point of the line joining **P1** and **P2** (see text).

Fig. 7.7b The distance **a** between the position of the **F₁** and the mathematical in-termediate **I**, can be resolved into two components, **b**, the perpendicular distance from **F₁** to the parental axis, and **c** the distance from this intercept to one parent. (After O'Higgins et al., 1988)

Overall, for the stretch of the vertebral column from C1 to T2 we found that crosses between inbred strains gave a value of 0.48 ± 0.15 for the **c** distance, compared with an estimate of 0.5 assuming either no dominance or dominance equally divided between parents. For DOM and REC the corresponding figure was 0.46 ± 0.19. Interestingly, the values of the **a**, **b**, and **c** distances were all quite variable from vertebra to vertebra. We were unsure whether this signified genetic control of vertebral shape at a local level or if we were making poor es-timates of a figure that had a mean of 0.5 and a large variance. Local action of genes at different vertebral levels is well documented, especially in major "skele-tal genes," which often affect, for example, the cervical or lumbar regions only. We were unsure whether or not this applied to less-pathological shape genes, and the numbers of mice used were rather small.

In a second study (Johnson et al., 1992) we looked at a larger sample of mice from two more inbred strains and their reciprocal **F₁**s, this time using a minimal sample size of 50 individuals. These studies confirmed our earlier findings. Reciprocal **F₁**s did not differ significantly from each other, showing that no dis-proportionate amount of shape information is carried by the X- chromosome, and no large maternal effects are discernible. Differences between vertebral levels were again apparent, reinforcing our conviction of pleiotropic gene action between vertebrae.

With a larger sample size and greater familiarity with the technique, we felt at liberty to extend our methodology a little. Theoretically, a change of shape from **a** to **b** should also change all Fourier coefficients. In practice we have found that this is not the case: the very fact that we can select some Fourier coefficients as contributing to discrimination (by having high F-ratios) had led us to suspect differences either in the mean or variance between components. When we looked systematically we found that a little more than 50% of Fourier coefficients varied significantly between strains in this comparison (CBA, C57BL). The pattern of differences was also vertebra specific, that is, different vertebrae differed from each other in different aspects of shape.

The F_1 is also of considerable interest. Whereas a Fourier coefficient differs in the parental strains, its value in the F_1 may be like the CBA parent (39%), like the C57BL parent (28%), intermediate (25%), or different from both parents (8%). The legitimacy of isolating individual Fourier coefficients in this way may be questioned, but it is, perhaps, justified since we know very little of how they behave in practice, except that changes are nonintuitive. It came as something of a surprise that the coefficients behaved in this way, almost as if they were inherited. But on reflection, we were not surprised that the shapes, which are, after all, being described with the sum of the coefficients, behaved in the same way.

The overall "potence ratio" (see Wigan, 1944, for discussion of this useful but neglected term) of 6:4 in favor of CBA is reflected in generalized distances between parental strains and the F_1. If we look at individual vertebrae, however, we find that the dominance of CBA over C57BL is clear in the series from C1 to C6 and T1, but reversed for C7 and T2, indicating a switch in dominance from CBA to C57BL in the shape of the upper thoracic spine, and reinforcing the concept of local action of shape genes at different segmental levels.

7.5.3 Single gene effects

The shape relationship between the inbred strains and their F_1s proved to be a complicated one. The mechanism of shape inheritance does not conform to a simple linear model of intermediacy. This is probably a function of the large number of shape genes that may be acting concurrently in such a cross. This complex situation can be formally simplified by looking at the effects of a single segregating gene, perhaps one whose effects are large and unsubtle, almost to the point of being pathological. We chose the undulated strain (Wright, 1947; Green, 1981), which has been shown to result from a point-mutation in the highly conserved Pax-1 paired box containing gene (Balling et al., 1988). Undulated mice have a short, kinked tail and vertebrae that are malformed throughout the spine. The various processes of the vertebrae are also reduced. This is particularly notable in the transverse processes, the *tuberculum anterius* of C6, and the spinous process of T2. This raises questions for the quantitative morphologist. That is, do we particularly notice the effects of undulated, which we know to occur all along the spine, on the C6 and T2 vertebrae because the well-defined features are affected? Or are the intervening vertebrae (C7 and T1) affected less or merely in a less spectacular fashion? Or is undulated an example of local pleiotropy, which we suspect exists at different vertebral levels?

In an endeavor to answer some of these questions, we looked at the C6 toT2 vertebrae from 23 un/un mice and 23 normal litter mates (Johnson et al., 1989). Using the first 20 pairs of sine and cosine components (so as to allow comparison between vertebrae rather than maximizing separation) we found (Figure 7.8a) a clear separation between C6 vertebrae, with un/un being a little more variable and giving a larger cloud size (often a feature of mutant phenotypes). We also found an acceptable (though less clear) separation for C7 (Figure 7.8b), and considerable overlap for T1 (Figure 7.8c). The T2 vertebrae showed a complication in that a clear separation was seen, but not by genotype (Figure 7.8d). Classification here was according to whether or not the T2 vertebra had a spinous process. The issue of undulated and the spinous process is complex (Grüneberg, 1963), as the gene, when in the C57BL stock, interacts with a factor in its genetic background, and only removes the spinous process in 50% of individuals. This suggests we should only compare the shapes of normal T2 vertebrae to those of the undulated strain with a spinous process. If this comparison is made (Table 7.3) a small but significant difference is seen.

It seems clear that, despite the fact that un or Pax-1 gene is active at all segmental levels, the effect is strongly pleiotropic over the short length of the vertebral column extending from C6 to T2, with a large effect on C6, a lesser one on C7, an insignificant effect on T1, and possibly a slightly larger one on T2.

7.5.4 Further studies of metameric variation

Metameric variation is obviously best studied along as much of the vertebral column as possible. Our studies on undulated were limited to a very short region. The studies using the techniques described so far were limited to the cervical and upper thoracic region (roughly the upper third of the spine) because they depend on the idea of expressing vertebral shape credibly as a two-dimensional outline. This is less easy to justify where a vertebra has a considerable antero-posterior dimension and its lateral view may contain as much shape information as its anterior projection.

In order to extend (1) the amount of the column sampled and (2) into the third dimension, we have modified our methodology. At the same time we have retained Fourier analysis and applied it in what we believe is a novel fashion. We focused on a number of variables (i.e., measurements taken between landmarks, not necessarily in any particular plane) captured from each presacral vertebra in

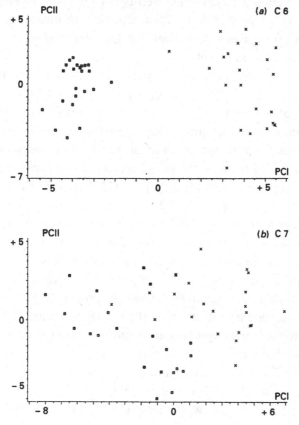

Fig. 7.8a through d Scores of individual normal and undulated vertebrae on first and second principal components. Shown are: C6 (plot a), C7 (plot b), T1 (plot c), and T2 (plot d). Squares = normal, x = undulated and stars indicate undulated without a spinous process (see text). (From Johnson and O'Higgins, 1989.)

the column, to examine changes in a particular feature from segment to segment (Johnson and O'Higgins, 1994). We used a frame grabber to save the images taken from a dissecting microscope via a TV camera. Employing a specially designed program to record the Cartesian coordinates of landmarks identified using a mouse, these coordinates were written to disc. Distances between landmarks can, thus, be collected fairly quickly and painlessly. Graphs of the magnitude of each feature against vertebral number can then be plotted (Figure 7.9). These may be used to visually assess metameric trends, the degree of within-group variation, and to compare patterns of metameric variation between variables. They also form a basis for the computation of average Euclidean distance (using 20 measurements per variable, one from each presacral vertebra). However, Euclidean distances, since they assess absolute distances between groups, take no account of within-

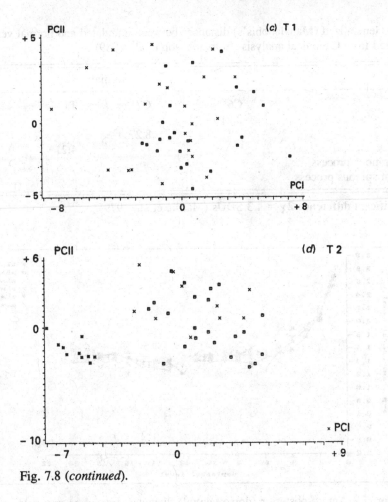

Fig. 7.8 (*continued*).

group relative to between-group variability; that is, they do not allow statistical significance to be precisely assessed. Mahalanobis's distance, on the other hand, does allow for a correct statistical appraisal. As a general rule, a reliable estimate of Mahalanobis's distance requires that the number of variables used must be considerably less than the number of individuals (see Section 7.4.3). For 20 variables (one from each presacral vertebrae) we would therefore need a sample of, say, 50 individuals. It is therefore desirable to summarize the original data in some way and express it using fewer variables.

We have applied the Fourier series, which is well adapted to modeling gently undulating shapes. In this case the gently undulating shape is not the vertebral outline but the graphical trace of a particular variable down the column. Since the Fourier series is best suited for circular or cyclical variation, and as we cannot guarantee that the magnitude of a variable, V, is similar for C1 and L3, we have

Table 7.3 Generalized (Mahalanobis's) distances between undulated and normal verte-brae derived from Canonical analysis (from Johnson et al., 1989).

Undulated	Normal			
	C6	C7	T1	T2
C6	22.16			
C7		8.27		
T1			4.21*	
T2 with spinous process				5.96
T2 without spinous process				9.72

* No significant difference $2\chi = 4.3$ SDUs (Mardia et al., 1979)

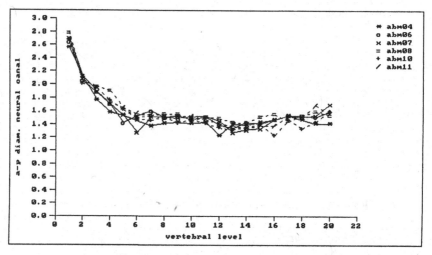

Fig. 7.9 Antero-posterior (dorso-ventral) diameter, treated as raw data, of the neural canal in the first 20 vertebrae of six C57BL mice (abm04 to abm11). (From Johnson and O'Higgins, 1993.)

"circularized" our dataset by reflecting it. Thus, the values of V (V_1 to V_{20}) along the column are represented by a series $V_1 \ldots V_{20} \ldots V_1$. The reflected data are then modeled by a portion of the Fourier series (Figure 7.10). In the illustration the original 20 variables are modeled quite accurately with six Fourier coefficients. Since the whole column is represented by 2π radians, and each vertebra by $2\pi/n$, different numbers of presacral vertebrae can also be accommodated.

Our studies using this technique are at present limited to a single study using a small number of mice from CBA and BALB/c inbred strains and the F_1s between them. However, this study has justified the methodology and provided some interesting facts about metameric variation. We measured only four dimensions from each vertebra, the antero-posterior (dorso-ventral) and transverse diameters of the neural canal, the height of the vertebral body, and the height of the spinous process. First of all, different dimensions showed different amounts of variability, with the least in the transverse diameter of the neural canal and the greatest in the spinous process. Second, the pattern of metameric variation seems to vary along the column. The antero-posterior diameter of the neural canal diminishes rapidly in the upper cervical region and again, more slowly, in the lower, whereas over the thoracic and lumbar regions, it is virtually constant. The transverse diameter of the neural canal follows a different pattern, increasing gradually through the cervical spine from C2 onwards, decreasing through the thorax, and then increasing again in the lumbar region. The antero-posterior height of the vertebral body and the spinous process each have their own different pattern. However, if we look at the first five cosine components, a_1 to a_5, (a_0 omitted because it is a scaling coefficient, and much larger than the others) we find strong correlations between most values. Negative correlations occurred between the spinous process length and antero-posterior diameter of vertebral body compared to the other dimensions. This implies that these dimensions vary inversely. If we invert the relevant curves (i.e., plot 1/size against vertebral number) we find (Figure 7.11) that all curves except that for dorso-ventral diameter of the neural canal are similar. The spinous process curve has a local perturbation corresponding to the long spine on T2 which we would argue, because of its inconsistency, is a specialization. This similarity is important. It means that a single descriptor (such as the concentration of a hypothetical morphogen along the column) could specify three of the four dimensions we measured. The significance of these find-

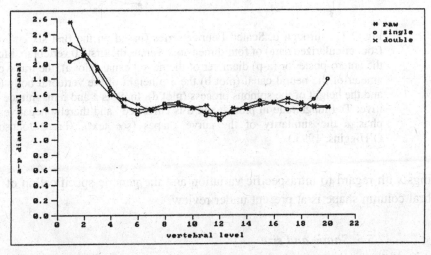

Fig. 7.10 The data of Figure 7.9 expressed as an arithmetical mean of the raw data (raw), with the first six components of a Fourier series based on the raw data (single) and with the first six components based on reflected (circularized) data (double). (From Johnson and O'Higgins, 1993.)

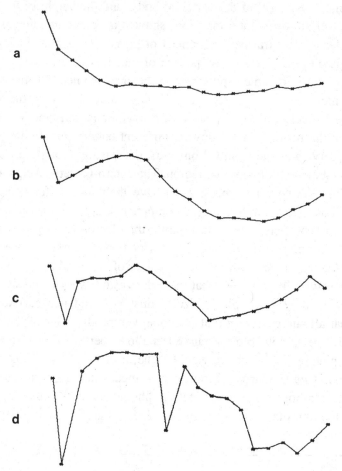

Fig. 7.11a through d. Scaled Fourier series (based on the first six components from circularized data) of four dimensions versus the first 20 vertebrae. Modeling the antero-posterior (a-p) diameter of the neural canal (plot a), the transverse diameter of the neural canal (plot b) the a-p height of the vertebral body (plot c) and the height of the spinous process (plot d). In plots a and b magnitude is positive: The magnitude in plots c and d is "inverted" and thereby negative to emphasize the similarity of the curve shapes (see text). (From Johnson and O'Higgins, 1993.)

ings with regard to intraspecific variation and the genetic specification of vertebral column shape is at present under review.

7.5.5 Shape and size

We were also interested in the change of shape associated with change of size. This size-related shape change is usually thought of as a result of mechanical influences. For example, engineering mechanics dictates that because of surface area:volume relations, a large cylinder must be proportionately thicker than a small one of the same strength. Although this approach is classically confined to interspecies comparisons, there is no reason why it should not also apply within a species. Shape should not vary isometrically between small and large mice — a larger mouse should have bones of a different shape. Selection for size, therefore, should constrain shape, and mice selected for size should have bones less variable than unselected controls. This raises two questions. Would only one shape fulfill the necessary mechanical constraints, or might a different solution arise from replicated selection? Also, would the processes of selection for larger and smaller size necessarily mirror each other?

We were fortunate in having access to material bred by Falconer (1973), allowing us to test these queries (Johnson et al., 1988a). He took a (very) outbred strain of mice, divided it into six replicates and selected for large and small six-week body weight in each line. We looked at the T1 and T2 vertebrae from six replicate large lines, six small lines, and six control lines. After 13–14 generations of selection, we asked ourselves if the increase (or decrease) in size led to measurable size-related shape change, and whether any shape change produced was similar among different replicates of the experiment.

An optimum correct classification of random tenths of the data was achieved for T1 and T2 by using the first 15 sine and first 15 cosine components from a sample of 15 males and 15 females from each replicate. We were able to classify 45–63% of T1 vertebrae correctly by replicate and 86–90% by size group. Canonical analysis of the first 30 components showed well-separated centroids of the large (QLA-QLF), control (QCA-QCF) and small (QSA-QSF) families of T1 vertebrae (see Johnson et al., 1988a). Clustering, by Ward's (1963) method, grouped replicates into large, control, and small families (Figure 7.12). T2 vertebrae classified less well (37–47% by replicate, 72–81% by group). A Canonical plot of T2 for the same three groups gave less well-separated centroids (see Johnson et al., 1988a). Ward's method showed two groups of controls clustering with small and large families, respectively (Figure 7.13). These proved to be the lightest and heaviest of the control replicates.

Clearly, the selection for size had also affected shape. We had no difficulty in splitting the large and control families for T1 and for T2 with the exceptions already noted. It was much more difficult, however, to classify replicates within groups, suggesting that the within-group shape was more similar than the between-group shape. Variability (roughly inverse to the difficulty of obtaining a correct classification, or proportional to the number misclassified) was least in small mice. The large mice showed the widest spread in the classification space. Taken together, these facts could be used to support the argument that shape (i.e., the most economical or efficient use of resources) was most limited by engineering constraints in small mice, whereas selection for large size allowed for more variation.

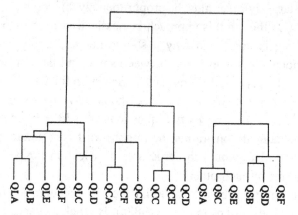

Fig. 7.12 Minimum spanning tree (Ward's method) for large (QLA-QLF), control (QCA-QCF) and small (QSA-QSF) families of T1 vertebrae. (From Johnson et al., 1988a.)

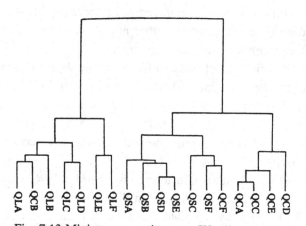

Fig. 7.13 Minimum spanning tree (Ward's method) for large (QLA-QLF), control (QCA-QCF) and small (QSA-QSF) families of T2 vertebrae. (From Johnson et al., 1988a.)

The distribution of replicates in classification space is also informative. Canonical axes allow us to visualize high dimensional space as a series of projections into two dimensions. The first Canonical axis is calculated to take up the largest between-group variance relative to the within-group variance. The second and subsequent orthogonal axes represent progressively smaller components of the variance. Working with primates, Ashton (1981) showed that in certain circumstances, biological significance can be attributed to the disposition of groups on these axes. Jolicoeur (1963) has suggested that the first principal component can be used for (allometric) size correction. (Canonical axes may be viewed as equivalent to principal components.)

In this study we removed size as a variable by equalizing vertebral areas. Any differences are, thus, not due to size *per se*, but could be due to the effect of size on shape. Viewing Canonical axis 1 for the T1 vertebrae, large and small families occupy opposite poles, and controls are intermediate (see Johnson et al., 1988a). On the second Canonical axis, small and large families occupy one pole and the controls the other. This is also largely true for T2 if we neglect the groups that are misclassified (see Johnson et al., 1988a). We suggest that selection for size produces vertebrae whose shape is size related (spread along axis 1); whereas controls, free from the constraint of selection, vary along axis 2, that is, in a non-size-related way, but are still arranged so that the largest (QCB) and smallest (QCF) tend toward the large and small poles of Canonical axis 1.

7.5.6 Another aspect of size — varying vertebral shape and age

Consideration of size-related shape change led us to consider another situation in which shape and size vary; namely, the process of growth and maturation. We looked at T1 and T2 vertebrae from an F_1 between BALB/c and CBA mice aged 25–60 days, both mathematically and using the scanning electron microscope (Johnson et al., 1988b).

We found it helpful to look at three types of plots: a Canonical plot of individual values derived from the first 20 sine and 20 cosine components, showing the first three Canonical axes (see Figures 4a and 4b in Johnson et al., 1988b); similar plots based on the *centroids* of each age group and shown here (Figures 7.14a, b); and two-dimensional plots of Canonical axis 1 versus Canonical axis 2 (Figures 7.15a, b). Plots of all individuals (not depicted here; see Johnson et al., 1988a) showed a large cloud of vertebrae aged 30–60 days, with most 25-day-old individuals well separated. This was even clearer from a plot of the group means (Figures 7.14a, b). The 25-day-old vertebrae were 7.63 (T1) and 6.51 (T2) SDUs away from their 35-day-old neighbors, whereas the average separation between other age groups was 3.65 SDUs (T1) and 3.98 SDUs (T2). Because the first two Canonical axes account for 75% (T1) and 83% (T2) of the variation, as opposed to 81% and 87% for the first three axes, little information is lost by simplifying the plot (Figures 7.15a, b). The pattern of shape change with age is now clarified. In two dimensions it is clear that the 25-day vertebrae are outliers and that the trend in the data is well represented by a regression line fitted through the remaining points.

This bears out the intuitive feeling that there should be a smooth seamless linear shape change underlying this kind of discontinuous pseudolongitudinal data. The 25-day mice proved to be an exception. In fact, 25 days was the youngest practicable age for this technique of bone preparation. Below this age vertebrae were incompletely ossified to such an extent that they tended to fall apart during preparation, and we might have expected a difference between vertebral mor-

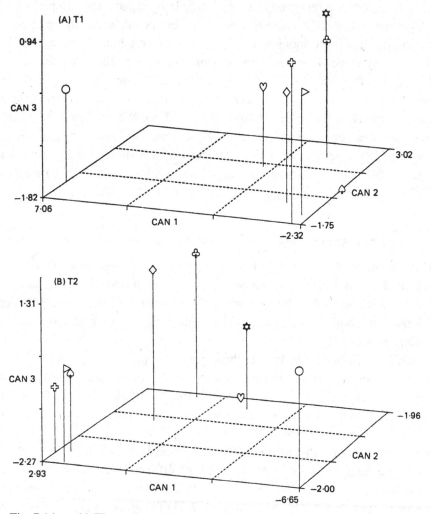

Fig. 7.14a and b Three-dimensional plot of the group centroids showing the shape relationships of T1 (plot A) and T2 (plot B) vertebrae at various ages. Balloon = 25d;   star = 30d;   heart = 35d;   club = 40d;   diamond = 45d;   spade = 50d; cross = 55d; flag = 60d . Based on the first 20 sine and 20 cosine components. The mirror imaging along axis 1 is artifactual and may be ignored. The first three axes account for 87.4% (T1) and 83.5% (T2) of the variance. (From Johnson et al., 1988b.)

phogenesis at younger ages and maturation of older bones. In fact, the rather slow, almost linear shape change over the period 30–60 days, when the vertebrae are increasing little in size, implies a constant pattern of remodeling. Although outside the scope of the present review, we were able to demonstrate this constant resorption pattern by scanning electron microscopy (O'Higgins et al., 1989).

7.5.7 *Other possible contributory factors — diet and virus infection*

The information contained in biological shape must represent the sum total of genetic and environmental factors acting upon an organism. So far we have concentrated upon genetically mediated shape change, because this is the largest determinant of the final shape reached by an adult bone. But the skeleton is known to be plastic, and to be closely integrated with musculature. The musculoskeletal

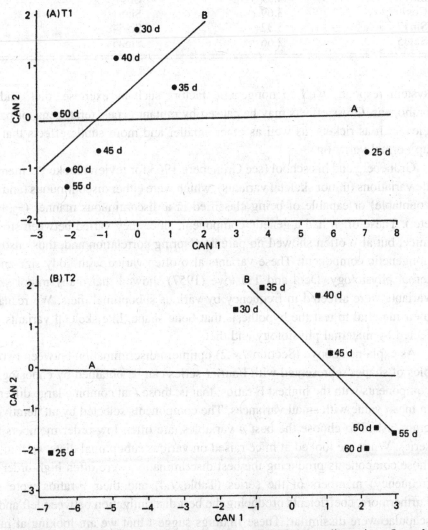

Fig. 7.15 The same data as in Figure 7.14 but reduced to two dimensions. In each case the trend line A is fitted to all data, whereas B is to data from mice aged over 30d. The first two Canonical axes account for 82% (T1) and 75% (T2) of the variance. (From Johnson et al., 1988b.)

Table 7.4 Best discriminators of shape (Variate = Fourier component number; in this notation even numbers are cosines, odd numbers are sines) for C1 and C2 vertebrae from mice raised on suboptimal diets (after Johnson et al., 1990).

C1(atlas)		C2(axis)	
Variate	F-ratio	Variate	F-ratio
Cos17	6.29	Cos3	9.51
Cos46	4.65	Cos12	5.44
Sin17	4.57	Cos4	3.84
Cos1	3.69	Sin57	3.59
Sin47	3.32	Sin54	4.30
Sin45	2.66	Sin25	3.26

system responds well to nongenetic factors such as exercise, diet, and even pathogens. Gross effects may be caused by mutant genes, teratogens, or pathologies such as rickets, as well as other smaller and more subtle effects that make up normal variation.

Grüneberg and his school (see Grüneberg 1963 for review) looked at many subtle variations (minor skeletal variants), which were either discontinuous (and hence countable) or capable of being classified in a discontinuous manner (−,+,++, etc.). These often had a genetic component, since they varied between strains of mice, but also often showed no parent/offspring correlation and, thus, also had a nongenetic component. These variants also often varied with body size and maternal physiology. Deol and Truslove (1957) showed that many minor skeletal variants were affected in frequency by various suboptimal diets. We reanalyzed their material to test the hypothesis that bone shape, like skeletal variants, is affected by maternal physiology and diet.

As explained above (Section 7.4.2) optimum discrimination between two samples of shapes represented with Fourier series can be obtained by choosing those components with the highest F-ratios; that is, those that combine large differences in mean value with small variances. The components selected by an iterative program, asked to choose the best n variables, are often low-order members in the series. When we looked at mice raised on various suboptimal diets we found that those components producing the best discrimination were often high-order (high frequency) members of the series (Table 7.4) and their F-ratios were small. Furthermore, coefficients producing the best discrimination between left and right scapulae were dissimilar. These findings suggest that we are looking at random variations and that there are, at best, only small shape differences between mice raised on different diets. Canonical analyses of the first ten Fourier coefficient pairs (see Figure 1b in Johnson et al., 1990), which allow for good shape reconstruction and contain most of the shape information in the samples, confirmed this (O'Higgins and Johnson, 1988). The dietary treatment did, however, produce significant reduction in femoral and humeral length between controls and dietarily challenged mice. Clearly, diet has an effect on limb length, and hence probably on overall size, but not on vertebral shape.

Very similar findings resulted from a study (Johnson, unpublished) of C57BL mice infected with the Bittner virus (Wickramaratne, 1974). We thought that the presence of this virus (which causes carcinoma of the breast in mice) might have a discernible effect on vertebral shape (following Grüneberg, 1970). Again, Fourier analysis merely showed a slight dispersion of cloud size in treated mice, rather than a clear change in shape or size. We must conclude that the type of changes that we have investigated are largely based upon genetic change, and that shape in mouse vertebrae is under very highly canalized genetic control.

References

Ashton, E. H. (1981). The Australopithecinae: Their biometrical study. *Symp. Zool. Soc. (Lond.)* **46**, 67–126.

Balling, R., Deutsch, U. & Gruss, P. (1988). Undulated, a mutation affecting the development of the mouse skeleton, has a point mutation in the paired box of Pax-1. *Cell* **55**, 531–5.

Bookstein, F. L., Strauss, R. E., Humphries, J. M., Chernoff, B., Elder, E. L. & Smith, G. R. (1982). A comment upon the use of Fourier methods in Systematics. *Sys. Zool.* **31**, 92–5.

Deol, M. S. & Truslove, G. M. (1957). Genetical studies on the skeleton of the mouse XX: Maternal physiology and variation in the skeleton of C57BL mice. *J. Genet.* **55**, 288–312.

Erlich, R., Baxter Pharr, R. & Healy-Williams, N.(1983). Comments on the validity of Fourier descriptors in systematics: a reply to Bookstein et al. *Sys. Zool.* **32**, 202–6.

Falconer, D. S. (1973). Replicated selection for body weight in mice. *Genet. Res.* **22**, 291–321.

Goodwin, B. (1993). Homology and a generative theory. *Acta Biotheoretica* **41**, 305–14.

Grüneberg, H. & McLaren, A. (1972). The skeletal phenotype of some mouse chimeras. *Proc. Roy. Soc. (Lond.-B)*. **182**, 9–23.

Green, M. C. (1981). *Genetic Variants and Strains of the Laboratory Mouse*. Stuttgart: Fischer Verlag.

Grüneberg, H. (1963). *The Pathology of Development*. Oxford: Blackwell.

Grüneberg, H. (1970). Is there a viral component in the genetic background? *Nature (Lond.)* **225**, 39–41.

Johnson, D. R. & O'Higgins, P. The inheritance of patterns of metameric variation in the mouse vertebral column. *J. Zool.* (in press).

Johnson, D. R., O'Higgins, P., McAndrew, T. J., Adams. L. M. & Flinn, R. M. (1985). Measurement of biological shape: a general method applied to mouse vertebrae. *J. Embryol. Exper. Morph.* **90**, 363–77.

Johnson, D. R., O'Higgins, P. & McAndrew, T. J. (1988a). The effect of replicated selection for body weight in mice on vertebral shape. *Genet. Res. (Cambridge)* **51**, 129–35.

Johnson, D. R., O'Higgins, P. & McAndrew, T. J. (1988b). The relationship between age, size and shape in the upper thoracic vertebrae of the mouse. *J. Anat.* **161**, 73–82.

Johnson, D. R., O'Higgins, P. & McAndrew, T. J. (1989). The effect of the gene undulated (un) on the shape of cervical and upper thoracic vertebrae in the house mouse (Mus musculus). *J. Anat.* **163**, 49–55.

Johnson, D. R., O'Higgins, P. & McAndrew, T. J. (1990). The effect of diet on bone shape in the mouse. *J. Anat.* **172**, 213–20.

Johnson, D. R., O'Higgins, P., McAndrew, T. J. & Kida, M.Y. (1992). The inheritance of vertebral shape in the mouse I: a study using Fourier analysis to examine patterns of inheritance in the morphology of cervical and upper thoracic vertebrae. *J. Anat.* **180**, 507–14.

Jolicoeur, P. (1963). The multivariate generalisation of the allometry equation. *Biometrics* **19**, 497–9.

Mardia, K. V., Kent, J. T. & Bibby, J. M. (1979). *Multivariate Analysis*. London: Academic Press.

Moore, W. J. & Mintz, B. (1972). Clonal model of vertebral column and skull development derived from genetically mosaic skeletons in allophenic mice. *Devel. Biol.* **27**, 55–70.

O'Higgins, P., Johnson, D. R. & McAndrew, T. J. (1986). The clonal model of vertebral column development: A reinvestigation of vertebral shape using Fourier analysis. *J. Embryol. Exp. Morph.* **96**, 171–82.

O'Higgins, P. & Williams, N. W. (1987). An investigation into the use of Fourier coefficients in characterising cranial shape in primates. *J. Zool. (Lond.).* **211**, 409–30.

O'Higgins, P. & Johnson, D. R. (1988). The quantitative description and comparison of biological forms. *CRC Crit. Rev. Anat. Sci.* **1**, 149–70.

O'Higgins, P., Johnson, D. R. & McAndrew, T. J. (1988). Mathematical and biological intermediacy in bone shape. Fourier analysis of cervical and upper thoracic vertebrae in the mouse. *J. Zool. (Lond.)* **214**, 373–81.

O'Higgins, P., Johnson, D. R. & Paxton, S. K. (1989). The relationship between age, size and shape of mouse thoracic vertebrae: a scanning electron microscopic study. *J. Anat.* **163**, 57–66.

Read, D. W. & Lestrel, P. E. (1986). A comment upon the uses of homologous-point measures in systematics: A reply to Bookstein et al. *Syst. Zool.* **33**, 241–53.

SAS Institute (1982). Users Guide: Statistics. Cary N.C.

Sneath, P. H. A. (1967). Trend surface analysis of transformation grids. *J. Zool. (Lond.)* **151**, 65–122.

Sneath, P. H. & Sokal, R. R. (1973). *Numerical Taxonomy*. San Francisco: W. H. Freeman.

Sokal, R. R. & Rohlf, F. J. (1969). *Biometry*. San Francisco: W. H. Freeman.

Steele, R. G. D. & Torrie, J. H. (1980). *Principles and Procedures of Statistics*. New York: McGraw Hill.

Ward, J. H., Jr. (1963). Hierarchical grouping to optimize an objective function. *J. Am. Stat. Assoc.* **58**, 236–244.

Wickramaratne, G. A. (1974). The role of latent viruses in subline differentiation in inbred strains of mice. *Genet. Res. (Cambridge.)* **24**, 11–17.

Wigan, L. G. (1944). Balance and potence in natural populations. *J. Genet.* **46**, 150–60.

Wright, M. E. (1947). Undulated: a new genetic factor in *Mus musculus* affecting spine and tail. *Hered.* **1**, 137–41.

Yasui, K. (1986). Method for analysing outlines with an application to recent Japanese crania. *Am. J. Phys. Anthrop.* **71**, 39–49.

8

Application of the Fourier Method in Genetic Studies of Dentofacial Morphology

LINDSAY C. RICHARDS, GRANT C. TOWNSEND AND
KAZUTAKA KASAI
The University of Adelaide
Nihon University School of Dentistry

8.1 Introduction

Quantifying the relative contributions of genetic and environmental influences on morphological variation in dental and facial structures is an important biological problem that continues to attract the interest of anatomists, physical anthropologists, geneticists, and clinicians in medicine and dentistry. Progress in this field has been limited for a number of reasons, including the: (1) complexity of craniofacial growth processes; (2) limited opportunities for experimental investigation in humans; (3) multifactorial mode of inheritance of dental and facial structures; and (4) significant methodological difficulties inherent in accurately describing the form of faces and teeth.

In the study of dentofacial variation, nonmetric, or qualitative, classifications may provide a satisfactory description of variations in shape, but these data cannot always be analyzed appropriately by the more powerful statistical methods. Conversely, metric methods based on combinations of linear and angular variables may provide reasonable representations of size, which are suitable for statistical analysis, but less appropriate for the description of shape variation.

Most genetic studies of dentofacial structures have relied on simple qualitative descriptors or on combinations of linear and angular variables to characterize morphology. These have been analyzed with the classical approaches of quantitative genetics, including twin and family studies, interpopulation comparisons, and the examination of individuals with chromosomal abnormalities.

Fourier descriptors provide a powerful numerical method to describe both the size and shape of complex structures (Lestrel, 1974; 1989). Once shapes have been quantified, statistical techniques can be applied in an attempt to detect underlying genetic factors that may not be apparent with less sophisticated methods.

To investigate the relative contributions of genetic and environmental factors

189

to the observed variation in dental and facial structures we have obtained data from several sources, including:

- dental casts and facial stereophotographs of more than 300 pairs of South Australian twins and their families enrolled in an ongoing study being undertaken in the Department of Dentistry at the University of Adelaide (Brown et al., 1987; Townsend et al., 1988).
- dental casts of healthy, young Japanese and Caucasian adults with no evidence of dental disease or malocclusion obtained from collections in the Department of Orthodontics, Nihon University School of Dentistry at Matsudo, and the Department of Dentistry, The University of Adelaide.
- cephalometric records obtained during a longitudinal study of the growth of Australian Aborigines conducted at Yuendumu in Central Australia (Brown and Barrett, 1973) and from the reference collection of the Department of Orthodontics, Nihon University School of Dentistry at Matsudo.

Using these records, the following investigations have been performed:

- comparisons of dental arch form in Japanese, Australian Aboriginals, and Caucasians.
- analysis of dental arch form in South Australian twins.
- comparisons of mandibular shape in Australian and Japanese subjects.
- morphometric analysis of facial profiles in twins and their families.

In each of these studies we have described the dental and facial morphology using Fourier functions to quantify size and shape with information derived from: (1) dental casts (Brown et al., 1983), (2) cephalometric radiographs (Richards, 1983); and (3) standardized photographs (Richards and Brown, 1981).

Standard reference points, or outlines established by evenly distributed radii constructed around the midpoint of a base line, were traced onto acetate drafting film (Figure 8.1). These data points were digitized to provide Cartesian coordinates with appropriate corrections for radiographic and photographic enlargements. The coordinates were then converted to polar form and Fourier coefficients computed to summarize size and shape using the methods described by Sekikawa (1986). The resultant Fourier functions were of the form:

$$y_i = \frac{a_0}{2} + \sum_{i=1}^{N} (\alpha_i \cos i\theta + \beta_i \sin i\theta), \tag{8.1}$$

where the constant a_0 represents the mean of the observed radii, the α_i and β_I are the Fourier coefficients and i represents the sequence of increasing frequencies included to generate the function. The amplitudes were computed from the coefficients in Eq. 8.1 as:

$$a_i = \sqrt{\alpha_i^2 + \beta_i^2} \tag{8.2}$$

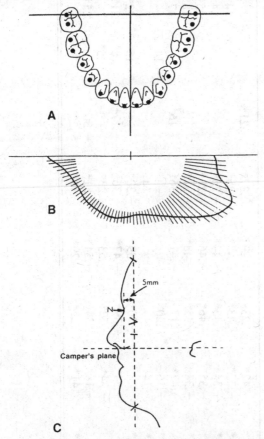

Fig. 8.1 Data acquisition methods for (A) dental arches with reference points indicated; (B) mandibular outlines illustrating the construction of radii to define outline points; and (C) facial profiles.

The number of terms required to satisfactorily describe the form depended on whether the aim was to summarize the shape in general or to specifically describe the finer detail. Generally, three amplitudes were sufficient to represent simple forms such as the dental arches, whereas eight to ten amplitudes were required for more complex structures such as facial profiles and mandibular outlines.

To interpret the Fourier amplitudes, we calculated correlations between each of the amplitudes and the radial distances from which they were generated. The pattern of these correlation coefficients indicated the anatomical regions described by each of the Fourier amplitudes. For example, Table 8.1 summarizes this information for facial profiles in South Australian twins, showing the correlations between selected radial lengths and the Fourier amplitudes. Based on this information the constant Fourier term (a_0) was interpreted as reflecting "size," with large positive correlations between this term and most radial distances. The first

Table 8.1 Significant (p < 0.05) correlations between selected radial lengths (dist01–dist61) and each of the Fourier amplitudes (a_0–a_9) describing facial profiles in South Australian twins. The anatomical region associated with each of distances is indicated. Nonsignificant correlations are not shown. Significance of p < 0.01 is indicated with an *.

	Region	a_0	a_1	a_2	a_3	a_4	a_5	a_6	a_7	a_8	a_9
dist01	forehead	.16	.34*	.25*	.42*	.23*	.29*	.37*	.21*		
dist06		.58*	.46*	−.44*	−.27*	−.43*	−.43*	−.25*			.14
dist11	nasion	.62*	.41*	−.44*	−.34*	−.45*	−.44*	−.25*			.14
dist16		.63*	.41*	−.43*	−.35*	−.46*	−.46*	−.28*			.15
dist21	nose	.65*	.38*	−.38*	−.38*	−.45*	−.48*	−.31*			.18*
dist26	nose	.61*			−.47*	−.17	−.41*	−.36*	−.16		.17
dist31		.21*	−.30*	.45*	−.16	.20*					
dist36		.16		.21*	.14			.14		−.14	
dist41	lips	.53*	.31*	−.19*		−.30*	−.39*	−.20*		.15	.22*
dist46	lips	.65*	.38*		−.34*	−.39*	−.47*	−.36*	−.17*	.18*	.21*
dist51		.60*	.45*		−.34*	−.52*	−.53*	−.23*		.25*	
dist56	chin	.62*	.52*		−.18*	−.34*	−.54*	−.51*	−.36*		.25*
dist61	chin	.19*	.32*	.29*	.41*	.25*	.29*	.38*	.21*		

Fourier amplitude (a_1) showed larger positive correlations with radial distances involving the forehead and chin regions. This amplitude was roughly interpreted as a "face height" factor. The subsequent amplitudes were associated with facial depth (a_2), nasal length (a_3), the nasolabial curve (a_4), and lip form (a_{5-6}). Studying the patterns of correlations between the Fourier amplitudes and the anatomical regions represented by the radii can provide important insights into variation in the described shapes. However, this information needs to be interpreted with caution as the correlations are commonly less than r = 0.70 suggesting that more than 50% of the variation remains unexplained. In addition, the presence of significant correlations between the Fourier amplitudes and radii does not establish any direct underlying biological relationships or controlling factors. Rather, the observed pattern of correlations provides a general guide to the regions represented by each of the amplitudes, making interpretation of differences between the Fourier functions derived for different individuals or populations more meaningful.

To determine the magnitude of errors involved in obtaining data in each of our studies, analyses of the differences between repeated measurements were undertaken. In general, the results indicated that mean differences were small and not statistically significant at the 5% probability level. In addition, error variances (S_E^2) were calculated from estimates of the standard deviations of single measures (S_E) as:

$$S_E = \sqrt{\frac{(x_i - x_2)^2}{2n}}, \qquad (8.3)$$

where the repeated measurements are X_1 and X_2 and n is the sample size (Dahlberg, 1940). For most variables the contribution of the error variance (S_E^2) to the total variance was small (less than 10%). For a few variables the error variance was relatively large, contributing more than 20% to observed variation. In these instances the results were interpreted with caution.

Although population comparisons can provide indirect evidence for the influence of genetic factors, twin studies provide a more direct insight into the relative contributions of genetic and environmental factors to the variation in size and shape.

Most dental and facial traits display continuous variation and are thought to be under polygenic control, with many genes exerting small influences. The classical model applied by quantitative geneticists (Falconer, 1989) describes the phenotypic expression of quantitative variables (P) as reflecting the combined influence of the genotype (G), which is the transmitted genetic information, and variation due to environmental influences (E), such as general health, diet, maternal health, climate, and other factors. This can be summarized by the equation P = G + E or expressed in terms of variances where $V_P = V_G + V_E$. Both the genetic and environmental contributions to phenotypic variance can be partitioned into a genetic variance (V_G) composed of an additive genetic component (V_A) representing the combined effects of all of the genes influencing a trait, a dominance component (V_D), which represents the effect of interactions between genes at a single locus, and an epistatic component (V_I) representing the results of interactions between genes at different loci. The relative contribution of the additive genetic factors (V_A) to the total observed variation (V_P) is referred to as the heritability (h^2). Estimates of heritability are population specific and depend on both the gene frequencies within populations and the nature and extent of the various environmental influences.

There are a number of methods for deriving estimates of heritability, each involving statistical and genetic assumptions which may or may not be realistic. Comparisons of traits in pairs of monozygous (MZ) and dizygous (DZ) twins provide a powerful method for partitioning the observed variation into genetic and environmental components and estimating heritabilities. The approach we adopted for the genetic analysis of Fourier amplitudes followed that of previous studies of occlusal variation in twins (Corruccini and Potter,1980; Sharma and Corruccini,1986; Townsend et al., 1988). First, some assumptions implicit in the traditional twin model were tested in the manner described by Christian et al., (1974) and Kang et al., (1977). A modified *t*-test, based on nested twin data in which the among-pair mean squares were used as the error term, was applied to test whether there were any significant differences in mean values of the Fourier constant and the computed amplitudes between MZ and DZ twins (Christian, 1979). Heterogeneity of total variances for MZ and DZ twins was then tested using a one-way ANOVA in which twin pairs were treated as groups of two, to provide among-pair mean squares (Christian et al.,1974). In addition, the ratio of the among-pair to the within-pair mean squares in the DZ twins was calculated to indirectly test whether environmental covariance in MZ twins might have inflated heritability estimates (Christian,1979).

Where there were no significant differences between the MZ and DZ means and total variances, heritability (h^2) was estimated from the intraclass correlation, r', by the path analysis method described by Lundstrom (1954) where the heritability is equal to:

$$h^2 = 2(r'_{MZ} - r'_{DZ}). \tag{8.4}$$

These estimates can, theoretically, range from 0 to 1, reflecting the proportion of observed phenotypic variation due to genetic factors.

8.2 Studies of dental arch form

Many methods have been developed to describe dental arch morphology, including simple subjective classifications (Hrdlicka, 1916), combinations of linear dimensions (Moorrees, 1959), complex curve-fitting methods (reviewed by Jones and Richmond, 1989) and the application of polynomials (Lu, 1966; Richards et al., 1990). The nonmetric methods and some of the simple techniques involving combinations of dimensions are easy to apply but are not very powerful descriptors of shape. The Fourier method has a number of advantages, most significantly its ability to separate the observed variation into components recognizable as "size" and "shape." In general, other mathematical methods tend to combine and thereby confound these two types of variation.

Comparisons of dental arch size and shape variation among populations provide indirect evidence of genetic differences between ethnic groups. To investigate these arch morphology differences, dental casts of 84 Australian Aborigines (35 male, 49 female), 48 Caucasians (36 male, 12 female) and 70 Japanese (38 male, 32 female) were selected and photographed. Tracings of the dental arches were made and the midpoints of the incisal edges and the buccal cusp tips of maxillary and mandibular teeth defined. The radial distances from a center at the midpoint of the line connecting the disto-buccal cusps of the second molars, and each of the identified dental reference points were used to calculate the Fourier functions. In theory, an uneven angular distribution of the radii constructed in this way can give an inappropriate weighting to some reference points. However, in the case of dental arches it can be argued that the angular distribution is not so uneven as to significantly bias the results, and Fourier amplitudes describing the maxillary and mandibular dental arches can be derived and interpreted by comparing Fourier amplitudes with combinations of linear arch dimensions.

Repeated measurement of a series of subjects indicated that the errors involved in deriving the Fourier amplitudes were small and did not make a significant contribution to the total observed variation (Table 8.2). In general, the constant term in the Fourier series (a_0) described arch size whereas the first amplitude (a_1) indicated the ratio of arch depth/width, and the second and third amplitudes (a_2 and a_3) described the relative taper of the arch.

The amplitudes describing the dental arches in males are shown in Table 8.3, and illustrate significant differences between groups. For example, in the mandibular arch, the constant or a_0 term, describing size, tended to be larger in Australian Aborigines than in either Caucasians or Japanese. Differences in the second amplitude, a_2, for Australian Aborigines were smaller than those for Caucasians and Japanese. For the third amplitude, a_3, Australian Aborigine values were larger than either Caucasian and Japanese. These differences in shape between groups suggested that Australians tended to have more parabolic arches. Figure 8.2 shows the arch forms described by Fourier descriptors.

Plots of the mean values for the Fourier amplitudes, reflecting the ratio of arch depth/width (a_1) and arch taper (a_2) in each of the populations (Figure 8.3), illustrated the differences in dental arch shape between the groups. In the maxilla the Australian Aboriginal males and females were characterized by a wider (larger

Table 8.2 Mean difference between 20 repeated determinations of Fourier amplitudes, standard errors of mean differences (S_E), and relative contributions of the measurement error (S_E) to total observed variance.

	Mean	S_E	Error %
a_0	−0.01	0.16	6.96
a_1	−0.02	0.20	0.96
a_2	0.06	0.04	4.48
a_3	0.02	0.03	8.48

Table 8.3 Fourier amplitudes describing maxillary and mandibular arch shape in Australian, Caucasian, and Japanese males.

	Australian aboriginal		Caucasian		Japanese	
	mean	s.d.	mean	s.d.	mean	s.d.
Maxilla						
a_0	37.32	1.67	35.58	1.52	35.68	1.45
a_1	7.47[j]	1.01	6.99	1.44	6.92	1.16
a_2	1.06	0.46	2.02[aj]	0.56	1.51[a]	0.44
a_3	0.43[c]	0.27	0.32	0.19	0.34	0.28
Mandible						
a_0	32.90	1.48	32.13	1.35	31.85	1.05
a_1	6.27	1.03	6.25	1.28	6.11	1.22
a_2	1.27	0.41	1.91[a]	0.57	1.76[a]	0.48
a_3	0.44[cj]	0.19	0.25	0.16	0.32	0.18

a—significantly larger than Australians ($p < 0.05$)
c—significantly larger than Caucasians ($p < 0.05$)
j—significantly larger than Japanese ($p < 0.05$)

mean values for a_1), less tapered (smaller values for a_2) arch compared with the other groups. In the mandible the differences in the arch depth/width ratio were less obvious, but Australian Aboriginal males and females and Japanese females tended to have more tapered arch shapes.

Although these population differences in arch size and shape might suggest an underlying genetic influence on dental arch form, a clearer indication of the relative contributions of genetic and environmental factors to the observed variation can be obtained from the comparison of Fourier amplitudes describing dental arch form in MZ and DZ twins. Table 8.4 shows the intrapair difference in the Fourier amplitudes for MZ and DZ twins, and the results of the variance analysis, which suggested the presence of significant genetic effects related to the constant describing size (a_o) and the first amplitude (a_1) describing arch shape. The intraclass

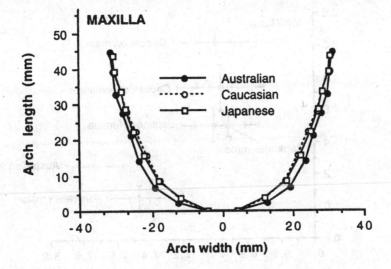

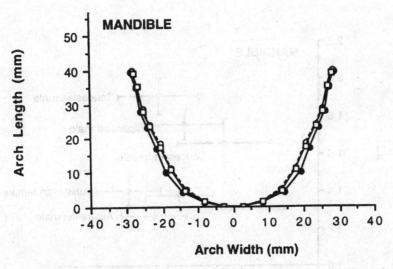

Fig. 8.2 Plots of maxillary and mandibular arches in Australian Aboriginal, Caucasian, and Japanese males.

correlation coefficients and heritability estimates (Table 8.5) indicated that genetic factors influenced maxillary arch shape, with significant positive correlations in MZ twins compared with the relatively lower correlations in DZ twin pairs.

The largest values of the intraclass correlation coefficients within MZ pairs were noted for the constant, a_0, (reflecting arch size) and for the first and second Fourier amplitudes, which described the ratio of arch depth/width and the rela-

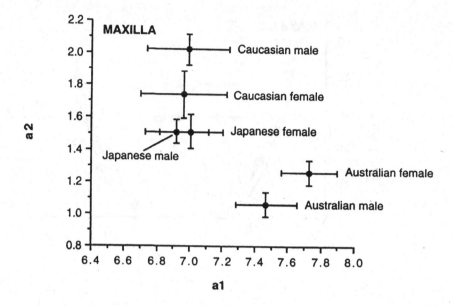

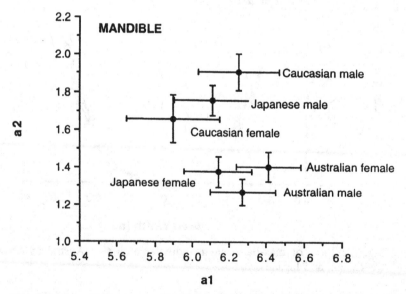

Fig. 8.3 Plots of Fourier amplitudes describing the dental arch depth/width ratio (a_1) and arch taper (a_2) in Caucasian, Australian Aboriginal, and Japanese males and females.

Table 8.4 Intrapair difference of Fourier Coefficients for maxillary and mandibular dental arches in MZ and DZ twins.

		Monozygous		Dizygous		F-ratio
		mean	variance	mean	variance	DZ / MZ
Maxilla	a_o	0.07	0.52	0.24	1.60	3.08**
	a_1	0.04	0.41	0.12	1.45	3.54**
	a_2	−0.01	0.10	0.09	0.14	1.40
	a_3	−0.02	0.02	−0.09	0.06	3.00**
Mandible	a_o	0.12	0.81	−0.08	1.83	2.26*
	a_1	−0.04	0.28	0.18	1.26	4.50**
	a_2	−0.09	0.15	0.10	0.13	0.87
	a_3	−0.02	0.03	−0.08	0.05	1.67

*$p < 0.05$
**$p < 0.01$, by one-tailed F-test

Table 8.5 Intraclass correlations and heritabilities for Fourier amplitudes describing maxillary and mandibular arch shapes.

		Intraclass correlation		Heritability
		MZ	DZ	h^2
Maxilla	a_0	0.87	0.62	0.50
	a_1	0.84	0.58	0.52
	a_2	0.81	0.77	0.08
	a_3	0.54	0.53	0.02
Mandible	a_0	0.74	0.43	0.62
	a_1	0.86	0.60	0.52
	a_2	0.72	0.74	−0.04
	a_3	0.30	0.10	0.40

tive taper of the arch, respectively, in both maxillary and mandibular arches. The largest correlations within DZ pairs were for the second term (a_2) in both the maxillary and mandibular arches with relatively lower correlations between other amplitudes. The fact that for these data some DZ correlations exceeded the theoretical maximum value (0.50) suggests that conclusions need to be drawn with care. Comparisons of correlation coefficients between the MZ and DZ groups indicated a significant genetic contribution to the variation in arch size (a_0) and arch depth/width ratio (a_1).

The results of both the interpopulation comparisons and the twin study confirm that the Fourier method provides an accurate, reproducible method for describing dental arch size and shape. The results suggest that genetic factors contribute mainly to the variation in arch size and to the arch depth/width ratio rather than to other aspects of the dental arch morphology.

When considered together, the results of our interpopulation comparisons and studies of twins suggest that differences in arch taper between ethnic groups might arise from environmental factors associated with functional differences, whereas variation in arch size and arch shape appear to have a larger genetic component.

8.3 Comparisons of mandibular shape

The form of the mandible is influenced by a number of environmental factors, the most significant of which is probably masticatory function. The fact that obvious differences exist between populations suggests, however, that genetic factors also influence mandibular form.

Fourier descriptors have been derived to describe mandibular outlines obtained from tracings of standardized lateral cephalometric radiographs of Australian Aborigines (20 males and 20 females) and Japanese (72 males and 71 females). For the Australian sample the Fourier constant (a_0) described 66% of the variation in mandibular form (Sergi et al., in press). The first amplitude (a_1) described 16% of the variation and was associated with the shape of the condylar process, ramus, and lower border of the mandible. The second amplitude described 9% of the variation and was also associated with condylar process and chin form. The third amplitude was associated with the form of the gonial angle. It accounted for less than 5% of the variation.

Differences in mandibular shape are summarized in Figure 8.4, which illustrates outlines for males generated from the mean values for the Fourier coefficients for the groups involved. In general, the Australian Aboriginal males showed relatively greater alveolar prognathism and more development of the gonial angle region. These differences in shape are reflected in the Fourier amplitudes (Figure 8.5), which were generally smaller in the Australian group than in the Japanese samples. This indicates that, excluding the influence of size, there was relatively less chin and condylar process development in the Australian group, which led to the impression that the alveolar and gonial regions are larger when size and shape are considered together.

8.4 Facial profile in twins and their families

Observed variation in facial morphology results from the influences of both environmental and genetic factors. Classical methods of analysis, based on the differences within pairs of MZ and DZ twins, have provided evidence of a strong

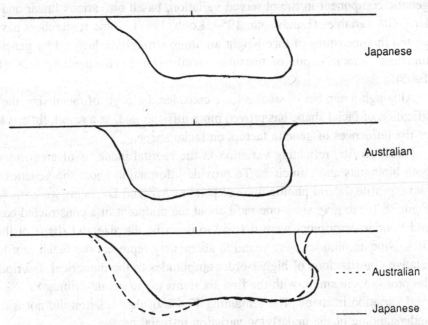

Fig. 8.4 Mandibular outlines generated from mean Fourier amplitudes for Japanese and Australian Aboriginal males.

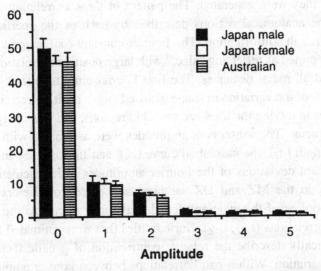

Fig. 8.5 Plots of Fourier amplitudes (with standard deviations) describing mandibular form for Japanese males, Japanese females, and Australian Aboriginal males and females combined.

genetic component in the observed variation, based on various linear and angular facial variables (Lundstrom, 1954; Lobb, 1987). Some researchers have suggested that measures of face height are more strongly influenced by genetic factors than are facial depths or measures directly related to the dentition (e.g., Hunter, 1965).

Although a number of studies have considered aspects of facial size, the quantification of facial shape has proved more difficult and, as a result, little is known of the influences of genetic factors on facial shape.

Facial profile, reflecting variation in the sagittal plane, is of great interest to both biologists and clinicians. To provide information about the genetics of the facial profile, lateral photographs of pairs of MZ and DZ twins were traced (see Figure 8.1c) to give sixty-one radii about the midpoint of a constructed baseline, and Fourier descriptors were derived to describe the size and shape of the profiles. Nine amplitudes were found to adequately represent the facial profile. The relative contributions of higher-order amplitudes to the numerical description of the profile were small, with the first six terms (a_1 to a_6) describing 82.6% of the total variation in shape. The remaining 17.4% of the variation did not add to an understanding of the underlying variation in facial profile.

To interpret the Fourier descriptors of the facial profile, we calculated correlations between each of the nine Fourier amplitudes and the sixty-one radial distances from which they were generated. The pattern of these correlation coefficients suggested the anatomical regions described by each of the coefficients. Table 8.1 summarizes this information. The Fourier constant (a_0), as previously discussed, was interpreted as reflecting "size," with large positive correlations between this term and all radial distances. The first Fourier amplitude (a_1), which accounted for 31.8% of the variation in shape, showed larger positive correlations with radial distances involving the forehead and chin regions, and was interpreted as a "face height" factor. The subsequent amplitudes were associated with facial depth (a_2), nasal length (a_3), the nasolabial curve (a_4), and lip form (a_5 and a_6).

Means and standard deviations of the Fourier amplitudes were calculated for males and females in the MZ and DZ samples, and correction factors were applied to enable pooling of the male and female data prior to calculating intrapair MZ and DZ correlations (r_{MZ}, r_{DZ}). Heritabilities (h^2) were estimated for all variables to numerically describe the relative contribution of genetic factors to the total observed variation. Within-pair correlations between Fourier amplitudes for MZ twin pairs were larger than those within DZ pairs (Table 8.6), suggesting a significant genetic contribution to variation in facial profile. The correlations within the DZ pairs were high for the constant term (a_0) and the first amplitude (a_1) but were otherwise relatively low (less than 0.32). Similarly, within the MZ pairs, the correlations were higher (greater than 0.60) for the first two

Table 8.6 Intraclass correlations in MZ and DZ twins and heritability estimates (h^2) for Fourier amplitudes (a_0 to a_9) describing the facial profile.

		Intraclass correlation		Heritability
		MZ	DZ	h^2
a_0	Size	0.62	0.48	0.28
a_1	Face height	0.76	0.47	0.58
a_2	Face depth	0.58	0.10	0.96
a_3	Nasal length	0.32	0.25	0.14
a_4	Lip form	0.44	0.32	0.24
a_5	Lip form	0.52	0.32	0.40
a_6	Lip form	0.41	0.08	0.46
a_7		0.28	—	—
a_8		0.21	—	—
a_9		0.17	0.01	0.32

terms, which reflected overall size (a_0) and face height (a_1), than for the other terms.

Interpretation of the pattern of correlations for the Fourier amplitudes describing face height (a_1), face depth (a_2), and nasal length (a_3) was complicated by the unreliability of the second amplitude (a_2) and by the very low correlation between MZ twin pairs. However, the data implicated genetic influences on variation in face height (a_1), with a relatively strong correlation in MZ pairs (r = 0.76), and a correlation approaching the theoretical maximum in DZ twins (r = 0.47).

Within-twin-pair correlations involving the amplitude that described face depth (a_2) were lower than might be expected, suggesting the presence of measurement errors associated with this variable. The differences between the correlation coefficients in MZ twins (r = 0.58) and in DZ pairs (r = 0.10) suggested a genetic contribution to variation in facial depth. Although the estimated heritability (h^2 = 0.96) for face depth was inappropriately high, due perhaps to the lower reliability of this variable, the overall pattern of correlations is consistent with other evidence suggesting that genetic factors make a significant contribution to the linear and angular variables describing facial convexity (Vanco et al., 1993).

The low correlation between scores for the third Fourier amplitude in MZ twins suggests that genetic factors were not the main determinants of variation in nose length. Similarly, the pattern of correlations in the MZ and DZ pairs did not imply a large genetic contribution to the Fourier amplitudes describing lip form.

To illustrate the pattern of variation in facial profile within family groups we obtained data for five families consisting of one pair of twins, their parents and

A

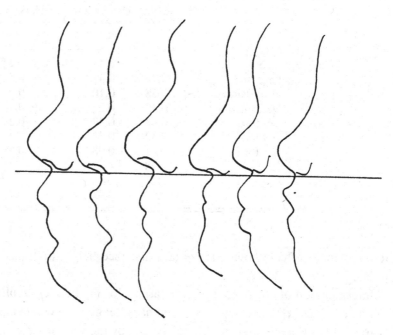

		Twin A	Twin B	Father	Mother	Brother	Sister
a_0	Size	23.03	21.12	29.17	23.68	20.31	24.46
a_1	Faceheight	11.86	10.05	15.52	12.45	12.40	10.86
a_2	Nasal depth	10.75	8.87	7.09	9.08	9.18	7.88
a_3	Nasal height	4.64	3.48	5.78	4.33	3.84	2.96
a_4	Lip form	3.11	3.54	0.65	1.02	1.45	2.95
a_5	Lip form	2.44	2.46	1.00	1.12	1.40	1.10

Fig. 8.6 Profile tracings of (a) MZ and (b) DZ twins and their families.

siblings (Pinkerton et al., in press). These are illustrated in Figures 8.6 and 8.7. As expected, Fourier amplitudes describing the size and shape of the face generally differed less between MZ twins than between family members, whereas the differences between DZ twins were of the same magnitude as those between other pairs of family members. Interestingly, the variation in size (reflected by the a_0 term) and nose length (a_3) was as great between mothers and fathers (genetically

B

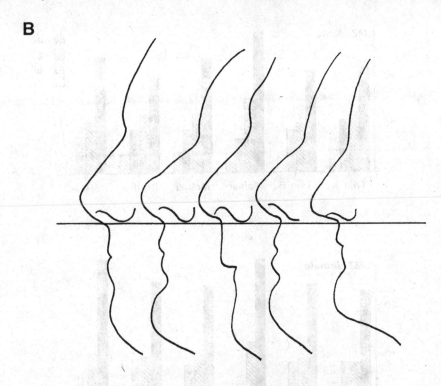

	Twin A	Twin B	Father	Mother	Sister
a_0 Size	24.89	30.21	26.34	27.50	21.26
a_1 Face height	9.49	11.03	10.92	11.61	7.61
a_2 Nasal depth	9.17	6.88	4.98	4.70	8.32
a_3 Nasal height	3.78	2.47	4.79	3.17	4.76
a_4 Lip form	2.99	0.74	2.27	0.32	2.52
a_5 Lip form	1.77	2.02	1.58	.22	1.72

Fig. 8.6 (continued)

unrelated) as between siblings (sharing 50% of genes), lending support to the low estimates of heritability found for these amplitudes.

8.5 Discussion

In each of our studies of dental and facial morphology, the use of Fourier functions to describe "size" and "shape" has provided new insights into the relative contributions of genetic and environmental influences to observed variation.

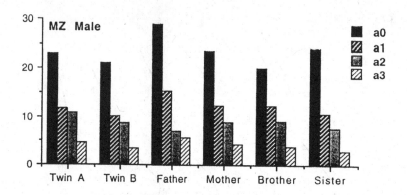

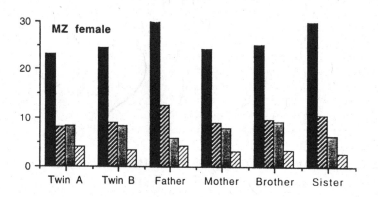

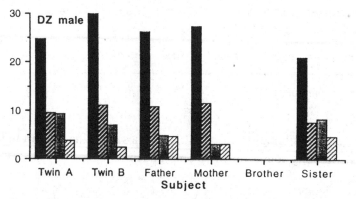

Fig. 8.7 Plots of Fourier amplitudes describing the facial profile for families.

Fourier descriptors have a number of advantages over more conventional metric approaches to study complex morphologies:

- In defining the structures to be studied, there is no need to identify homologous points. This makes it possible to study and compare the complex outlines, and removes the bias that can be introduced by the selection of reference points.
- The computation of the descriptors is straightforward and can be carried out using personal computers.
- Fourier descriptors provide for significant data reduction. Even the more complex structures we studied required fewer than ten harmonics to accurately describe outlines consisting of sixty-one original data points.
- The differentiation between "size" and "shape" terms makes it possible to consider these two types of variation independently.
- Computation of the Fourier functions facilitates comparisons between groups. For example, calculating sample means for each of the Fourier terms, summing and then plotting the curve described by series, makes visual comparisons between groups possible.
- In many cases the Fourier descriptors can be interpreted in terms of the underlying biological variation they describe. This is achieved by considering the pattern of correlations between each of the coefficients and the radial distances from which they have been derived. This important advantage makes it possible to subject Fourier amplitudes to relatively powerful statistical analyses in genetic studies and, as a result, to draw conclusions about the influence of genetic factors on the variation represented by each of the amplitudes.

In our studies of dentofacial morphology, the application of Fourier functions to compare dental arch form, facial profiles, and mandibular shape among different human populations and in twins, has provided new information about genetic and environmental influences on the variation in these structures. This includes:

- *Differences in dental arch size and shape among different human populations provided indirect evidence of the underlying genetic differences between the groups.* Whereas Australian Aborigines tended to have larger maxillary and mandibular dental arches than either Japanese or Caucasians, the most significant differences were aspects of arch shape. Caucasian and Japanese subjects displayed more-tapered maxillary and mandibular arches than Australians. In addition, comparisons of Fourier amplitudes describing dental arches in MZ and DZ twins provided evidence of the influence of genetic factors on arch size and the ratio of arch depth/width. Considered together these studies suggest that the observed differences in arch taper between the Australian Aboriginals and the other groups might arise from environmental factors, whereas variation in arch size and the arch depth/width ratio might have a larger genetic component.

- *Comparisons of mandibular form between an Australian Aboriginal sample and Japanese subjects indicated significant differences in both size and shape.* The evidence suggests this is a profitable area for further genetic studies.
- *Comparisons of facial profiles in twins and their families provided evidence of significant genetic contributions.* The results of these studies suggest a genetic component to the variation in face height, with little evidence of genetic factors influencing nose and lip form.

8.6 Acknowledgments

The assistance of Dr. M. Sekikawa, Dr. C. Vanco, Dr. R. Sergi, Sandra Pinkerton, Professor E. Kanazawa, Professor T. Ozaki, Professor T. Iwasawa, George Travan, Tracey Parish, and Pina Mangiarelli is appreciated.

Aspects of this research were supported by grants from The National Health and Medical Research Council of Australia, the Australian Academy of Science, the Japan Society for the Promotion of Science, Nihon University, The University of Adelaide, the South Australian Foundation for Dental Research and Education, and the Dental Board of South Australia.

References

Brown, T., Abbott, A. H. & Burgess, V. B. (1983). Age changes in dental arch dimensions of Australian Aboriginals. *Am. J. Phys. Anthrop.* **62**, 291–303.

Brown, T. & Barrett, M. J. (1973). Dental and craniofacial growth studies of Australian Aborigines. In *The Human Biology of Aborigines in Cape York*, ed. R. L. Kirk. Canberra: Australian Institute of Aboriginal Studies.

Brown, T., Townsend, G. C., Richards, L. C. & Travan, G. R. (1987). A study of dentofacial morphology in South Australian twins. *Austral. Dent. J.* **32**, 81–90.

Christian, J. C. (1979). Testing twin means and estimating genetic variance: Basic methodology for the analysis of quantitative twin data. *Acta Genet. Med. et Gemmellologiae* **28**, 35–40.

Christian, J. C., Kang, K. W. & Norton, J. A. (1974). Choice of an estimate of genetic variance from twin data. *Am. J. Hum. Genet.* **26**, 154–61.

Corruccini, R. S. & Potter, R. Y. P. (1980). A study of occlusion and arch widths in families. *Am. J. Orthod.* **78**, 140–54.

Dahlberg, G. (1940). *Statistical Methods for Medical and Biological Students.* London: George Allen and Unwin.

Falconer, D. S. (1989). *Introduction to Quantitative Genetics.* New York: Longman.

Hrdlicka, A. (1916). Contribution to the anthropology of Central and Smith Sound Eskimos. *Ann. Pap. Am. Mus. Nat. Hist.* **5**, 177–285.

Hunter, W. S. (1965). A study of the inheritance of craniofacial characteristics as seen in lateral cephalograms of 72 like-sexed twins. *Trans. Europ. Orthod. Soc.* **41**, 59–76.

Jones, M. L. & Richmond, S. (1989). An assessment of the fit of a parabolic curve to pre- and post-treatment dental arches. *Brit. J. Orthod.* **16**, 85–93.

Kang, K. W., Corey, L. A., Evans, M. M., Christian, J. C. & Norton , J. A. (1977).

Dominance and environmental variances: Their effect on heritabilities estimated from twin data. *Hum. Hered.* **27,** 9–21.

Lestrel, P. E. (1974). Some problems in the assessment of morphological size and shape differences. *Yearbk. Phys. Anthrop.* **18,** 140–62.

Lestrel, P. E. (1989). Some approaches toward the mathematical modeling of the craniofacial complex. *J. Cran. Genet. Develop. Biol.* **9,** 77–91.

Lobb, W. K. (1987). Craniofacial morphology and occlusal variation in monozygous and dizygous twins. *Angle Orthod.* **57,** 219–33.

Lu, K. H. (1966). An orthogonal analysis of the form, symmetry, and asymmetry of the dental arch. *Arch. Oral Biol.* **11,** 1057–69.

Lundstrom, A. (1954). Nature versus nurture in dento-facial variation. *Europ. J. Orthodont.* **6,** 77–91.

Moorrees, C. F. A. (1959). *The Dentition of the Growing Child.* Cambridge: Harvard University Press.

Pinkerton, S. K., Richards, L. C., Townsend, G. C. & Kasai, K. Facial profiles in South Australian twins and their families. *J. Dent. Res.* (in press).

Richards, L. C. (1983). Adaptation in the masticatory system. Descriptive and correlative studies of a precontemporary population. Ph.D. thesis, The University of Adelaide.

Richards, L. C. & Brown, T. (1981). Dental attrition and age relationships in Australian Aboriginals. *Archaeol. Phys. Anthrop. Oceania* **16,** 94–8.

Richards, L. C., Townsend, G. C., Brown, T. & Burgess, V. B. (1990). Dental arch morphology in South Australian twins. *Arch. Oral Biol.* **35,** 983–9.

Sekikawa, M. (1986). Fourier analysis of the dental arch form. *Jap. J. Oral Biol.* **28,** 43–61.

Sergi, R., Richards, L. C., Kasai, K. & Townsend, G. C. Age changes in mandibular shape. *J. Dent. Res.* (in press).

Sharma, K. & Corruccini, R. S. (1986). Genetic basis of dental occlusal variation in northwest Indian twins. *Europ. J. Orthod.* **8,** 91–7.

Townsend, G. C., Richards, L. C., Brown, T. & Burgess, V. B. (1988). Twin zygosity determination of the basis of dental morphology. *J. Foren. Odonto-Stomomatol.* **6,** 1–16.

Vanco, C., Kasai, K., Richards, L. C., Sergi, R. & Townsend, G. C. (1993). Genetic and environmental influences on facial profile. *Austral. Dent. J.* **38,** 329–30.

9

Fourier Analysis of Size and Shape Changes in Japanese Skulls

FUMIO OHTSUKI, TERUO UETAKE, KAZUTAKA ADACHI,
PETE E. LESTREL AND KAZURO HANIHARA
Tokyo Metropolitan University
Tokyo University of Agriculture and Technology
The University of Tokyo
School of Dentistry, UCLA &
International Institute for Advanced Studies

9.1 Introduction

Morphological changes in Japanese cranial and facial form over time were described by Suzuki (1967). He examined a sample ($n = 542$) from the Jomon (Neolithic) through the Early Modern (Historic) period. These cranial materials were identified as adult males free from any congenital malformations or artificial deformations.

Suzuki described the face of a typical protohistoric Japanese as long headed, with a broad face, wide and flat nasal roots, and a fairly strong prognathism. These morphological aspects gradually changed becoming round-headed, displaying a more constricted face, and narrower and higher nasal roots, and less prognathous jaws. Nevertheless, two characteristics associated with the protohistoric period, dolichocephaly and prognathism, remained quite pronounced.

Some of the contributing factors influencing these chronological changes in skull morphology were hypothesized to be socioeconomic and environmental (Suzuki, 1967). Another presumed factor was the choice of mate. Finally, he also suggested that differences in Japanese physiognomy (as seen in these craniofacial or physical characteristics) from prehistoric to modern times were due to diverse origins. His studies were based on metrical and nonmetrical observations as well as physical measurements taken from the Martin and Saller (1957) handbook.

Traditionally, skull form has been described with ratios or indices as introduced by Martin and Saller (1957). However, only a small percentage of the visual information contained in any biological form can be ascertained with these biometrical measures. Lestrel established the use of Fourier analysis for describing biological forms, especially in the field of physical anthropology (Lestrel, 1974; Lestrel and Brown, 1976; Lestrel and Roche, 1976; Lestrel et al., 1977). The

Fourier method was shown to be more effective than conventional measurements in the analysis of complex shapes (Lestrel, 1975).

The particular Fourier series utilized here consists of a finite series in theta composed of sine and cosine terms where the measurements are in polar coordinates (r,θ) over the interval from 0 to 2π:

$$f(\theta) = a_0 + \sum_{i=1}^{n} a_i \cos i\theta + \sum_{i=1}^{n} b_i \sin i\theta. \tag{9.1}$$

The data consist of equally spaced vectors as radii from an internal center to the boundary outline. The a_0 term, or constant, is the mean of all the radial measurements, the a_i and b_i are the ith Fourier coefficients for each harmonic determined by least squares, and n is the total number of harmonics (see Chapter 2). If the function to be fitted by the Fourier series is symmetrical, or if symmetry can be deliberately imposed, then the sine terms will become zero and Equation 9.1 reduces to Equation 9.2 (Lestrel, 1974; Lestrel and Roche, 1976):

$$f(\theta) = a_0 + \sum_{i=1}^{n} a_i \cos i\theta. \tag{9.2}$$

Expanding the series to include the first three harmonics yields Equation 9.3:

$$f(\theta) = a_0 + a_1 \cos \theta + a_2 \cos 2\theta + \ldots + a_n \cos n\theta. \tag{9.3}$$

Three particularly advantageous characteristics of the Fourier series are:

- The coefficients are orthogonal and hence can be treated as independent variables, which makes it possible to separately analyze the contributions that each harmonic makes to the total form.
- The capability exists, under certain constraints, of relating numerical differences in the Fourier coefficients to actual differences in the observed form.
- The size and shape components can be separated from each other (Needham, 1950).

9.2 Chronological changes in Japanese skull shape: previous work

9.2.1 The cranial outline (lateral view)

We initially attempted to assess chronological size and shape changes in the Japanese skull in *Norma lateralis*, or lateral view (Ohtsuki et al., 1983). Materials used were adult male skulls covering four age periods: Jomon, Kamakura, Muromachi, and Edo (these groups are discussed in detail in Section 9.3).

Mean raw data values (vector distances or radii; see Section 9.3.4) containing both size and shape information were compared for the four age periods. Fourier analysis was applied to detect chronological changes in skull shape after standardization of size. Statistical results obtained from the GLM procedure (SAS, 1979) disclosed shape differences as well as size differences. In particular, the Jomon showed the most divergence in skull form among the four groups according to Duncan's multiple range test (Woolf, 1980). These results were not unexpected since the Jomon represent the earliest population stratum, whereas the inhabitants of Kamakura, Muromachi, and Edo age periods appeared considerably later.

These changes in lateral skull shape outline appear to be a decrease or flattening at the parieto-occipital margin accompanied by an overall expansion at the frontal aspect and inferior aspect of the occipital region. These studies were then extended to the antero-posterior (*Norma occipitalis*) and vertical (*Norma verticalis*) views.

9.2.2 The cranial outline (occipital view)

In the occipital view, the mean vector values of the Jomon specimens are considerably different from the other three groups. This trend was particularly clear at the left parietal margin. However, once the effect of size was removed, the occipital shape of the Jomon skulls was found to be similar to those of the more recent age periods. In other words, chronological shape changes in Japanese skull form viewed from the occipital direction could not be detected, although size differences were present (Ohtsuki et al., 1993a).

9.2.3 The cranial outline (vertical view)

The analysis of size and shape changes from the vertical view (*Norma verticalis*) was also initiated and differences in the vectors to the parietal margin could be demonstrated. Analysis by the Fourier technique indicated the presence of chronological skull shape changes after the removal of size. Differences were found in the size-standardized vectors, again for the left parietal margin (Uetake et al., 1993) and differences were also found for the occipital region (Ohtsuki et al., 1993a). It is of interest that differences between the four age periods seem to be focused on the left side, both for the vertical and occipital views. Analysis of Japanese skulls by Moiré contourography also found the left side more to be frequently affected (Takayama, 1984; Takeuchi et al., 1985). The analysis was recently extended to include Modern age period specimens, which allowed the comparison of skull shape changes over five instead of four age periods (Ohtsuki et al., 1995).

9.3 Methodology associated with the present study

The skull materials of the Modern age period are from the 1st Department of Anatomy, the Jikei University School of Medicine. The adult male skull material from the earlier four age periods are from the collection at the University Museum, The University of Tokyo.

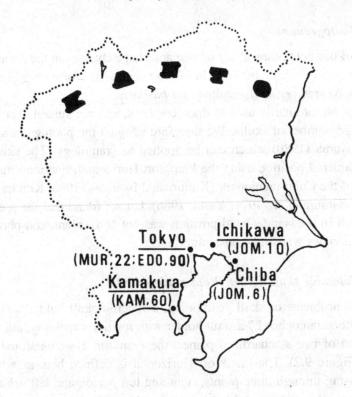

JOM.Jomon;KAM.Kamakura
MUR.Muromachi & EDO.Edo

Fig. 9.1 Number of specimens excavated from each site. They cover four age periods. Only the location of the Modern age period specimens ($n = 65$) is not indicated in this figure. The location of all materials was restricted to the Kanto District.

The Kanto District of Japan (in which Tokyo is located) is the source of all the excavated materials (Figure 9.1). The materials cover five age periods: Jomon (2000–1000 BC), Kamakura (14^C AD), Muromachi (15^C AD), Edo (18^C AD), and Modern (19^C and 20^C AD). The sample sizes, as well as the abbreviation for each group, are shown in parentheses. For the Jomon period (JOM) there were 10 skulls excavated from Ichikawa and 6 from Chiba. For the Kamakura period (KAM) there were 60 skulls. For the Muromachi period (MUR) there were 22 skulls. For the Edo period (EDO) there were 90 skulls, all from Tokyo, and for the Modern period, 65 skulls, also from the Kanto District. Thus, all excavated materials were from the Kanto District and confined to a range of 50 km from Tokyo. To further control the geographical variability, the birthplaces of the Modern age period subjects were also limited to the Kanto District.

9.3.1 Photogrammetry

For this aspect of our continuing study of size and shape changes in the Japanese skull, we examined the skull outlines in *Norma frontalis*, *Norma occipitalis*, *Norma lateralis*, *Norma verticalis*, and *Norma basilaris*.

The Dioptrograph, ordinarily used to draw contours, was not suitable here because of the large number of skulls. We therefore adopted the photogrammetric method of Takayama (1980) which can be applied to craniology. The skull is fixed in a standardized position using the Frankfort Horizontal. Previous studies have made use of the Cubuscraniophor (Kimura and Iwamoto, 1968; Kimura and Iwamoto, 1969; Kimura et al., 1971; Yasui, 1986), but we found that the precise fixation of a skull in the standardized position was not convenient, and photography of the five views was time-consuming.

9.3.2 Rotatable skull holder (Rotacraniophor I)[1]

To resolve these problems of skull positioning, a specific skull holder was developed. This "Rotacraniophor I," a skull holder with rotation capability, allowed the determination of two standardized planes, the Frankfort Horizontal and the sagittal plane (Figure 9.2). The Frankfort Horizontal is defined here as a horizontal plane passing through three points, right and left *porion* and left *orbitale*. The horizontal aspect is determined by the upper ridge of both aural supporting rods and the upper ridge of the *orbitale* pointer, with the skull held horizontally. *Porion* is then located ca. 0 to 3 mm above the upper ridge of the aural supporting rod, providing an estimate for the Frankfort Horizontal. When the skull is held vertically, *porion* is positioned ca. 0 to 3 mm anterior to the frontal ridge of the aural supporting rod. The sagittal plane is defined as a plane passing through *nasion* and perpendicular to a line connecting the center of both (left and right) external auditory meata. Thus, this line is the axis of the aural supporting rods. Further details are described elsewhere (Adachi et al., 1989).

Although skull photographs were taken in all the five views, the lateral view is the subject of the present study. The results for the five age periods from the other views are still undergoing analysis.

9.3.3 Photography

Photographs (*Norma lateralis*) of the skulls were taken using a NIKKOR ED with a 800m F8.0 Lens. The object-camera distance was fixed at 12 meters yielding a field-of-view of the image close to actual size. The magnification error was found

---

[1] The Rotacraniophor I is available from the Riken Photoelasticity Institute. Contact the senior author for details.

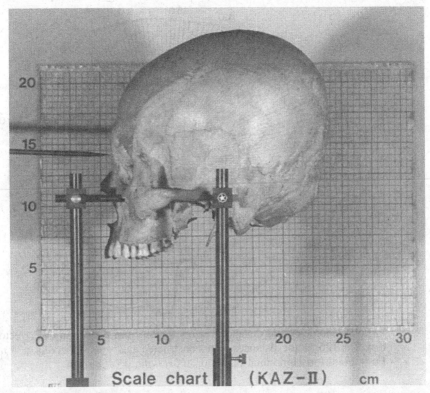

Fig. 9.2 Photograph of the skull in the lateral view (*Norma lateralis*). The quadrille graph paper in the background was merely used as background and does not indicate the actual size. (Reproduced from Adachi et al., 1989, with permission of the publisher.)

to be less than 1% (Takayama, 1980). ILFOSPEED semi-matte photographic paper was used to provide a stable image, that is, to reduce object distortion.

9.3.4 Geometry and vectors

The photographs of the skull images were traced onto dimensionally stable acetate sheets, and the radial vectors, numbered 1 to 126, were drawn (described below). The vector center (VC) was determined as follows (Figure 9.3):

1. A line was drawn (Frankfort plane) from *or* (*orbitale*) to *po* (*porion*), the superior limit of the *external auditory meatus*, *eam*.
2. A line was then constructed connecting *na* (*nasion*) and the superior limit of *po*. This line was then extended postero-inferiorly until it intersected the occipital aspect.

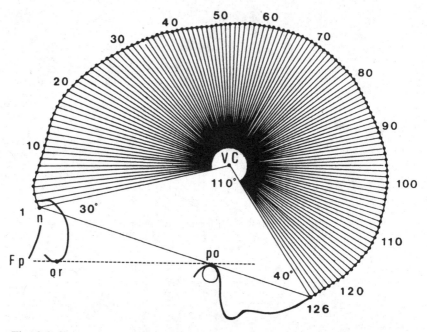

Fig. 9.3 Geometry and distance vectors drawn from the origin (VC, vector center). The 126 vector distances were constructed at 2° equal intervals. n = *nasion*; Fp = Frankfort Horizontal plane; or = *orbitale*; po = *porion*.

3. At *nasion*, a 30° angle was constructed as illustrated. At the intersection with the occiput, a 40° angle was constructed . The 30° and 40° lines were extended until they intersected inside the skull. This became the center for the vectors or distances to the cranial vault (VC). The choice of this geometry was dictated by goodness-of-fit considerations. That is, it insured that the fit was not unduly influenced by the presence of "sharp corners" at *nasion* and the occipital aspect (Lestrel, 1975).

4. Vectors or radii were constructed, at equal 2° intervals, using the **na**-VC line as the starting vector and ending with the VC-occipital aspect line as vector 126 as shown in Figure 9.3.

Once constructed, the radial distances, 1 to 126, were measured using a digitizing device which, under software control, also allowed the simultaneous display of these distances on a monitor. The precise details are available from the senior author.

These distances from the VC to the cranial vault were then submitted to a specially written Fourier analysis program (Parnell and Lestrel, 1977). This program uses the vector distances, from 1 to 126, at equal 2° spacing over the interval of

[0 to π], that is, 180°.The data are then reflected, creating a mirror image (with a new set of vector distances 126 to 1) yielding a total of 251 distances (the first distance being the same as the last) now over the required interval of [0 to 2π]. This reflection allows for a 50% reduction in coefficients, as the sine terms will now vanish because of the imposed symmetry (Eq. 9.3). The program also computes the centroid of the figure.

The initial vector distances to the vault outline, between distance 1 and distance 126, subtend an angle of 252° (not 180°). This results in a distortion, or more precisely, a topological transformation. These topologically transformed vector distances are then used to compute the Fourier function. The predicted distances derived from the Fourier function are then re-created back into an undistorted figure and visually displayed (Figures 9.4, 9.5, and 9.6). This procedure can be considered as an "inverse" topological transformation. The details involved in this application of the Fourier method to the cranial vault, as well as to other skeletal data, are reported elsewhere (Lestrel, 1974, 1975; Lestrel et al., 1977).

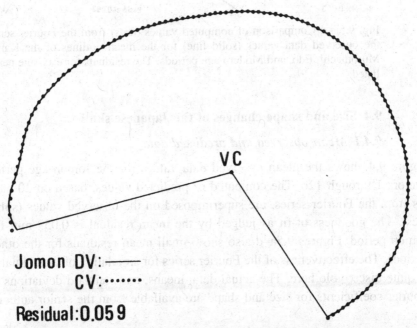

Fig. 9.4 Drawings showing the goodness-of-fit of the Fourier series for the skulls of the Jomon age period. DV denotes the observed data values as shown by a bold line. CV is the expected or computed value, which is shown by dots. The goodness-of-fit of the Fourier series was judged by the mean residual or difference between the observed data (actual curve) and the predicted values provided by the Fourier series.

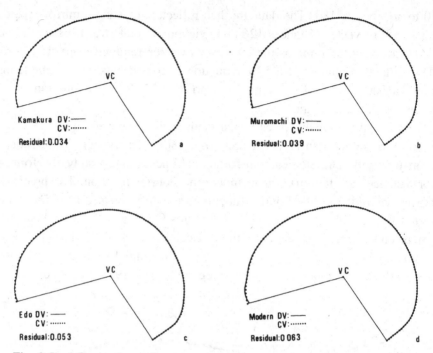

Fig. 9.5a–d Comparison of computed values (dots) from the Fourier series and the observed data values (solid line) for the mean outlines of the Kamakura, Muromachi, Edo, and Modern age periods. The residuals for each age period are listed.

9.4 Size and shape changes of the Japanese skull

9.4.1 Mean observed and predicted data

Figure 9.4 shows the mean observed data values for the Jomon age period for vectors 1 through 126. The computed or predicted values, based on 20 harmonics from the Fourier series, are superimposed on the observed values (solid circles). The goodness-of-fit as judged by the mean residual is 0.059 mm for the Jomon period. Figures 9.5a–d also show small mean residuals for the other age periods. The effectiveness of the Fourier series for simulation of the cranial shape is quite discernible here. The actual data, means, and standard deviations of the Fourier coefficients for size and shape are available from the senior author.

9.4.2 Area standardization

Although form and shape are often used synonymously, it has been suggested that Form = Size + Shape is a more appropriate relationship (Needham, 1950). By standardizing the area within the bounded outline for each case, the effect of size can be greatly minimized or entirely removed. Each case was scaled up or down to an arbitrary standard area of $10,000\pi$ mm^2. In other words, each term in the series was multiplied by the ratio between the standard area divided by the actual area. Means and standard deviations of the Fourier coefficients after correction for size differences are also available from the senior author.

9.4.3 Attaching biological meaning to the Fourier coefficients

Only for reflected data, as described here, does the a_2 coefficient measure the antero-posterior extent of the cranial outline (Lestrel, 1974; Lestrel and Roche, 1976). This coefficient is rather small for the Jomon (11.42), but tends to increase as the skull becomes more recent, the exception being the Modern group. The values are: 15.40 for the Kamakura, 15.46 for the Muromachi, and 16.46 for the Edo group. The value drops to 12.99 for the Modern group. From the a_2 term one

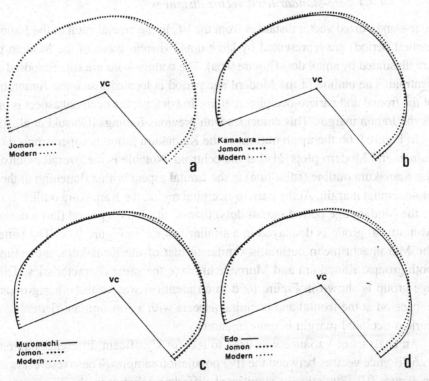

Fig. 9.6a Comparison of the mean size-standardized values of Jomon and Modern age periods. The mean values for the Jomon age period are represented by closed circles, and those of the Modern, by closed circles and a bold line. Fig. 9.6b–d. The mean size-standardized values of the Kamakura (6b), Muromachi (6c), and Edo (6d) age periods are plotted as closed circles. The mean outline of the Jomon skull is represented by a thin line and the Modern skull by a bold line.

may infer that the cranial shape for the Modern age period displays a tendency toward a more rounded shape. On the other hand, the a_3 coefficient tends to decrease with more recent samples: 6.43 for the Jomon, 5.20 for the Kamakura, 5.15 for the Muromachi, 4.55 for the Edo, and 4.11 in the Modern group. This coefficient is associated with a more vertical extent of the cranial outline (Lestrel and Roche, 1976). From these results one may further infer that the cranial shape is flattening near the vertex as a function of age. The Jomon group illustrates the largest magnitude in this coefficient, which is associated with the least flattening of the vault near the vertex. Although the coefficients associated with the higher harmonics (smaller in magnitude) also show trends with descending age, they are increasingly difficult to relate to changes in skull shape. Nevertheless, the size and shape changes shown here are not nearly as dramatic as those displayed by Lestrel (1974) between *H. neanderthalensis* and *H. sapiens*.

9.4.4 Size-standardized vector distances

Size-standardized vector distances from the VC to the cranial vault of the Jomon (the earliest period) are represented by large dots, whereas those of the Modern period are illustrated by small dots (Figure 9.6a). The comparisons are superimposed on the centroid. The outline of the Modern age period is located inside the Jomon outline at the frontal and parieto-occipital regions. In contrast, the occipital aspect is outside of the Jomon margin. This coincides with previous findings (Ohtsuki et al., 1983).

In Figure 9.6b the mean outline of the Kamakura group is superimposed on the Jomon and Modern plots, also at the centroid. Notable is the overall extension of the Kamakura outline (solid line) at the frontal aspect with a flattening at the parieto-occipital margin. At the parieto-occipital region the Kamakura outline is closer to the Modern age period (small dots) than to the Jomon period (large dots). The Muromachi group is displayed in a similar fashion in Figure 9.6c. The pattern of the Muromachi mean outline is similar to that of the Kamakura, indicating that both groups, Kamakura and Muromachi, have the same characteristics. The Edo age group is shown in Figure 9.6d in an identical way to the other groups. The expansion at the frontal and occipital aspects with a concomitant flattening at the parieto-occipital margin is quite apparent.

An analysis of variance was used to test for significant differences among the 126 distance vectors between the five population samples. These results are shown in Figure 9.7. Statistically significant differences between the five age periods were found for a majority of the distances along the skull margin at $p < 0.01$ (open circles) and at the occipital aspect $p < 0.05$ (solid circles).[2]

---

[2] Although these results certainly appear valid, technically, the probability of a Type I error is greater than indicated by the α-level, a consequence of the large number of variables. Also, ANOVA procedures do not take into account the correlation (potentially high) between adjacent vectors.

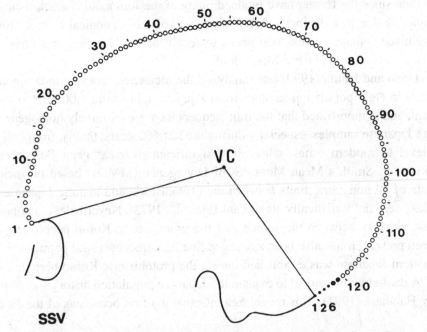

Fig. 9.7 Illustration of the statistically significant results for the vector distances among the five groups. An open circle signifies p < 0.01 and a filled circle, p < 0.05.

Other statistical analyses were carried out to numerically classify the five population groups. Q-mode correlation coefficients among the samples based on the VC distances were computed pairwise and a two-dimensional scattergram drawn (Figure 9.8). The 2-D scattergram was derived from the correlation matrix using Torgerson's multidimensional scaling method (Torgerson, 1965). Finally, the VC distances were assessed using cluster analysis to determine population affinities. The group-averaging method was utilized and the results displayed in a dendrogram (Figure 9.9).

9.5 Discussion

In Figures 9.8 and 9.9, the close affinity between the Kamakura, Muromachi, and Edo groups is demonstrated. The "shape distances," from the Jomon age period on, are increasing as age decreases. These results agree with previous findings based on more conventional anthropological measurements, and seem to reflect the micro-evolution of the Japanese population. However, the shape distance between the Jomon and Modern is comparatively small in contrast to the distances at the other age periods. This is in contradiction to the findings of Hanihara (1985) and Dodo (1986). Those authors both inferred that the Ainu originated from the

Jomon since the former have retained many of the Jomonoid characteristics, pre-sumably due to a distinctly different process of morphological change from the mainland Japanese. Also, one needs to recall that the Jomon people have been classified as Proto- or Pre-Mongoloid.

Dodo and Ishida (1990) later analyzed the incidence data of cranial nonmetric traits in eight population samples from Japan, ranging from 4,000 BP to the present, and demonstrated that the trait frequencies were extremely homogeneous in the Japanese samples, especially during the last 600 years, that is, from early medieval to modern times, when little significant overseas gene flow occurred. However, Smith's Mean Measures of Divergence (MMDs) based on incidence data of 20 nonmetric traits *between* the protohistoric and historic Japanese samples, were not statistically significant (Sjøvold, 1973). Nevertheless, a population discontinuity between the Jomon and the protohistoric Kofun people may have been present. It has also been suggested that the direct ancestral population of the modern Japanese was established during the protohistoric Kofun period.

A dual-structure model to explain the Japanese population history was proposed by Hanihara (1991). This model assumes that the first occupants of the Japanese

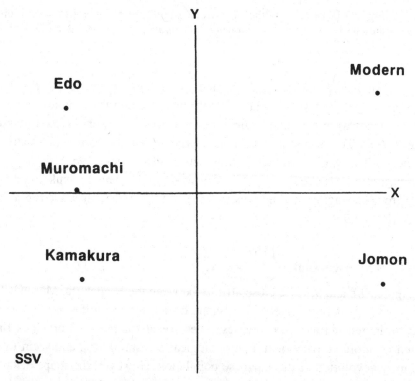

Fig. 9.8 The two-dimensional scattergram derived from the correlation matrix using Torgerson's (1965) multidimensional scaling method.

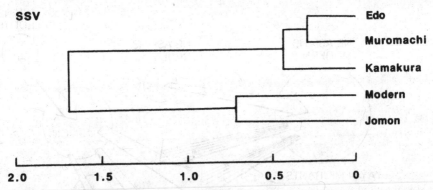

Fig. 9.9 A Dendrogram that shows "shape distances" between the mean skull outlines of the five populations assessed by cluster analysis. The group-averaging method of cluster analysis was used. Numbers under each line refer to the distance converted from the Q-mode coefficients.

Archipelago from which the Neolithic Jomon were derived came from somewhere in South Asia during the Upper Paleolithic. A second wave of migrants from northeast Asia then arrived during, and continued after, the Aeneolithic Yayoi age. Populations of both lineages gradually mixed with each other (Figure 9.11, in Hanihara, 1991). There seems to be no contradiction between the findings by Dodo and Ishida (1990), and the dual-structure model proposed by Hanihara (1991). Yasui (1987) however, basing his observations on skull shape, reached a different conclusion, suggesting that only the midsagittal outline (*Norma lateralis*) of the Sakhalin Ainu females displayed a close similarity to the Jomon (Tsukumo) people. The other Ainu groups tended to resemble the Kofun and Kamakura people rather than the Jomon.

9.6 Conclusions

Shape changes of the Japanese lateral (*Norma lateralis*) skull outline, over five age periods, have now been numerically described. Photographs of the earlier four age groups from other views; that is, the sagittal (*Norma occipitalis*) and vertical views (*Norma verticalis*), fitted with the Fourier series to assess shape changes over time, have been previously described (Ohtsuki et al.,1983; Ohtsuki et al., 1993a; 1993b). These shape changes over the four age periods (Jomon, Kamakura, Muromachi, and Edo) were examined in the three anatomical views. The results indicated that the skull shape of the Jomon was different from three descendant populations. This is, generally, in agreement with the findings of Dodo and Ishida (1990) and Hanihara (1991).

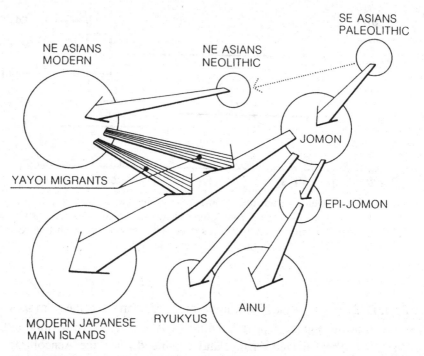

Fig. 9.10 Schematic diagram illustrating the history of the Japanese. Population groups are scattered according to the 1st and 2nd principal component scores obtained from 9 cranial measurements in males (reproduced from Hanihara, 1991, with permission of the publisher).

The present results suggest that the "shape distance" from the Jomon population to the Modern group is rather moderate in contrast to the Jomon distance to the other three age groups. These results differ from the previous findings described above. Q-mode correlation coefficients among the populations, based solely on shape, also established the close relationship of the Jomon and Modern groups.

Scrutinizing the z-scores of the 126 vector distances, one finds that the scores for the Modern sample are larger than those for the Jomon. Specifically, from the frontal to the anterior aspect of the parietal, and vice versa for the posterior aspect of the parietal to the occipital. Accordingly, the Modern skulls are more protruded at the fronto-parietal aspect and tend to be depressed at the parieto-occipital aspect. This indicates that the skull shape of the Modern population is becoming more rounded. However, the overall similarities in shape in the lateral view again suggests a close affinity between both groups as previously seen from the Q-mode correlation coefficients.

Although significant skull shape changes have been found in the lateral view, further analyses of the micro-evolution from the Jomon to Modern populations will need to be carried out, perhaps with detailed comparisons of the individual cranial components, such as the parietal, occipital, and so on. Finally, chronological age and geography could also be important factors, and these may be particularly influenced by size.

The causative factors involved in the microevolution of the physical traits seen in the modern Japanese have been the main focus of our work, and the cranial vault clearly displays the presence of such systematic forces over time. It seems increasingly apparent that these changes in skull shape, initially characterized by a strong protrusion of the supraorbital ridges and a bun-shaped occipital aspect, seem to have inexorably led toward a more rounded shape with the slow passage of evolutionary time.

References

Adachi, K., Ohtsuki, F. & Hattori, M. (1989). Development of rotatable skull holder "Rotacraniophor I." *J. Anthrop. Soc. Nippon* **97**, 393–405.

Dodo, Y. (1986). Metrical and nonmetrical analyses of Jomon crania from Eastern Japan. In *Prehistoric Hunter-Gatherers in Japan — New Research Methods,* ed. T. Akazawa & C. Aikens. *Univ. Mus., Univ. Tokyo, Bull.* no.27.

Dodo, Y. & Ishida, H. (1990). Population history of Japan as viewed from cranial nonmetric variation. *J. Anthropol. Soc. Nippon* **98**, 269–87.

Hanihara, K. (1985). Origins and affinities of Japanese as viewed from cranial measurements. In *Out of Asia: Peopling the Americas and the Pacific*, ed. R. Kirk & E. Szathmary. Canberra: The Journal of Pacific History.

Hanihara, K. (1991). Dual-structure model for the population history of the Japanese. *Japan Review* **2**, 1–33.

Kimura, K. & Iwamoto, S. (1968). Photometric studies of the location of some measuring points at the lateral view of the Japanese skulls. *Bull. Fac. Phys. Educ., Tokyo Univ. of Educ.*, **7**, 69–78 (in Japanese).

Kimura, K. & Iwamoto, S. (1969). Photometric studies of the chronological changes of the lateral view in the Japanese skulls. *Bull. Fac. Phys. Educ., Tokyo Univ. of Educ.* **8**, 133–40 (in Japanese).

Kimura, K., Takahashi, A. & Iwamoto, S. (1971). Photogrammetric studies on the sex difference of the modern Japanese skull. *Bul. Fac. Phys. Educ., Tokyo Univ. of Educ.* **10**, 77–86 (in Japanese).

Lestrel, P. E. (1974). Some problems in the assessment of morphological size and shape differences. *Yearbk. Phys. Anthropol.* **18**, 140–62.

Lestrel, P. E. (1975). Fourier analysis of size and shape of the human cranium: A longitudinal study from four to eighteen years of age. Ph.D. diss., University of California, Los Angeles.

Lestrel, P. E. & Brown H. D. (1976). Fourier analysis of adolescent growth of the cranial vault: A longitudinal study. *Hum. Biol.* **48**, 517–28.

Lestrel, P. E. & Roche, A. F. (1976). Fourier analysis of the cranium in trisomy 21. *Growth* **40**, 385–98.

Lestrel, P. E., Kimbel, W. H., Prior, F. W. & Fleischman, M. L. (1977). Size and shape of the hominoid distal femur: Fourier analysis. *Am. J. Phys. Anthropol.* **46,** 281–90.

Martin, R. & Saller, K. (1928). *Lehrbuch der Anthropologie.* (3rd ed.) Stuttgart: G. Fischer.

Needham, A. E. (1950). The form-transformation of the abdomen of the female pea-crab, *Pinnotheres pisum. Proc. Roy. Soc.(Lond.) B.* **137,** 115–36.

Ohtsuki, F., Lestrel, P. E., Adachi, K. & Hanihara, K. (1983). Shape changes in the Japanese skull: Fourier analysis. *XIth Internat. Congr. Anthropol. Ethnol. Sci.* D-304-00. Vancouver, Canada.

Ohtsuki, F., Uetake, T., Adachi, K., Lestrel, P. E. & Hanihara, K. (1993a). Fourier analysis of shape changes in the Japanese skull: Sagittal view. *Okajimas Folia Anat. (Jpn.).* **70,** 127–38.

Ohtsuki, F., Uetake, T., Adachi, K., Lestrel, P. E. & Hanihara, K. (1993b). Fourier analysis of shape changes in the Japanese skull: Vertical view. *Forma* **8,** 297–307.

Ohtsuki, F., Uetake, T., Adachi, K., Kato, S., Lestrel, P. E. & Hanihara, K. (1995). Fourier analysis of shape changes in the Japanese skull: Lateral view. *Anthropol. Sci.* **103,** 182.

Parnell, J. N. & Lestrel, P. E. (1977). A computer program for comparing irregular two-dimensional forms. *Comput. Prog. Biomed.* **7,** 145–61.

SAS Institute (1979). SAS User's Guide. 1979 ed. North Carolina: SAS Institute, Inc.

Sjøvold, T. (1973). The occurrence of minor nonmetrical variants in the skeleton and their quantitative treatment for population comparisons. *Homo* **24,** 204–33.

Suzuki, H. (1967). Microevolutional changes in the Japanese population from the prehistoric age to the present day. *J. Fac. Sci. Univ. Tokyo* **3,** 279–309.

Takayama, H. (1980). An examination of photographic measurement in craniology. *J. Anthrop. Soc. Nippon* **88,** 249–68.

Takayama, H. (1984). Morphological analysis of modern Japanese facial skeleton by Moiré contourography. *J. Anthrop. Soc. Nippon* **92,** 125–6.

Takeuchi, S., Kato, S. & Tokudome, M. (1985). Observations on the occipital region of the skull of the Edo-period people excavated at Tokyo with the aid of Moiré contourography. *Acta Anat. Nippon* **60,** 414.

Torgerson, W. S. (1965). Multidimensional scaling similarity. *Psychometrika,* **20,** 379–93.

Uetake, T., Adachi, K., Lestrel, P. E., Hanihara, K. & Ohtsuki, F. (1993). Chronological changes in the form of the Japanese skull: Vertical view. *Bull. Fac. Gen. Ed., Tokyo Univ. Agri. Tech.* **30,** 149–67.

Woolf, C. M. (1980). *Statistics for Biologists — Principles of Biometry.* Princeton: D Van Nostrand Company, Inc.

Yasui, K. (1986). Method for analyzing outlines with an application to recent Japanese crania. *Am. J. Phys. Anthropol.* **71,** 39–49.

Yasui, K. (1987). Morphological transition of the Japanese cranium: An analysis of the cross-sectional outlines. *Mem. Fac. Sci. Kyoto Univ. (Ser. Biol.)* **12,** 11–88.

10

Craniofacial Variability in the Hominoidea

BURKHARD JACOBSHAGEN

Anthropologisches Institut im FB Biologie, der Justus-Liebeg-Universität Gießen

10.1 Introduction: The measurement of cranial shape

This chapter surveys some approaches to craniometry, each of them designed to represent the surface or internal morphology in 3-D. The reason for using 3-D measurements is to prevent an irreversible loss of data in the first phase of numerical description. Simplifications leading to 2-D data, or other processes to reduce the number of descriptors, can be carried out subsequently. In contrast to the fixed sequence of conventional caliper measurements, such sets of variables can be redefined later (Moyers and Bookstein, 1979). Moreover, new types of variables, for example, spatial measures, might be added. Furthermore, landmarks can be included that are not necessarily defined by conventional anatomical descriptions. An additional advantage with such 3-D measures, besides an increase in efficiency, is that problems of biomechanical function may be correlated with morphology and other biological approaches can be analyzed more adequately (Jacobshagen, 1985).

The range of 3-D measuring techniques includes mechanical contact devices, such as 3-D digitizers (Menk, 1978), and noncontact methods (see Table 10.1). The latter methods utilize visible light, X-rays, electron spin resonance (ESR), nuclear magnetic resonance (NMR), and other techniques. In the morphometry of bony structures, those methods, like ESR and NMR, that are primarily used to measure physiological parameters (soft tissue), have no particular advantages. Moreover, tomographic techniques, including X-ray-based computerized tomography (CT), also suffer from poor 3-D resolution, making them undesirable for precise morphometric purposes.

Holographic documentation of 3-D objects for morphometry has been minimally reported (Sheinberg and Anderson, 1976). Classical techniques such as stereophotogrammetry utilizing slightly differing views between two camera locations, based on Moiré patterns, have also been used. For generally smooth surfaces, this is an adequate solution, especially since charge-coupled device (CCD)

Table 10.1 A survey of some of the three-dimensional craniometric approaches available (adapted from Jacobshagen, 1985).

Measurement technique	Properties	Equipment	Method of operation*	Results
Mechanical 3-D-measurement	direct contact with landmarks	special measuring device with computer	semi-automatic	discrete points (surface descriptive)
Stereophotogrammetry	optical triangulation	2 cameras, pass-point system, stereocomparator, computer	semi-automatic or fully-automatic	defined or undefined discrete points (in non-analytic technique: contour lines)
Laser-Stereovideogrammetry	optical triangulation	laser scanner, 2 CCD-cameras, computer	fully-automatic	undefined discrete points (surface descriptive)
Moiré technique	grid-distortion	grid-projector, 1 camera, evaluation unit	non-automatic or semi-automatic	discrete points or contour lines (surface descriptive)
X-ray-stereophotogrammetry	optical triangulation	stereo-x-ray unit, pass-point system, stereocomparator, computer	semi-automatic	discrete points or contour lines (surface and volume descriptive)
Computerized tomography	x-ray absorption	computer-tomograph with integrated computer	fully-automatic	gridded cross-sections
NMR-tomography	nuclear-magnetic resonance	NMR-tomograph with integrated computer	fully-automatic	gridded cross-sections
Holography	optical intereference	laser, photo plates, special evaluation unit	no details at present	no details at present

*non-automatic, the whole measurement procedure is carried out manually and visually; semi-automatic, the selection of landmarks or profiles is carried out manually, data transfer and storage is automated; and fully automatic, the complete process of measuring is automated.

cameras of sufficient resolution are now available and the measurement procedure is largely done under computer control (Ashizawa et al., 1983; 1985; Frobin and Hierholzer, 1978; Tereda and Kanazawa, 1974; Okoshi, 1976). For industrial purposes, automatic laser-scanners for 3-D measurements have also been developed (e. g., a device by Cyberware in Germany).

Stereophotogrammetry, already mentioned, has been used for different anthropometric purposes such as craniometrics (Creel, 1976), body measurements (Herron, 1972), as well as for morphometrics on teeth and face (Savara, 1965). In the following example, a variation of stereophotogrammetry, visible light was utilized for morphometric measurements on hominoid skulls. The author used the Kellner-System (Kellner, 1980), which has the advantage of allowing the use of a wide range of cameras, in contrast to the special stereophotogrammetric equipment usually required (Jacobshagen, 1980; 1981a; 1981b; Jacobshagen et al., 1988). In this application two simple Rollei 35 mm rangefinder cameras were used. A specially designed calibration procedure allowed for compensation of lens distortion to be instituted at the final steps of data processing. Photographs were made using a reference point system device to enable an exact reconstruction of the skull in various views (Figure 10.1). To make the surface features more visually apparent for photographic purposes, a random-pattern slide (composed of black irregular dots on a white background — details available from the author) was projected onto the skull.

The focus distance was approximately 1.2 m. The fairly large base length of about 39 cm between the Rollei 35 mm cameras provided good stereoscopic separation of images with correspondingly good depth resolution. The overall precision of the point measurements was found to lie within a sphere of 0.3 mm^3 diameter using test objects. As the reference point system was mounted on a turntable (Figure 10.1), it was possible to photograph the skull in the standardized anatomical positions to obtain four views: one frontal, two lateral, and one occipital. Three additional views of the vertical and basal skull areas and the mandible required nonstandardized orientations.

The 3-D landmark measurements were generated using an analytic stereocomparator interfaced with a computer terminal (Jacobshagen, 1980; 1981b). The measurement sequence was done via computer and allowed for a rapid and flawless succession of points to be entered. Approximately 300 points were measured on each skull. Figures 10.2a and b illustrate the density of landmarks for a frontal and a lateral projection. Besides a number of anatomically defined landmarks, others were used to represent curvatures such as the orbital circumference and the midsagittal profile.

All measurements were related to the Frankfort Horizontal. The extension to 3-D was defined by the auricular points and the inferior orbital point of the left

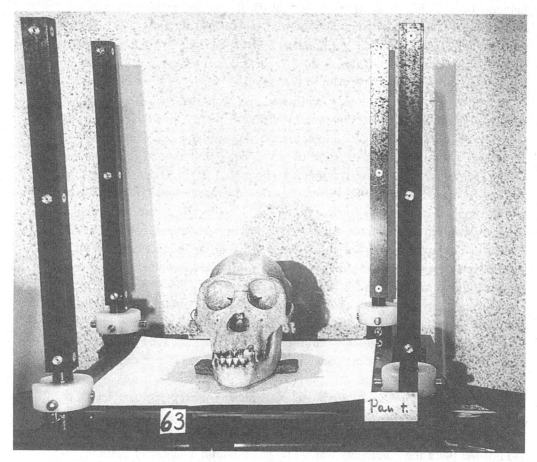

Fig. 10.1 A chimpanzee skull placed onto a reference point system device with turntable.

orbit (Figure 10.3). Whereas the anatomically defined landmarks were utilized to calculate conventional distance measures from point pairs, the other points were used for the quantification of 2-D profiles using Fourier analysis.

10.2 Harmonic analysis of orbital and sagittal contours

Fourier analysis[1] was chosen as a tool to quantify craniometrical data of five hominoid samples: *Pan troglodytes* (male, n = 25), *Gorilla gorilla* (male, n = 30; female, n = 29), *Pongo pygmaeus* (male, n = 30), *Homo sapiens sapiens* (most of

---

[1] The preparation of data used for Fourier analysis was carried out according to Lestrel, (1975; 1980; 1982) and Lestrel and Roche, (1986), which I have called the "vector method." An alternative approach, the "tangent-angle method," is described by Zahn and Roskies (1972) and was applied by Pasian and Santin (1985) as well as by Pesce-Delfino and Ricco (1983). See also Jacobshagen (1992).

them probably male, n = 25). For description and sources of this material see Jacobshagen (1981a). Morphological structures used for harmonic analysis were the contour of the left orbit and the midsagittal outline of the whole skull. In the initial data acquisition step (stereophotogrammetry) a good representation of the orbital contour was maintained by a set of closely sampled points under visual control. The density of the sampling of points was adapted to the actual curvature. This procedure was found to be adequate for representing the morphology with sufficient resolution and little distortion.

10.2.1 Orbital data

To prepare the orbital data for input to the Fourier analysis program, additional steps were necessary:

1. Conversion from 3-D to 2-D was undertaken by projecting the coordinates into a plane that closely parallels the major aspects of the individual orbit, re-

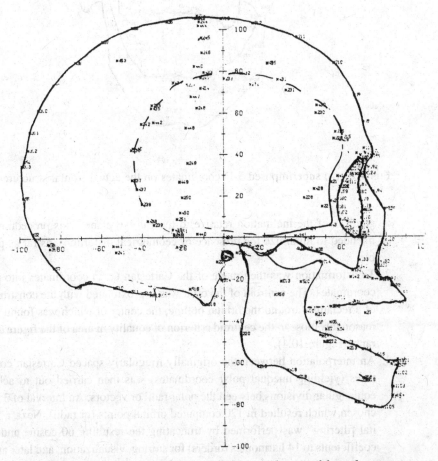

Fig. 10.2 (a) Superimposed 3-D coordinates on the actual lateral structures.

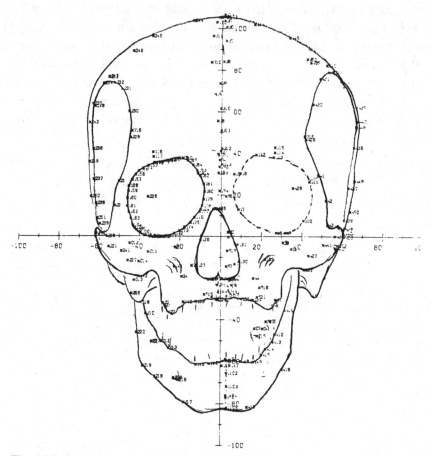

Fig. 10.2 (b) superimposed 3-D coordinates on the actual frontal structures.

gardless of the inclination relative to the ear-eye plane. This procedure was intended to minimize the presence of geometrical distortions, as far as possible.

2. Transformation was then made of the Cartesian (x, y) coordinates into polar coordinates. The centroid of the orbit was approximated with the construction of a rectangle around the orbital outline, the center of which was found to be reasonably close to the centroid criterion of equality in area of the figure quadrants (Figure 10.4).

3. An interpolation between the originally irregularly spaced Cartesian coordinates (yielding unequal polar coordinates) was then carried out to achieve equiangular divisions between the polar radii or vectors. An interval of 3° was chosen which resulted in 120 computed orbital points (or radii). Next, a "digital filtering" was performed by truncating the resulting 60 cosine and sine coefficients to 14 harmonics (orders) for storing, visualization, and later analy-

ses. The data reduction was dependent on the goodness-of-fit between the original data points and the Fourier curves. A maximum deviation of 0.6 mm was accepted — approximately equal to the error arising from the measuring process used to determine the original orbital contour (Figure 10.5). The starting point for the Fourier analysis was the most distal point (from the centroid) on the left eye contour. Sampling was carried out in a clockwise direction with respect to the frontal view.

10.2.2 Sagittal profiles

The quantification of the midsagittal outlines was carried out to test the ability of Fourier analysis to document the overall morphology. The sagittal outline was described with 46 landmarks (Apes) or 47 landmarks (Homo). The sampling was, relatively speaking, not as detailed as the orbital aperture, resulting in a slightly

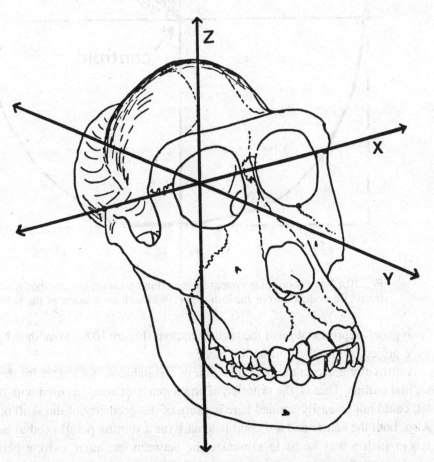

Fig. 10.3 The 3-D coordinate system (see text). Reproduced from Jacobshagen (1982) with permission of the publisher.

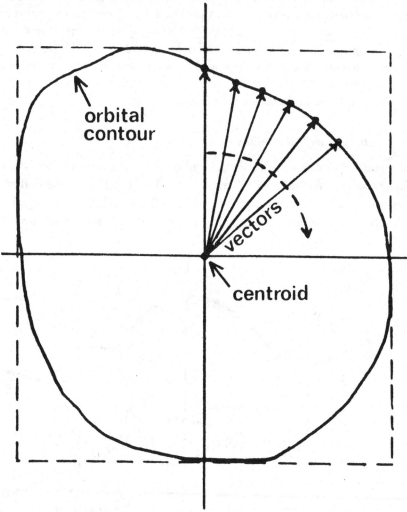

Fig. 10.4 Polar coordinate system for the orbital contour; vectors shown schematically. Reproduced from Jacobshagen (1982) with permission of the publisher.

"polygonal" representation of the sagittal contour (Figure 10.6). Mandibular points were projected into the midsagittal plane.

A difficulty arose in the determination of a suitable vector center for the midsagittal outline. That is, the criterion of equal quadrant areas, as used with the orbit, could not be easily applied here because of the geometry of the skull outline. Also, both the center and a second landmark (as a starting point) need to be chosen in such a way so as to minimize the between-specimen and the between-species variation. For superimpositions of different primate species the cranial base is considered to be a relatively comparable and stable structure (Hofer, 1965).

Abbie (1953) recommended the use of the landmark *hormion* as a substitute for the endocranial landmark *sella*. Since a center at *hormion* would exclude portions of the occipital region, a projection of *hormion* onto the ear-eye plane was chosen instead, which is even closer to the endocranial surface.

The second requirement, the choice of starting point for the Fourier analysis, also had to be defined. The chosen landmark *condylion laterale* proved to be far from optimum, as its small distance from the center caused considerable angular variation.

Using the same approach as with the orbit, the original landmarks were interpolated to generate 365 points (resulting in sampling intervals of 0.986°) for Fourier analysis. The resulting Fourier series was then truncated at the 70th order (harmonic). The energy (or power), s^2, above this order was only around 0.01% computed as a percentage of the total from Equation 10.1:

$$\sum_{i=1}^{n} s_i^2 = \sum_{i=1}^{n} (a_i^2 + b_i^2), \tag{10.1}$$

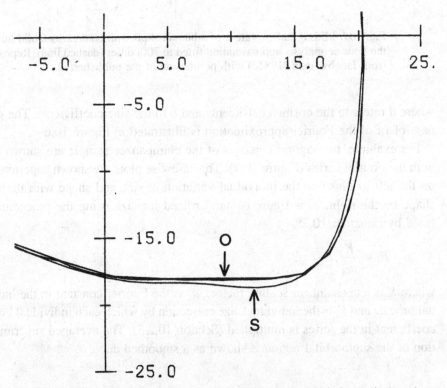

Fig. 10.5 Maximum deviation of 0.6 mm between the orbital margin based on original measurements ("O") and curve generated from the Fourier synthesis ("S"). Reproduced from Jacobshagen (1982) with permission of the publisher.

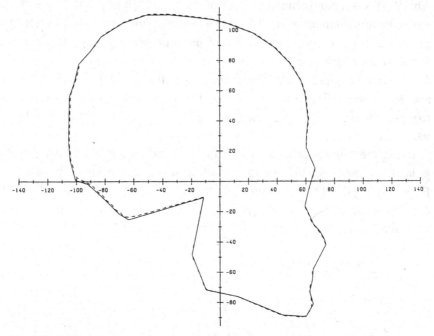

Fig. 10.6 Midsagittal curvature of a human skull with landmarks (solid line) and the Fourier analysis approximation fitted to 70th order (dashed line). Reproduced from Jacobshagen (1981c) with permission of the publisher.

where *a* refers to the cosine coefficients and *b* to the sine coefficients. The goodness-of-fit of the Fourier approximation is illustrated in Figure 10.6.

For example, the sagittal contours of the chimpanzee sample are shown fitted with the Fourier series (Figure 10.7). The casewise plots are shown superimposed on the left to illustrate the individual variation in size and shape with the mean shape on the right. This figure is standardized for size using the procedure defined by Equation 10.2:

$$F_i = \frac{K}{a_{0i}},$$

(10.2)

where K is a constant (or scaling factor), a_0 is the Fourier constant in the individual series, i and F_i is the individual size correction by which each individual Fourier coefficient in the series is multiplied (Section 10.2.3). The averaged superimposition of the supraorbital region is shown as a smoothed curve.

10.2.3 Individual variation of the orbital contour

To visualize the variation in orbital contour, the individual outline of each specimen was plotted in two ways. First, "anatomically," that is, visually, as the Fourier

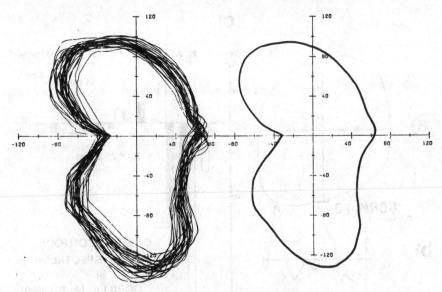

Fig. 10.7 Superimposed midsagittal contours of chimpanzee (left) and the mean or average contour (right).

representation of the orbital outline (Figure 10.8a). Second, the polar coordinates were replotted in a diagram (Figure 10.8b), where the y-axis reflects the length of the vectors as absolute deviations (or differences) from the mean or a_0 term (in mm). The x-axis shows the angle scaled from [0 to 2π]. This type of diagram clearly displays the deviations of an individual orbital contour from a circle.[2] For example, these two types of diagrams with both the corresponding Fourier coefficients, for the first 14 harmonics (Figure 10.8c), and the corresponding energy or power spectrum[3] (Figure 10.8d) are shown for an individual Gorilla specimen.

The intraspecific variation present in each of the five species utilizing the above procedure (Figure 10.8), is not illustrated here because of space constraints (refer to Jacobshagen, 1982). Considerable size and shape differences are present.

10.2.4 Evaluation of the Fourier series and its visualization

In addition to the display of individual variation, statistical processing of the Fourier data was also carried out. As the constant term, a_0, is a size parameter, the average size and standard deviations of the constant for the orbital and midsagittal contours can be calculated (Figure 10.9). Also shown are the correlation

---

[2] Editor's note: The mean of all vectors, the a_0 term or constant, can be visualized as a circle in polar coordinates and as a horizontal line in Cartesian (Figure 10.8b).

[3] Editor's note: The energy spectrum here is slightly different from the usual power spectrum, which is computed as $s_i^2 = \dfrac{a_i^2 + b_i^2}{2}$ (see Chapter 2).

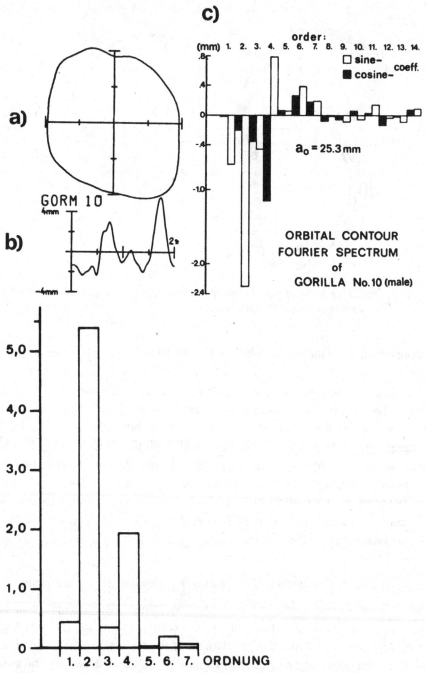

Fig. 10.8 Orbital contour of a male gorilla with four different representations (see text): (a) "anatomical" view in polar coordinates; (b) deviations from the mean or a_0 constant plotted along the *y*-axis with the angular interval [0 to 2π] plotted along the *x*-axis; (c) the Fourier sine and cosine coefficients (to 14th order); and (d) the corresponding energy spectrum. Reproduced from Jacobshagen (1982; a–c) and from Jacobshagen (1992; d), with permission of the publisher.

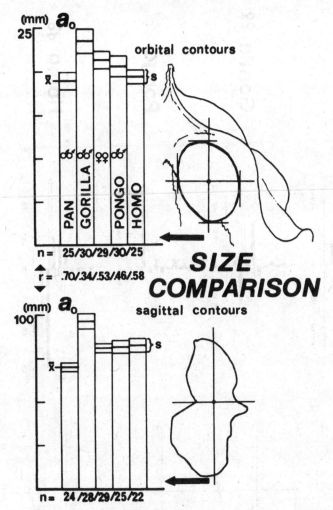

Fig. 10.9 Size comparisons based on the mean, or a_0 constant, (with their means and standard deviations) for the orbital contour (top) and the midsagittal contour (bottom). Also shown are the correlations between these structures (see text). Modified from Jacobshagen (1982).

coefficients of the a_0 term for the orbit versus the midsagittal contour. There seems to be a stronger relationship in the small skulls (chimpanzees), whereas a lower correlation was observed, when sagittal cresting occurs (male gorillas). Humans fit within the range of female gorillas and orangutan.

Using the individual size parameter a_0, and multiplying the Fourier coefficients by Equation 10.2, the coefficients can readily be corrected for size, and utilized for shape statistics[4] (Lestrel, 1974). Plots superimposing all the size-normalized

---

[4] Editor's note: Although the a_0 term is rapidly computed and has been used as an approximate estimate for size, a more precise estimate is the actual bounded area.

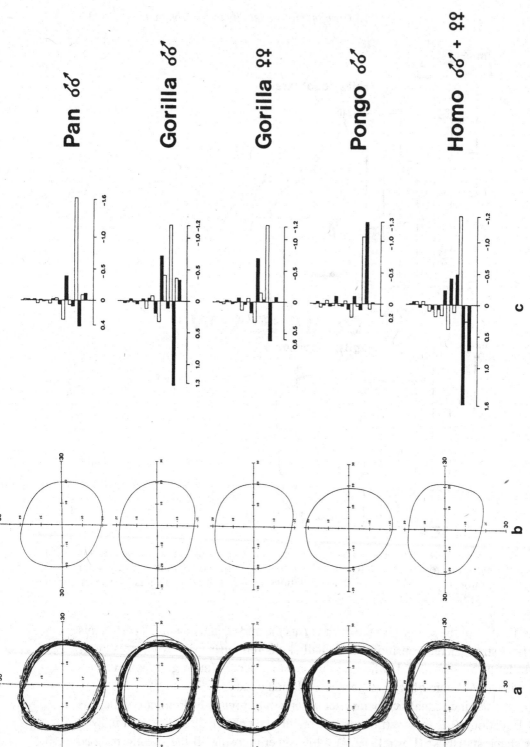

Fig. 10.10 Intraspecific variation of the orbital contours for the five primate samples: (a) superimposed individual curves standardized for size; (b) the mean or averaged orbital shapes; and (c) the corresponding Fourier sine and cosine coefficients are shown up to 10th order.

240

Table 10.2 Number of factors and the proportion of the variation explained (only factors with eigenvalues equal to, or greater
than one, are shown here).

	using $a_0 + 14$ energy scores
Pan	6 (77.1%)
Gorilla, males	6 (70.9%)
Gorilla, females	6 (73.2%)
Pongo	7 (74.0%)
Homo	6 (79.4%)

individual contours give a visual impression of the typical amount of shape variation in each species (Figure 10.10a). The individual contour data were then averaged to produce a "typical" mean orbit (Figure 10.10b). Finally, the corresponding Fourier coefficients, up to the 10th order, are shown in Figure 10.10c.
These demonstrate that Fourier analysis is an efficient tool for the detailed description of shape, with a minimum of parameters. Visually, the average contours
of male and female gorillas display only slight differences, whereas the Fourier
coefficients show marked differences (Jacobshagen, 1992).

10.3 Factor analysis of the Fourier coefficients

Although the Fourier coefficients are mathematically orthogonal descriptors (i.e.,
independent), correlations may, nevertheless, arise in the data subsequent to
Fourier analysis. This suggests the presence of biological relationships between
the different structures.

To gain further insight into some of these structural relationships, factor analysis was applied to the orbital data of the primate samples. Prior to factor analysis, internal testing, instituted to check the adequacy of the correlation matrices,
led to negative eigenvalues and a nonpositive definite matrix or otherwise "ill-
conditioned" correlation matrices, as well as poor results with the Kaiser-Meyer-
Olkin measure of sampling adequacy. This was resolved when the energy scores
(spectra) were used as data in the factor analysis instead of the individual Fourier
coefficients. Table 10.2 displays the factor analysis results.

Viewing the energy scores, some similarities can be found between the chimpanzee and human samples. In fact, coefficient loadings are associated more often with a common factor than with the other primate comparisons (not shown
here). Nevertheless, a pronounced morphogenetic similarity in the orbits of the

species Pan and Homo cannot be simply assumed. Factor analyses of different samples are difficult to compare beyond an overall structural similarity in the factor patterns. One reason for this has to do with the orientation of the factor matrices within the mathematical space: rotation criteria are designed to optimize parameters and not necessarily to achieve comparability between different analyses. So subtle changes in correlation patterns may lead to major deviations in the resulting pattern of factor loadings.

Comparison of the factor loadings of the constant, a_0, with loadings of energy scores of male gorillas, orangutan, and the human sample, (in which individual size was eliminated using Equation 10.2) suggest some allometric dependancies

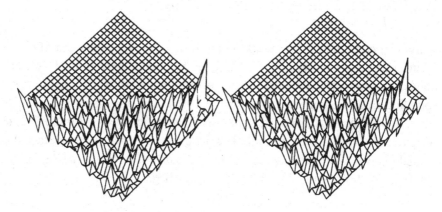

Fig. 10.11 Stereogram of the correlation matrix based on the complete series of Fourier coefficients of male gorillas. First row and first column meet in the right corner, diagonal values and empty fields (the "flat" triangle) are set to zero (the highest positive correlation — at the right — is r = 0.77; the highest negative correlation — hidden — is r = −0.81).

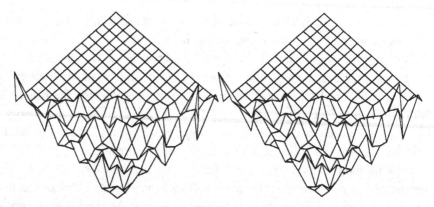

Fig. 10.12 Stereogram of the correlation matrix based on the energy spectra of male gorillas (the highest correlation — at the right — is r = 0.55; the highest negative correlation — hidden — is r = −0.53).

Table 10.3 Classification of sexes using stepwise-discriminant anaylsis and using a *set* of variables.

Character	Discriminating power (%)
Orbital breadth	93.3
a_0	88.2
Orbital height	76.3
Orbital index (Martin No. I 42)	66.1

between size and shape. This phenomenon can be evaluated by looking at the underlying correlations. The highest negative and positive correlations found between the a_0 term and single Fourier coefficients are: $r = -0.32$ and $r = 0.38$ for *Pan* (4x > 0.3), $r = -0.55$ and $r = 0.21$ for male *Gorilla* (3x > 0.3), $r = -0.26$ and $r = 0.31$ for female *Gorilla* (1x > 0.3), $r = -0.42$ and $r = 0.41$ for *Pongo* (4x > 0.3), and $r = -0.39$ and $r = 0.37$ for *Homo* (5x > 0.3). In each species the same coefficients were never found to correlate with the constant a_0 (see also Section 10.5).

Because of problems interpreting the factor analytic results, an attempt was made to visualize the complete correlation patterns. As an example, the patterns of male gorillas are shown in Figures 10.11 and 10.12. Although only shown here for male gorillas, they look very similar in all four species and show a "noisy" pattern.

10.4 Sex and species discrimination

With the application of stepwise-discriminant analysis, the sexual dimorphism in orbital shape of male and female gorillas can be evaluated. For a comparison with univariate analyses, conventional height, breadth, and index values were used (Table 10.3). As expected, the best discrimination is given by the complete Fourier series (Table 10.4). Although size information is retained in the constant, a_0, of the Fourier series, when the coefficients are normalized for size (Equation 10.2), there results a decrease in discriminatory power. Interestingly, the conventionally measured orbital breadth seems to have exactly the same diagnostic power for sexual discrimination as the latter variable set. The energy spectrum is as efficient as the constant, a_0, for sexual discrimination. All other combinations of characteristics are less adequate for this purpose.

Looking at the diagrams of the averaged energy spectra (Figure 10.13) of the orbital circumference of the male and female gorillas, one can now clearly recognize the seemingly indistinguishable differences seen in Figure 10.10.

For species discrimination, in contrast to sexual differences, stepwise-discriminant analysis was also applied. The stepwise procedure selected 21 coefficients. The seven excluded were: coefficients 2, 14, 18, 23, 26, 27, and 28. The dis-

Table 10.4 Classification of sexes using stepwise-discriminant analysis and using *single* variables.

Input characters	Orders included in stepwise procedure*	Discriminating power (%)
a_0 + 28 Coefficients	a_0 + 8 Coefficients (2, 9, 3, 8, 3, 4, 11, 1)	97.1
a_0 + 28 Coefficients (normalized for size)	a_0 + 12 Coefficients (2, 9, 3, 8, 3, 4, 10, 11, 10, 13, 13)	93.3
a_0 + 14 Energy Scores (normalized for size)	a_0 + 4 Energy Scores (2, 3, 4, 5)	88.2
a_0 + Coefficients up to 5th order	a_0 + Coefficients (2, 3, 4, 3, 4)	86.7
28 Coefficients (normalized for size)	10 Coefficients (1, 2, 1, 10, 11, 5, 4)	81.3
14 Energy Scores	6 Energy Scores (1, 12, 13, 2, 7, 11)	78.1

*The proportion of cosine and sine coefficients was always ~50%.

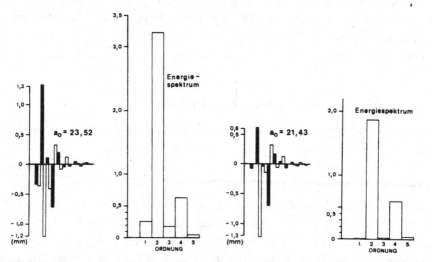

Fig. 10.13 Averaged Fourier coefficients and their corresponding energy spectra, for male (left) and female (right) gorillas. Reproduced from Jacobshagen (1992) with permission of the publisher.

crimination between species was 100%, whereas sex discrimination using the same analysis led to 93.3% correctly classified cases in male and 96.6% in female gorillas. Thus, despite the large individual diversity in the size and shape, the intraspecific variation is less than the interspecific one.

A numerical evaluation of Penrose's shape-distances was also computed (Penrose, 1954). Figure 10.14 shows two representations of the results for the

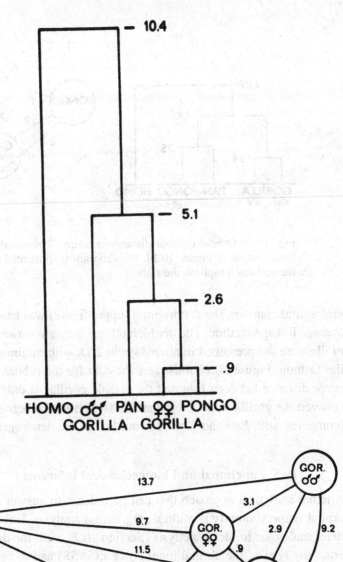

Fig. 10.14 Midsagittal contour distance measures using Penrose's shape distance. Data based on the Fourier series of all five primate samples. Dendrogram based on the unweighted average linkage method (top) and the nonhierarchical graph, where the distances are exactly reflected by the length of the lines connecting the circles (bottom).

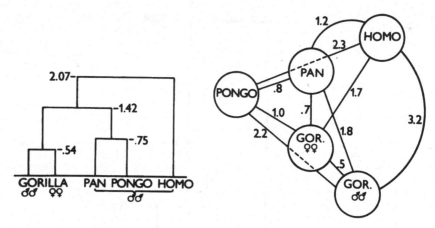

Fig. 10.15 Orbital contour distance measures computed in an identical fashion those shown in Figure 10.14. Dendrogram is illustrated on the left and the non-hierarchical graph on the right.

midsagittal contour. The dendrogram (upper figure) was based on the unweighted average linkage method. The nonhierarchical diagram (lower figure) is an attempt to illustrate the presumed relationships in 2-D, with minimal distortion. In a similar fashion, Figure 10.15 illustrates the data for the orbital margin. Here the average distance between *Pan* and the female gorillas is only slightly greater than between the gorilla sexes. Note that the large distance between *Pongo* and *Homo* (compared with *Pan* and *Homo*) is masked in the dendrogram.

10.5 Functional and biomechanical inferences

Another analytical approach that can be utilized to answer questions concerning orbital shape allometry is multiple regression analysis. However, problems arose here, analogous to factor analysis (Section 10.2), with the data structure. To generate any results, the normal input criteria (SPSS) had to be broadened considerably, especially for the chimpanzee sample. Although these data have to be treated with caution, some of the parameters are listed in Table 10.5.

These data suggest the presence of a strong, but complex, dependency between size and shape. In a manner similar to that of the whole skull, the orbital margin is formed by a variety of different influences based on genetics as well as functional constraints during ontogenesis. Important factors may be: (1) the relative positions of braincase, orbits, and facial skeleton; (2) the strength of the masticatory apparatus; and (3) the robusticity of the supraorbital region (Vogel, 1966). The volume of the eyeball is of minor significance with respect to the orbital volume (Schultz, 1940).

Table 10.5 Survey of multiple regression results based on the Fourier coefficients corrected for individual size. The SPSS Input criteria for the variables were: PIN = 0.1, POUT = 0.2, FIN = 1, FOUT = 0.5. The squared multiple R equals the proportion of explained variation from the regression equation. The significance of F shows the reliability of R^2 (see text).

	Pan	*Gor.* (m)	*Gor.* (f.)	*Pongo*	*Homo*
multiple R	0.61	0.94	0.92	0.81	0.96
mult. R^2	0.37	0.89	0.84	0.66	0.93
Sign. F	0.04	0.0007	0.0019	0.0029	0.0002
coeff. included	25,4,7,5	7,9,8,1, 24,12,5, 10,15,6,2, 14,11,3, 19,17	14,7,3,17, 22,16,18, 9,11,28,5, 21,19,20	3,18,17,28 10,7,9,5, 24	9,25,8,12, 20,11,15, 13,27,23, 24,17,26

Due to this complex network of functional interdependencies, the possibility of statistical analyses directly implicating one or more of these functional relationships seems remote. Moreover, it also seems likely that the same may be true for the pattern of correlations between the sagittal and orbital Fourier data, as well as between the Fourier series and conventional distance and angular measurements. As discussed earlier, the correlation patterns were different for each of the skull samples analyzed here. The only common inference is a correlation between those characteristics that are strongly affected by size, for example, as seen in the the a_0 term for both the orbital and sagittal curvatures, which are correlated with length, breadth, and height distances.

A possible exception, suggested by the correlation patterns, can be found in the 3-D coordinates of the skulls in this study. The attachment of the masseter muscle seems to be a distinct causative factor in the variation of the vertical aspect of part of the zygomatic arch (Jacobshagen 1981a; 1981c).

These results do not support the view of a simple relationship between the various categories of traits examined here. Although this conclusion may seem "uncomfortable," it is a consequence of the utilization of methods that allow for more complete numerical descriptions. Moreover, although these methods open up the possibility of additional research opportunities, they do not necessarily *lead to easier interpretations*.

In order to reduce some of these inherent problems in the analysis of functional and biomechanical mechanisms, it is, perhaps, more efficient and easier to focus on smaller anatomical areas with limited functional complexity. Consequently, to arrive at more precise interpretations, especially of functional complexes, more simulations of the biomechanical forces and their interactions (on bony structures) over small anatomical areas are needed. On the other hand, metrical analyses using Fourier analytic methods could assist in the building and testing of hypotheses, as a first step toward the biologically significant modeling of functional complexes (Jacobshagen, 1985). Subsequently, the application of 3-D Fourier analysis, and other undiscovered methods, can be considered the next step toward a much more exhaustive evaluation of the rich morphological information that is readily apparent visually, but remains largely unattainable (Lestrel, 1975). Although technical problems will increase, it seems increasingly inevitable, given the 3-D character of the majority of biological organisms, that this is the research direction of the future.

References

Abbie, A. A. (1953). The cranial centre. *Z. Morph. Anthrop.* **53**, 6–11.

Ashizawa, K., Kuki, T. & Kusumoto, A. (1985). Fringe patterns and measurements on dorsal Moiré topography in Japanese children, aged 13 and 14. *Am. J. Phys. Anthrop.* **68**, 359–65.

Ashizawa, K., Tsutsumi, E., Kurihara, S. & Yoshizawa, T. (1983). Installation of Moiré cameras for somatometrical use. *Bull. Faculty Domestic Sci.* **19**, 49–61.

Creel, N. (1976). A stereophotogrammetric system for biological applications (abstract). *Am. Assoc. Phys. Anthrop.* **45**, 173.

Frobin, W. & Hierholzer, E. (1978). A stereometric method for the measurement of body surface using a projected grid. *Soc. Photo-Opt. Instr. Engr. (SPIE)* **166**, 39–44.

Herron, R. E. (1972). Biostereometric measurement of body form. *Yearbook Phys. Anthrop.* **16**, 80–121.

Hofer, H. (1965). Die morphologische analyse des schädels des menschen. In *Menschliche Abstammungslehre. Fortschritte der Anthropogenie*, ed. G. Heberer. Stuttgart: Gustav Fischer.

Jacobshagen, B. (1980). Grenzen konventioneller techniken und möglichkeiten alternativer ansätze in der anthropometrie. Mit einem beispiel für den einsatz der biophotogrammetrie in der schädelmeßtechnik. *Z. Morph. Anthrop.* **71**, 306–21.

Jacobshagen, B. (1981a). Die variabilität des schädels pongider und hominider primaten. Vergleichende funktionell-morphologische untersuchungen unter besonderer berücksichtigung der vom kauapparat belasteten Strukturen. Ph.D. diss., University of Hamburg.

Jacobshagen, B. (1981b). The limits of conventional techniques in anthropometry and the potential of alternative approaches. *J. Hum. Evol.* **10**, 633–7.

Jacobshagen, B. (1981c). Comparison of morphological factors in the cranial variation of the great apes and man. In *Proceedings in Life Sciences*. Vol. A: *Primate Evolutionary Biology*, ed. B. Chiarelli & R. S. Corrucini. Heidelberg: Springer.

Jacobshagen, B. (1982). Variations in size and shape of the orbital contour. A comparison between man and the greater apes using Fourier analysis. *Anthropos* **21**, 113–30.

Jacobshagen, B. (1985). Quantification of form — possibilities and problems for the functional approach. In *Evolution and Morphogenesis*, ed. J. Mlíkovsky & V. J. A. Novák. Praha: Academia.

Jacobshagen, B. (1986). Size and shape of the orbital outline: A multivariate comparison and analysis of intraspecific variation in four hominoid species. In *Primate Evolution. Proceedings of the 10th Congress of the International Primatological Society*, Vol.1, ed. J. G. Else & P. C. Lee. Cambridge: Cambridge University Press.

Jacobshagen, B., Berghaus, G., Knußmann, R., Sperwien, A. & Zeltner, H. (1988). Fotogrammetrische methoden. In *Anthropologie. Handbuch der Vergleichenden Biologie des Menschen*, Vol. I/1, ed. R. Knußmann. Stuttgart, New York: Gustav Fischer Verlag.

Jacobshagen, B. (1992). Fourieranalyse. In *Anthropologie. Handbuch der Vergleichenden Biologie des Menschen*. Vol. I/2, ed. R. Knußmann. Stuttgart, New York: Gustav Fischer Verlag.

Kellner, H. (1980). Verfahren und Vorrichtung zur photogrammetrischen Vermessung von räumlichen objekten insbesondere nach Verkehrsunfällen. *Patentschrift 2631226*, Deutsches Patentamt.

Lestrel, P. E. (1974). Some problems in the assessment of morphological size and shape differences. *Yearbk. Phys. Anthrop.* **18**, 140–62

Lestrel, P. E. (1975). Fourier analysis of size and shape of the human cranium: a longitudinal study from four to eighteen years of age. Ph.D. diss. University of California, Los Angeles.

Lestrel, P. E. (1980). A quantitative approach to skeletal morphology: Fourier analysis. *Soc. Photo-Opt. Instr. Engr. (SPIE)* **166**, 80–93.

Lestrel, P. E. (1982). A Fourier analytic procedure to describe complex morphological shapes. In *Factors and Mechanisms Influencing Bone Growth*, ed. A. D. Dixon & B. Sarnat. New York: Alan R. Liss.

Lestrel, P. E. & Roche, A. F. (1986). Cranial base shape variation with age: A longitudinal study of shape using Fourier analysis. *Hum. Biol.* **58**, 527–40.

Menk, R. (1978). Stéréométrie crânienne: réalisation d'un vieux rêve de l'anthropologie. *Arch. Suisses d'anthrop. Générale* **42**, 23–30.

Moyers, R. E. & Bookstein, F. L. (1979). The inappropriateness of conventional cephalometrics. *Amer. J. Orthod.* **75**, 599–617.

Okoshi, T. (1976). *Three-Dimensional Imaging Techniques*. New York: Academic Press.

Pasian, F. & Santin, P. (1985). Morphological analysis of extended objects. In *Data Analysis in Astronomy*, ed. L. Scarsi & V. Di Gesu. New York: Plenum Press.

Penrose, L. S. (1954). Distance, size and shape. *Ann. Eugen.* **18**, 337–43.

Pesce-Delfino, V. & Ricco, R. (1983). Remarks on analytic morphometry in biology: Procedure and software illustration. *Acta Stereol.* **2**, 458–68.

Savara, B. S. (1965). Application of photogrammetry for quantitative study of tooth and face morphology. *Amer. J. Phys. Anthrop.* **23**, 427–34.

Sheinberg, B. M. & Anderson, W. L. (1976). Possible applications of holography to physical anthropology (abstract). *Amer. J. Phys. Anthrop.* **45**, 206.

Schultz, A. H. (1940). The size of the orbit and of the eye in primates. *Am. J. Phys. Anthrop.* **26**, 389–408.

Tereda, H. & Kanazawa, E. (1974). The position of Euryon on the human skull analyzed three-dimensionally by Moiré contourography. *J. Anthrop. Soc. Nippon* **82**, 10–19.

Vogel, C. (1966). Morphologische studien am gesichtsschädel catarrhiner primaten. *Bibl. Primatol.*, Fasc. 4. Basel: S. Karger.

Zahn, C. T. & Roskies, R. Z. (1972). Fourier descriptors for plane closed curves. *IEEE Trans. Comput.* **C-21**, 269–81.

11

The Heuristic Adequacy of Fourier Descriptors: Methodological Aspects and Morphological Applications

VITTORIO PESCE DELFINO, TERESA LETTINI
AND ELIGIO VACCA
Consorzio di Ricerca DIGAMMA, Bari, Italy

11.1 Setting the Scene

In 1917 and in 1942, D'Arcy Thompson posed the question of size and shape relationships in terms of "Growth and Form" by linking it to two fundamental concepts: (1) basing morphological classifications on a nondimensional type of mathematics (analytical functions); and (2) verifying the "principle of discontinuity," particularly as this principle occurs in morphogenesis (e.g., chaos/catastrophic theory). It is this rethinking in *Systema Naturae* terms that Thompson should be remembered for, not just for his well-known method of coordinate transformations. To clarify this we begin by quoting from Thompson:

> When we begin to draw comparisons between our algebraic curves and attempt to transform one into another, we find ourselves limited by the very nature of the case to curves having some tangible degree of relation to one another; and these "degrees of relationship" imply a classification of mathematical forms, analogous to the classification of plants or animals in another part of the *Systema Naturae*. An algebraic curve has its fundamental formula, which defines the family to which it belongs; and its parameters, whose quantitative variation admits of infinite variety within the limits which the formula prescribes (D'Arcy Thompson, 1942:1093–94).

Continuing:

> We never think of "transforming" a helicoid into an ellipsoid, ... [and] ... So it is with the forms of animals. We cannot *transform* an invertebrate into a vertebrate, nor a coelenterate into a worm, by any simple and legitimate deformation, nor by anything short of reduction to elementary principles. A "principle of discontinuity," then, is inherent in all our classifications, whether mathematical, physical or biological (1942:1094; italics in original).

The problems of allometry and shape measurement have become particularly important. There has been increasing awareness that differential growth rates are responsible for allometric changes; that is, as fundamental mechanisms for shape differentiation in biological structures. This problem of allometry was recognized by Holloway who indicated that: "measurements such as length, width, height, whether in chords or arcs, only describe space . . . and further run into the abyss of allometric correction . . . If additional information (shape?) to size is expected, some methods of allometric correction must be used" (Holloway, 1981:158).

Measures of developmental length, as distances, together with angles, ratios, and areas, represent the primary data for the evaluation of size. The attempt to describe shapes utilizing ratios and other simple measures has yielded questionable results.

Poor results may also arise from stereological methods based on probabilistic approaches that are widely used in histomorphometry. As Aherne and Dunnill indicated: "morphometric methods which have been devised for randomly disposed or isotropic elements cannot be validly applied to anisotropic elements or tissue components" (Aherne and Dunnill, 1982:22). Note also the inadequacy of this approach with respect to the anisotropy of biological structures: "a cursory knowledge of anatomy will at once reveal that this (isotropy) is a property possessed by very few structures and tissues" (Aherne and Dunnill, 1982:22). In contrast to isotropic distributions, the distribution of biological components arises from numerous, diverse morphogenetic factors (structural, functional, active, passive, stable, provisory, etc.).

Consequently, this raises the issue of whether analytical morphometric techniques can overcome the inherent limitations in discrete measurements and still be effective for describing shapes; and moreover, whether the above approach can even be applied in a totally different context. That is, can we emulate the visual information that resides in the human physiognomy for identification purposes? We base this recognition on aspects such as dimensions, color, and, above all, the shape of constituent elements (nose, forehead, profile, etc.). The change in anatomical profile (it's shape) undoubtedly contains more information than any other factor (e.g., color or isolated dimensions). For example, any measure connecting landmarks such as *glabella*, the nasal spine, and the extremity of the nasal bones, will not tell us anything about the shape of the nose (i.e., its contours). Yet, it is the profile between those points that is of physiognomic value and forms the basis for identification because our visual perception is quite sensitive to differences in shape. Put differently, the brain is an effective shape recognizer but not describer.

The eye detects subtle differences by looking for contrasts, the greater the contrast between comparisons, the more readily can differences be distinguished. Nevertheless, the exact physiochemical processes involved in the reception, recognition and integration of visual stimuli, remain largely unknown. In contrast to the human capability to identify rapidly and process the information present in complex forms, at least in a qualitative sense, the mathematical replication of visual information has been slow in forthcoming (Lestrel, 1982:393).

In contrast, methods to quantify morphology have been based on relatively simple procedures for dimensional evaluations, which are, however, limited for the analysis of shape. Since the shape (outline) of an anatomical region needs to be reduced to quantifiable characteristics, one approach is to consider outlines or profiles as curves, leading to descriptions based on analytic geometry.

Fourier analysis has been suggested as the best analytical method to describe irregular biological shapes (Diaz et al., 1990; Jacobshagen, 1982; Johnson et al., 1986; Lestrel, 1974; 1982; Masashi et al., 1992; O'Higgins and Williams, 1987).

Other proposed methods include biorthogonal grids using homologous-point landmarks (Bookstein, 1978); finite element methods (Lestrel, 1989); and fractal geometry (Weibel, 1994). Lestrel provided a critical review of the first two methods together with elliptical Fourier functions. It is instructive to read his conclusions — with which we are in agreement:

> As descriptive methods of shape and shape changes, all three methods reviewed can be considered as major innovative improvements when compared to conventional metrical approaches. But as ultimate numerical models, they remain incomplete. Each method has constraints that limit its application to certain classes of shapes or aspects of those shapes. At the moment there is no single method that extracts a majority of the information that is present in all morphological forms. This visual information, so readily observable, has remained elusive to quantification (Lestrel, 1989:89).

Methods that use discrete landmarks exclude information about shape between landmarks. Moreover, the biological homology of these landmarks is often arbitrary. Thus, in the words of K. Z. Lorenz, from his Nobel Prize lecture (Stockholm, 12 December 1973): "Strictly speaking, the term homologous can only be applied to characters and not to organs . . . As a pupil of . . . Ferdinand Hochstetter, I had the benefit of a very thorough instruction in the methodological procedure of distinguishing similarities caused by common descent (homology) from those due to parallel adaptation. In fact, the making of this distinction forms a great part of the comparative evolutionist's [*morphologist's*] daily work." It is apparent that the criteria of point homology is in contradiction with the concept of homology as discussed above. Homology has to be based on similarity of one or more parts to be considered in an evolutionary sense. This implies that the evaluation of homologous similarities first requires the numerical description of shape.[1]

Consequently, we shift to a discussion of quantification of this visual information. While in agreement with Lestrel's earlier point about the "elusiveness of

---

[1] Editor's note: Reference should also be made to Chapters 1 and 4 for discussions pertaining to homology.

the quantification of visual information," we feel that a mathematical approach to model visual perception is feasible.

This modeling can be accomplished via analysis of natural language used as a formal, if imperfect, description of visual information: "The study of form may be descriptive merely, or it may become analytical. We begin by describing the shape of an object in simple words of common speech: we end by defining in the precise language of mathematics" (Thompson, cited in Lestrel, 1982:393).

We can safely assume that Thompson did not imply that from common speech one could directly proceed to the language of mathematics, but rather, to arrive at a mathematical description of form it is necessary to start with a precise understanding of the meaning of words used to describe shapes.

11.2 A model of visual information derived from natural language

Several years ago we attempted to implement Thompson's approach with an analysis of the natural language used by morphologists to describe shape. We were motivated by an anthropological controversy.

At the beginning of this century, numerical description of the cranial shapes was vigorously opposed by Sergi (1900; 1911) who deplored "the cabalistic abuse" and severely denounced the "inadequacy" of such a numerical approach for description and classification. He supported his view with examples that showed that regularly shaped figures were indistinguishable on the basis of indices, even when the shapes were very dissimilar.

To correct this situation, Sergi proposed the use of a binomial classification of skulls based on morphological description. However, his solution to the craniometrical inadequacy of measures and indices can also be legitimately criticized, in that the method lacked objectivity which only a numerical description could provide. Sergi was, however, aware that a binomial system of nomenclature was incomplete in terms of the variability in shape.

The following terms display the inadequacy of such descriptive attempts to describe skull shape: *ellipsoides eucampylus, ellipsoides parallelepipedoides, ooides planus, pentagonoides subtilis, pentagonoides latus, beloides aegyptiacus, byrsoides americanus, sphaeroides Arkansas, sphenoides parvus, rhomboides floridensis*. This large number of terms led researchers to begin collecting words to build a technical dictionary of terms describing the shape of objects (cells, nuclei, particles, tissues, organs, etc.). These terms are routinely used in medical diagnosis. Examination of these terms led to the detection of three major categories. The first group refers to words that are indicative, in a generic way, of shape: *monomorphic, bizarre, regular, irregular, more or less regular, dysmorphic, polymorphic, pleomorphic, anaplastic*. The second category includes words that are more specific and involve evaluation based on symmetry, and so on: *elongated, elliptic, stick-like, hand mirror-like, egg like, distorted, asymmetric, round, oval, pear-shaped, reniform, horseshoe-shaped, spindly, oat-shaped.* The third category comprises words that are, more or less, descriptive of marked irregularities of the contour: *polygonal, full of corners, lobulated, multilobulated, hyperlobulated (clover leaf, flower cells, popcorn cells), stellate, indented, cleaved, cerebriform, convoluted.*

These morphological descriptions suggest that shapes contain elements of two distinct categories: (1) distortions of symmetry; and (2) perturbations of contour. In the absence of quantification it is not surprising that the morphologist is forced to use such terms in diagnosis to somehow distinguish between these two categories. In fact, actual objects display an intermixing of the categories, as well as others involving magnitude, scaling, and so on, requiring a more complex evaluation. Partitioning shapes into these categories is not new and its roots can be traced to medieval times as recognized by the Gestalt philosophers (Ungerer, 1966). Consider M.C. Escher's *Circle Limit IV* woodcut which is depicted in Figure 11.1 (Locher, 1971). Our visual perception is able to discriminate angels (top figure) from devils (bottom figure) by the outline. We will return to these outlines subsequently. Using a "gestalt" approach, the *sharply cut, pointed, hooked,* and *hard* features can be referred to the devil (referred to as "takete," an imaginary term for bad, malignant), whereas *rounded, soft, smooth, gently sloped* features refer to angels ("maluma," another term signifying kind, benign). See Lazotti and Tarantino (1984) for further details.

Consequently, we felt that any analytic description of actual shapes, no matter how correct, must be considered inadequate if the numerical description does not deal with *symmetry* and *contour perturbations.* Indeed, if these two parameters are not effectively separated, it is likely that approximations, (e.g., a Fourier series) allowing separation (orthogonality) and reduction in number of variables, may be of limited advantage, although still useful for such procedures as multivariate discriminant analyses. However, even this advantage may be minimal if the correspondence between the parameters (e.g., Fourier coefficients) and the morphological characteristics is not adequately understood. Quoting Dullemeijer (1974):

> Mathematical tools are methods; they serve for the analysis, the discrimination or the generalization, but the explanation is typological, causal or historical rather than mathematical. A true mathematical idea implies a formulation of the relations of shape and form to other phenomena in terms of exactly defined mathematical parameters.

Thus, a working hypothesis is that *symmetry* and *contour perturbations* must be related in some way to functional considerations. For example, cellular distortions

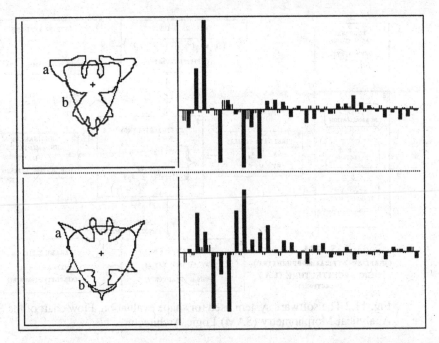

Fig. 11.1 The two intertwining figures (angels and devils) from M. C. Escher's *Circle limit IV*, a 1960 woodcut: (a) original contour; (b) the fundamental shape and Fourier spectra.

may be related to cytoplasm contents or nucleus position, whereas perturbations of the contour could be induced by surface membrane arrangements. Likewise for skulls; the first category could refer to morphologic adaptations of the braincase as a response to evolutionary changes influencing the brain, whereas the second category could potentially lead to the description of subspecific or individual local features resulting from chewing stress and strain adaptations, ontogenetic variations, and so on.

Moreover, the heuristic power of a global description of form by itself is considerably lessened when compared to the additional information derived from symmetry and contour perturbations as discussed above. However, it is not possible to achieve this goal with single, isolated methods. What is needed is an algorithmic architecture composed of multiple procedures, considered jointly, allowing a general solution to facilitate descriptions, comparisons, systematics, and so forth, as well as subsequent multivariate statistical analysis.

11.3 The SAM logical architecture

The SAM (Shape Analytical Morphometry) approach provides for the extraction of descriptive shape measures using procedures that generate numerical parame-

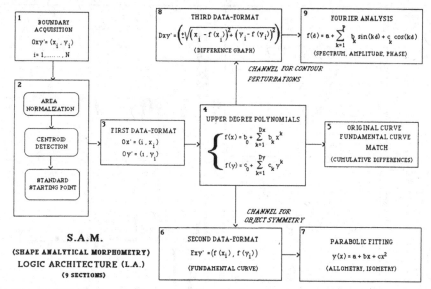

Fig. 11.2 The software system used for shape evaluation. Flow chart of the Shape Analytical Morphometry (SAM) Logic Architecture.

ters that can be uniquely linked to morphological characteristics (Pesce Delfino and Ricco, 1983; Pesce Delfino, et al., 1986a; Pesce Delfino, et al., 1987a; Pesce Delfino, et al., 1990a).

The logic of the SAM architecture assumes that the objects are 2-D closed or open curves viewed in terms of distortions (asymmetries) and irregularities of the outline. The latter characteristics may be pronounced indentations, extroversions, and offshoots, which can have a high degree of repetition.

The crucial step involves partitioning the total information content into two separate channels consisting of dedicated procedures, one dealing with symmetry and the other with contour. The SAM Logical Architecture is displayed in Figure 11.2. This figure shows the logical flow and algorithms in nine sections describing different operations: Box 1 — digital acquisition of boundaries; Box 2 — operations of area normalization, centroid detection, and standardization of starting point; Boxes 3, 6, and 8 — rearrangement of data into suitable formats; and Boxes 4, 5, 7, and 9 — computation of the major analytical algorithms. The steps involved in partitioning the total shape information start with Box 4, once it is in the proper data format (Box 3, see Section 11.3.2). The channel evaluating the symmetry of an object is derived from Box 6 which is then fitted with a parabolic curve (Box 7). The channel for evaluating perturbations of the contour is computed from Box 8 which is then used to compute the Fourier analysis (Box 9). An illustration of each computation available from the SAM. is shown in Figure 11.3.

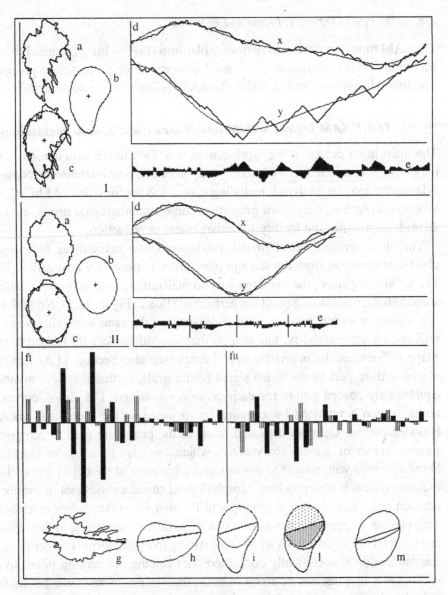

Fig. 11.3 Shape evaluation of tumor nodes of the breast: (a) original outlines of two tumor nodes (I, II) (Box 1, Figure 11.2); (b) closed curve that results from merging the two separate (one for x- and other for y-values) polynomials generating the fundamental shape (FS) (Box 6); (c) superimposition of the FS onto the original shape; (d) plot of abscissa values (x) and ordinate values (y) together with the separately derived polynomials (Boxes, 3, 4, and 5); (e) graph of the differences between the original outline and the FS after rectification of the FS curve (zero line) (Box 8); (f) Fourier spectra for first 15 harmonics (Box 9); (g) identification of asymmetry minimum axis on the original outline of irregular node I (Box 7); (h) corresponding minimum asymmetry axis located on the fundamental shape to identify the starting point on the boundary (Box 7); (i) maximum asymmetry evaluation for node 1 (Box 7); (l) fine dots: allometric area; sparse dots: isometric area and symmetric model of the figure (see Figure 11.4); and (m) maximum asymmetry evaluation for the more regular node II (Box 7). (Refer to Figure 11.2 for Box numbers.)

The SAM runs on a dedicated hardware platform (ISP — Image Signal Processor, Motorola CPU) with specially designed equipment for video signal processing. The following subsections describe the SAM architecture in some detail.

11.3.1 SAM logical architecture (boxes 1 and 2, data acquisition)

The main input device of the work-station is a TV camera with a zoom lens. It may also be connected to a microscope or other imaging instruments. Numerous 2-D objects may be analyzed, including curves and profiles obtained by other devices (craniograms, maximum projection curves, photographic prints, etc.). The zoom lens provides the facility for enlargement or reduction.

The object profiles are digitized, producing (x, y) coordinates in a counter-clockwise direction from the starting point (Box 1, Boundary acquisition, Figure 11.2). Once digitized, the operations of normalization, centroid computation and standardization of starting point are performed (Box 2, Figure 11.2). Normalization is a scaling procedure to insure that all objects are the same size. This is not only required for comparability, but also so that size differences do not mask subtle shape differences. To normalize open curves (see also Section 11.3.4), they are placed with respect to the video signal (raster grid), so that the *same* number of equidistantly spaced points for each curve are obtained. For closed curves, the normalization is obtained by a scaling factor based on the actual area under the bounded outline. Centroid detection refers to the procedure used to compute the figure's center of gravity (barycenter) which can also be used for orientation. Standardization with respect to position and orientation of the object within the co-ordinate system is also required. Morphological characteristics that determine orientation (e.g., craniometrical points) will be used for curves (open or closed) of such objects. In general, open curves can be positioned according to their first and last points. For closed curves without overriding morphological considerations, the standardization is specifically concerned with finding the starting point. In fact, with respect to a nucleus or a cell contour, there is no intrinsic rule for finding a unique point as a standard first point. However, the starting point here, although arbitrarily determined, is nevertheless, strictly defined (see Section 11.3.3).

11.3.2 SAM logical architecture (boxes 3, 4, 5, and 6, fundamental curve)

The term "fundamental shape" refers here to a new curve which approximates the original profile, although by introducing a smoothing effect. This fundamental shape, largely free of contour perturbations, retains the original distortions of symmetry.

This procedure generates parameters of contour irregularities in terms of the relationship between the fundamental curve and the original profile. Initially, the abscissa and ordinate coordinates (x,y) are separated into two plots, one of x versus i (i being a new horizontal axis) and the other y versus i (Box 3). The i-axis consists of a series of positive integers with an increment of one which ranges from 1 to N, where N equals the number of points into which the profile has been subdivided. Separate polynomial functions are then fitted to each of these plots (Box 4).

The coefficients of the two upper-degree[2] polynomials (one for the x- and the other for the y-coordinate) are now calculated using least squares. This procedure has been adopted because one can always obtain two series of values[3] which are unique with respect to the coordinate system.

The maximum order (degree) of each polynomial is not of interest here because the purpose is only to approximate (smoothing) the original profile, not to provide a precise fit. Thus, each polynomial (in x or y) is truncated well before the maximum degree is attained, that is, prior to the appearance of anomalous peaks that are related to localized contour perturbations. Prior studies have found degrees of 6, 7, or 8 acceptable. Finally, after *rounding the values* from the two polynomial functions, respectively $f(x)$ and $f(y)$, to integers, they are merged together to produce a new series of x, y coordinates that represent the *fundamental shape* (vector Fxy, Box 6).

The original curve and its fundamental shape are then compared and the differences evaluated with various measures. These included square root of mean square error (RMS), total and mean error, absolute total and mean error, positive error, and negative error.

The original figure and the function curves have the same number of points. However, since the fundamental curve is a smoothed approximation of the original curve it will contain a certain number of repeated points with the same coordinates (due to rounding to integers). The ratio between the number of repeated points and the original number of points is another parameter of difference between these two curves. After this parameter has been computed, the repeated points are removed (in effect a second smoothing) to make the fundamental curve free of any residual information about contour perturbations. The fundamental curve is then ready for further analysis (Box 7).

11.3.3 SAM logical architecture (box 7, symmetry)

The procedure for the evaluation of symmetry consists of repeatedly calculating the coefficients of a parabola using a least squares fit to the fundamental shape. The procedure involves the manipulation of the fundamental shape by a stepwise rotation through 180° with small angular steps (10° being typical). The rotation

---

[2] Editor's note: Upper degree refers to the maximum order of the polynomial.
[3] Editor's note: This parametric procedure avoids the complications arising with multivalued functions.

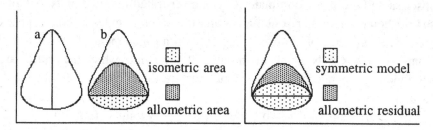

Fig. 11.4 Analysis of symmetry based on the fundamental shape of a triangle: (a) condition of minimum asymmetry (about a vertical symmetry axis) and (b) condition of maximum asymmetry (about a horizontal symmetry axis). Figure on the left shows a model in which the duplicated isometric area generates a symmetric figure while the allometric area is viewed as a "residual" or the fraction which is not accounted for by the condition of symmetry.

is performed with respect to the barycenter (centroid). For each rotational step, a parabola is computed, then an arc/chord index is constructed using the trace of the parabola (arc) and the straight line segment (chord) joining the ends of the parabolic arc where they cross the boundary of the fundamental shape. The trace of the square term of the parabola (cx^2, Box 7), may be more or less convex (Figures 11.3 and 11.4). Note that this procedure is independent of the coordinate system.

For each rotational step the area between the parabolic arc and its chord is determined (allometric area) as well as the area defined by the space between the chord and the profile of the fundamental shape (isometric area). This latter measure is determined from the side opposite to the one which displays the convexity of the parabola (see Figures 11.3l and 11.4).

At each step, the arc/chord locations corresponding to the maximum perpendicular distance between arc and chord (maximum asymmetry) and the minimum difference (minimum asymmetry) are also found and these distances are recorded (Figure 11.3h,i). The minimum difference corresponds to the most symmetry in the figure; when the symmetry is perfect the coefficient of the square term of the parabola becomes zero and the arc of the parabola flattens to a straight line coinciding with the chord that determines the axis of symmetry. When the chord and arc coincide, the allometric area will also be equal to zero and the domains on the two sides of the chord will have equivalent area. In any other case, the sum of the isometric and the allometric areas will be equal to one-half the total area of the fundamental shape. Finally, since the computation of a parabola contains three degrees of freedom, it is also possible to represent the allometry in terms of a physical vector with its *modulus*, *direction*, and *versus*. This yielded the following three measures:

(1) the arc/chord ratio is the *modulus* of the vector; the *direction* is given by the perpendicular to the chord passing through the maximum convexity of the parabola, and the *versus* of the vector is indicated by the side toward which the convexity of the parabola is oriented;

(2) the allometric area (allometry %), constitutes the difference between the parabola arc and its related chord, expressed as a percentage of one-half of the total area of the figure as given by the fundamental curve (see Figures 11.31 and 11.4); and

(3) the isometric area (isometric difference), is that domain which does not contain the convexity of the parabola. Moreover, this domain, if duplicated on the opposite side of the chord, generates a symmetrical figure.

The ability to view allometry/symmetry relationships (Box 7) in physical vector terms provides a useful procedure because allometry is the result of local differential growth. The analysis of a whole figure requires 18 steps and yields a family of 18 vectors.

This procedure, independent of the coordinate system, is also used to define a standardized starting point on the boundary of the original closed curve required for the other SAM sections. Required for Fourier analysis (see Section 11.3.4), this point is located at the intersection of the profile with the chord extended at the point of most symmetry (Figure 11.3g). As there are two intersections, the one nearest the barycenter is chosen as the starting point. The barycenter is located within the allometric area and, if this area is equal to zero, the barycenter is on the resulting straight line.

The same approach using allometric and isometric shape differences applies to open curves. In this case a new figure is constructed in which the two profiles to be compared are redrawn after a lateral reflection (inversion) of the second profile. For example, in the case of frontofacial profiles, a two-faced figure will be constructed so that the two profiles can be compared. This two-faced figure is named "Janus" after the ancient Latin God (Figure 11.5). Here the coefficients of the parabola are calculated only once; because the procedure is applied to open curves placed in a standardized position, the Janus construction is not submitted to rotation. Asymmetries resulting from differences between the two compared profiles can thus be evaluated. In this case, the convexity of the parabolic arc depends on the trend of the outline composed of the two profiles. If only one profile is involved, the Janus construction will substitute a straight line for the second profile.

11.3.4 SAM logical architecture (boxes 8 and 9, Fourier analysis)

The Fourier harmonic analysis aspect of SAM is based on a trigonometric polynomial (series) from which are calculated a set of Fourier sine/cosine coefficients

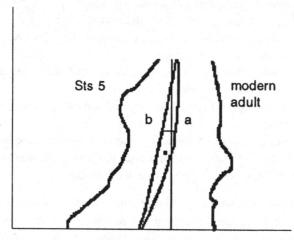

Fig. 11.5 A "Janus" like construction of the fronto-facial profile of (STS 5) and *H. sapiens* (adult). A parabolic fit was utilized to compare the two profiles in terms of asymmetry: (a) arc of the parabola; (b) straight line segment (chord) joining the extremities of the arc. The bounded segment defined by the arc and chord, is the allometric area and contains the barycenter (small solid square). The convexity of the parabola, its inclination and positioning of the arc/chord complex, describe the comparison of the two profiles in terms of isometric differences and the prevalence allometric considerations.

for a finite number of sinusoidal harmonics. The number of harmonics cannot exceed one-half the number of points on the boundary if the number is even; if odd, the number of harmonics is equal to one-half the number of points minus one.[4] From the coefficients, amplitude and phase values are calculated for each harmonic. The amplitude is related to the absolute size of coefficients, and the phase is based on the ratio of the sine/cosine coefficients (see Chapter 2).

This is an extremely powerful procedure which provides a precise representation of any form with very low residuals. One approximates such forms by simply adding harmonics.

Further, comparisons of different profiles can be graphically described by superimposing these sinusoidal contributors,[5] either singly or as summations of subsets of these contributors (demonstrated subsequently).

In our opinion, two major problems need to be resolved to make harmonic analysis more amenable to the analysis of outlines of the type of interest here and to provide useful statistical results. The first problem is theoretical and suggests that Fourier analysis should be not used to describe the shape of a given object

---

[4] Editor's note: Known as the Nyquist frequency criterion.
[5] Editor's note: The authors intended usage of "contributors" is broadly defined to include both sine/cosine coefficients as well as phase and amplitude computed from the sine/cosine coefficients.

in its totality (i.e., including detailed outline irregularities), but rather should be limited to an analysis of contour perturbations. Hence the need, as discussed previously, to partition the form into symmetry and contour perturbation components.

The second problem is more technical and is related to the structure of the data submitted to SAM. Whereas open curves can be positioned using a local coordinate system determined by homologous points, it is also possible to view them as a series of distances from each point on the curve to the corresponding point on a straight reference line passing through some suitable point, for instance, one of the homologous points on the curve. Further, to place the (open) profile into a standardized position, it may be necessary to make this reference line parallel to one of the coordinate axes. This displacement should be designed to pass through the selected (homologous) point.

However, results using the above procedure will depend on the chosen standardization. To free the profile from the constraints of positioning, an alternative is to compute a localized reference line based on a first-order fit (a straight line) to the points on the profile (Figure 11.6). This approach is suitable for open curves which do not present recurrent outlines (multivalued). The distances between each point on the profile and the corresponding point on the reference line are then ordered from the starting point to the ending point. This series of distances derived from such open curves corresponds to the "difference graph" for closed curves and is the data submitted to Fourier analysis.

Other open and closed curves can be placed into the polar coordinate system, where the distances to the points on the profile are defined as lengths of radii vectors from an origin (e.g., the barycenter). However, this approach cannot be generalized; it is limited to figures without recurrent outlines or marked contour convexities (e.g., profile indentations in the form of an "S"), that is, are single-valued.

Elliptical Fourier analysis, where subsequent points on the outline are connected with chords, in its different versions (Kuhl and Giardina, 1982; Moellering and Raynor, 1982) and applications (Diaz et al., 1990), overcomes the technical limitations arising with recurrent outlines.

For Fourier analysis, the format of the dataset (Box 8) must be completely generalized; that is, not restricted to a particular class of shapes. As we have previously seen, the polynomials are calculated separately for the abscissa and the ordinate to define the fundamental shape, which eliminates ambiguities such as multivalued functions.

The fundamental curve, given by the two polynomials, is then used (Box 8) as a reference line in the Cartesian system. This requires that the fundamental curve be rectified.[6] At the same time, the points of the original curve are displaced to

---

[6] Editor's note: The sense of rectification in this context is one of smoothing; i.e., the "fundamental curve" is subject to a geometric linearization resulting in a reference line.

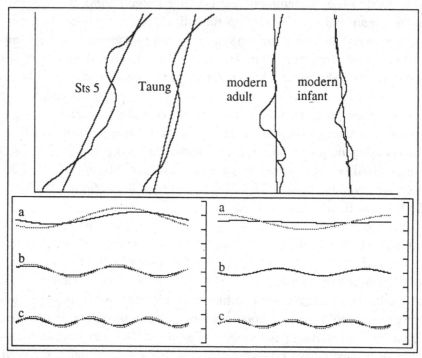

Fig. 11.6 Fourier analysis applied to the specimens illustrated in Figure 11.8 after a detrending procedure was imposed by linear regression: (a) superimposition of the first-order harmonic; (b) second-order harmonic; and (c) third-order harmonic for (left) Sts5 and Taung 1; ((right) Modern adult and Modern infant (solid line infant specimens and dotted line adult specimens).

a new position in 2-D space where they correspond to the points on the fundamental curve serving as the reference line. The procedure connects each point of the original curve with the point of the fundamental curve. The distance between each displaced original curve point and the corresponding point of the rectilinear fundamental curve is then computed[7] (Box 8). Each distance will be considered positive if the point on the original curve is located outside (above) the fundamental curve and negative if it is located inside (below) the curve. If two points coincide (where the two curves cross), the distance value will be zero. In this way we will have a series of positive and negative values with respect to the rectilinearized fundamental curve (Figures 11.7g, 11.8g, 11.3e). Thus, we have a way of obtaining an oriented unidimensional series of values with an irregular but periodic trend, suitable for graphical representation (a difference graph), and which

---

[7] Editor's note: The end result is that the fundamental curve becomes a straight line and the information retained is the corresponding point by point distances (differences) between the original curve and the fundamental curve prior to rectification.

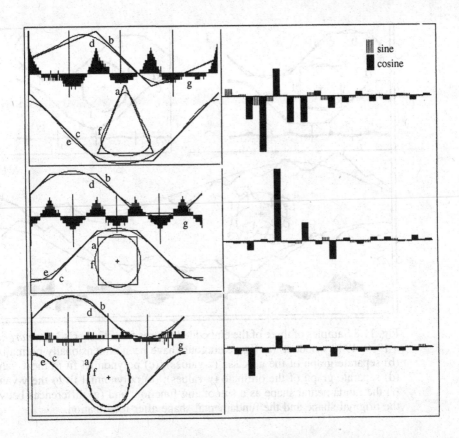

Fig. 11.7 Determination of the fundamental shape of a triangle, a rectangle, and an ellipse with their respective Fourier spectra: (a) figure contours have been dimensionally normalized; (b) separate graph of the abscissa (*x*-values); (c) polynomial fit to the *x*-values; (d) separate graph of the ordinate (*y*-values); (e) polynomial fit to the *y*-values; (f) the fundamental shape as a smoothing function; and (g) differences between the original shape and the fundamental shape after rectification (see text).

can be submitted to Fourier analysis. From the above discussion it should be apparent that for a regular polygon with the number of sides assumed to be infinite (for which the fundamental curve is a circle), the distance values become zero, yielding two identical superimposed circles with no significant information to be submitted to Fourier analysis.

To obtain a series of coefficients that allow for the reconstruction of the original profile, it is necessary to perform harmonic analysis twice. That is, the differences are calculated using abscissa and ordinate values respectively, on both the original series (Box 3) and the polynomial functions (Box 4). Once the computed SAM harmonic solution is obtained (Box 9), a reconstruction of the origi-

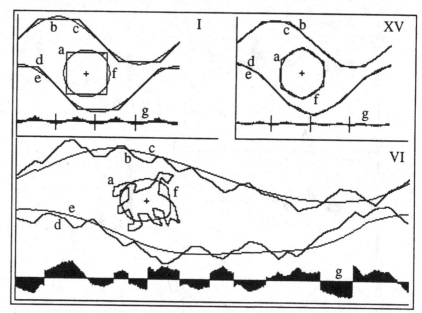

Fig. 11.8 Samples of three of the embedded figures from Escher's *Metamorphose*, a 1939–1940 woodcut: (a) figure contours have been dimensionally normalized; (b) separate graph of the abscissa (*x*-values); (c) polynomial fit to the *x*-values; (d) separate graph of the ordinate (*y*-values); (e) polynomial fit to the *y*-values; (f) the fundamental shape as a smoothing function; and (g) differences between the original shape and the fundamental shape after rectification.

nal curve can be made to graphically verify the fit of the reconstructed curve with the original curve. The reconstruction is made as follows: all the Fourier terms (up to the maximum order as determined from the number of points) are summed (Figures 11.6, 11.9, and 11.10). This generates a difference graph (recalling that this operation is performed twice, separately for the *x* and *y* values). Further, these reconstructed difference graphs are then added to the polynomial function curves (again, separately for the *x* and *y* values), obtaining separate abscissa and ordinate original profiles, which are finally merged to re-create the original curve.

From a classificatory point of view (for closed curves), the above analysis is carried out once using the resulting merging of the difference values obtained from the abscissa and ordinate data. In this case, the sign will be defined by the position of each original profile point with respect to the fundamental curve, positive if the point is outside, negative if inside, and zero at crossings on the boundary.

The result of such a harmonic analysis is typically represented by the Fourier spectrum (NB: not a power spectrum). This can be illustrated with a bar graph where the harmonic contributors are arranged in increasing harmonic order from left to right. The sine/cosine coefficients with positive values are displayed above the horizontal axis, the negative values below the axis. Striped bars identify the sine coefficients, solid bars the cosine coefficients.

Unfortunately, Fourier coefficients used directly are not amenable to easy statistical handling and may generate ambiguous results because they contain amplitude and phase information.

From the SAM Fourier analysis logic (Box 9) the following parameters are generated from the sine/cosine coefficients: (1) amplitude values; (2) phase values; (3) harmonic amplitudes ordered by harmonic number; (4) sum and mean of amplitudes for the whole set or for selected subsets of harmonics; (5) amplitude, percent with respect to the sum; (6) amplitude, percent with respect to the maximum amplitude value; (7) reordered series of harmonic amplitudes according to decreasing magnitude; and (8) roughness factor (i.e., the integral of the spectrum).

The generation of a one-dimensional series from the 2-D domain (the closed curve re-expressed as a difference graph) results in a small first-order harmonic. The maximum amplitude value is generally determined by a harmonic order other than the first. The harmonic with maximum amplitude plays the role of a "dri-

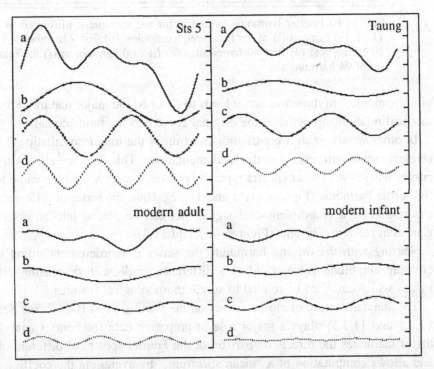

Fig. 11.9 The separate harmonics and their summation (harmonic synthesis) for the specimens illustrated in Figure 11.11: (a) summation of the first three harmonics; (b) first-order harmonic; (c) second-order harmonic; and (d) third-order harmonic.

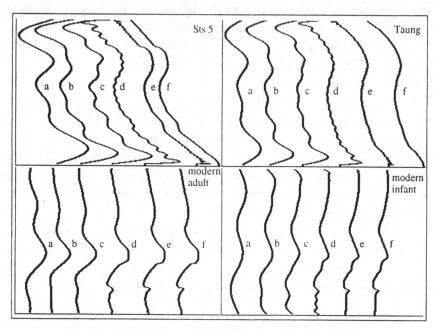

Fig. 11.10 Fourier harmonic synthesis for the specimens illustrated in Figure 11.11: (a) summation of the first three harmonics; (b) first 7 harmonics; (c) first 20 harmonics; (d) first 40 harmonics; (e) first 60 harmonics; (f) the maximum set of 94 harmonics.

ving harmonic," in the sense that it tends to describe the major feature of the geometrical relationship between the original curve and the fundamental.

In other words, such a contributor determines the major "oscillations" in the original curve with respect to the fundamental one. This does not arise with open curves where the Fourier spectra typically possess a large amplitude value for the first-order harmonic (Figures 11.11 and 11.12). Thus, the series of difference values resembles a one-dimensional signal that can be placed into an orthogonal Cartesian reference system (Figures 11.3 and 11.7).

Starting with the driving harmonic, the series is reordered according to decreasing amplitude value, yielding a difference graph with decreasing distance values which can then be related to subtle morphological features.

The standardization of closed curves on the starting point (Box 2, see Sections 11.3.1 and 11.3.3) plays a major role in preparing data for Fourier analysis, in that it facilitates the direct comparison of the Fourier spectra of different objects and allows computation of a "mean spectrum," by averaging the coefficients of same order for a sample.

The crucial difference between the Fourier analysis procedure advocated here, and elliptical Fourier functions, is that whereas the sum of first few elliptical contributors may roughly resemble the shape of the object in a way similar to the fundamental curve approach, they do not allow for the partitioning into the symmetry and contour perturbations components. These are computed with an independent algorithm (Box 4) and do not make use of Fourier analysis.

11.4 Analysis of artificial shapes

Using SAM, three regularly shaped figures were investigated: (1) a triangle; (2) a rectangle; and (3) an ellipse (Figures 11.4 and 11.7) based on dimensions proposed by Sergi (1900). Table 11.1 displays the amplitude values of the first 7 Fourier harmonics as a percentage of the total sum of the first 30 harmonics as well as their phase angles. Also shown is the summation of the amplitudes.

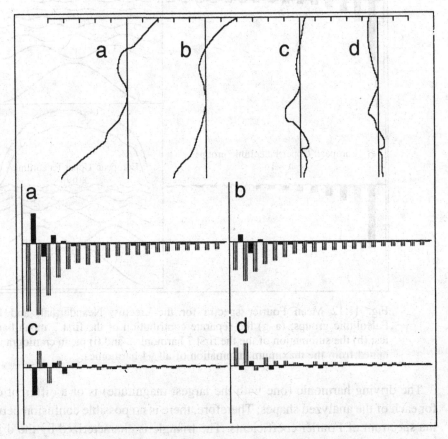

Fig. 11.11 A standardized vertical reference line to assess the profile of: (a) STS 5; (b) Taung 1; (c) adult *H. Sapiens;* and (d) infant *H. sapiens*. Also shown are their associated Fourier spectra.

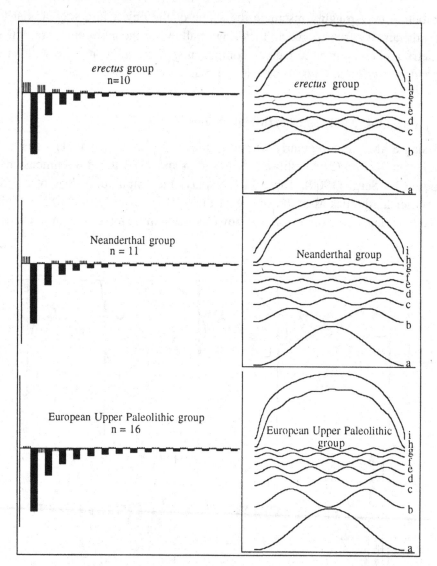

Fig. 11.12 Mean Fourier spectra for the erectus, Neanderthal, and Upper Paleolithic groups; (a–g) the separate contribution of the first 7 mean harmonics; (h) the summation of the the first 7 harmonics; and (i) mean craniograms obtained from the maximum summation of all 94 harmonics.

The driving harmonic (one with the largest magnitude) is of a different order for each of the analyzed shapes. Therefore, there is no possible confusion between the spectrum of Fourier coefficients. The triangle is characterized by the driving harmonic of third order, the rectangle by the fourth order, and the ellipse by the second order. The information on shape differences viewed as different ratios between the sides (for triangles and rectangles), or as diameters (for ellipses), is ex-

Table 11.1 Harmonic analysis applied to a triangle, a rectangle, and an ellipse. Amplitude % (upper row) and phase angle values in degrees (lower row) are shown for the first seven harmonics. The amplitude values are computed as a percentage of the total consisting of 30 harmonics. Also listed on the left is the summation of the seven amplitudes.

Harmonic order >	1°	2°	3°	4°	5°	6°	7°	Amplitude sum
triangle	3	10	24	12	11	11	2	13493
	185°	93°	82°	280°	90°	89°	242°	
rectangle	0.7	5	0.6	44	0.8	11	2	9322
	90°	81°	268°	270°	64°	276°	94°	
ellipse	5	28	4	12	4	2	2	5638
	108°	87°	97°	271°	78°	242°	104°	

pressed by the amplitude and phase values of the other harmonics. Thus, only for the equilateral triangle does Fourier analysis recognize the presence of a third-order harmonic and only for the square, the actual presence of a fourth-order harmonic.

Asymmetry (Box 7), is very pronounced in the triangle (Figure 11.4), where there is high allometry compared to the rectangle and ellipse, which have similar fundamental shapes and, therefore, have the same allometric rate for the same rotational step.

Another set of figures with different shapes ("L", "I", "T"), but with the same perimeter and area are illustrated in Figure 11.13. The SAM provides "fingerprints" of the three figures by using a combination of parameters (Boxes 5, 7, and 9). The "L"-shaped figure shows a Fourier spectrum driven by the second order harmonic and displays pronounced values of minimum and maximum asymmetry (Figure 11.13b, c). The "I"-shaped figure shows a Fourier spectrum also driven by the second order harmonic but shows very low values for maximum and minimum asymmetry (Figure 11.13b, c). The "T"-shaped figure shows a Fourier spectrum driven by the third order harmonic and an absence of asymmetry (Figure 11.13b) but with high asymmetry after a 90° rotation (Figure 11.13c).

Returning to Escher, as discussed in the introduction, we can now use SAM to analyze the *Circle limit IV* and *Metamorphose* drawings. *Circle limit IV* can be considered as an example of the global view of morphometric analysis, as applied to artificial figures. The problem, thus, consists of discriminating angels from devils; a taxonomic goal.

The *Circle limit IV* values of perimeter, area, abscissa projection, ordinate

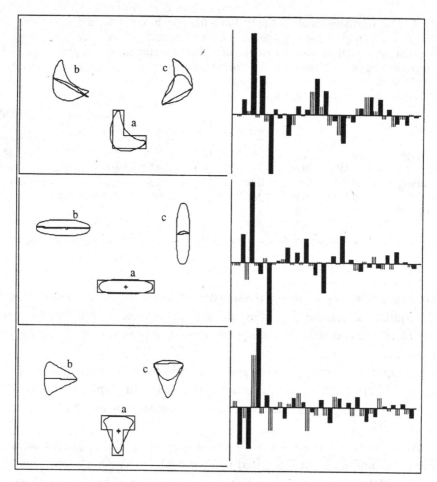

Fig. 11.13 Analysis of artificial L-, I-, and T-shaped objects having the same perimeter and area: (a) original with its fundamental curve; (b) minimum asymmetry; (c) maximum asymmetry and, on the right, their Fourier spectra.

projection, and roundness factor are all equal for the two figures, and thus, completely unable to distinguish angels from devils in spite of clear shape differences. The SAM system discriminated well between the angel and devil forms using the parameters from Box 5. Specifically, the angel was driven by the third-order harmonic and the devil by the eighth-order harmonic (Figure 11.1).

Escher's *Metamorphose* could be simulated by using a "morphogenetic" approach, which computed step-by-step changes of a regular figure (square) into another one (hexagon). Escher's drawing consists of a continuous strip in which the morphological transformation ends with a wasp flying out from a hexagon (where, according to Thompson, a morphological discontinuity takes place). This discontinuity is logically resolved here. The SAM system computes the *Metamorphose* transformations giving unambiguous fingerprints for each figure (Figure 11.8). The results of the first *Metamorphose* feature, the square, is easy to understand. The fundamental curve (Figure 11.8f) is a circle, and we see that the four corners of the square are outside it and the four sides are crossing it. The two computed polynomials, the separately plotted x and y values, show the primary relationships between the given fundamental and original curves; the difference graph (Figure 11.8g) shows four peaks of positive values and four peaks of negative ones. Positive peaks correspond to the four corners of the square and negative peaks correspond to its four sides. The Fourier spectrum is dominated by the fourth-order driving harmonic. The hexagon figure is also easy to understand (Figure 11.8). There are now six positive and six negative peaks. Comparing the height of the corresponding peaks of the square and of the hexagon, we see that the peaks of the hexagon are lower than for the square, due to the higher number of sides (making it closer in form to a circle). The Fourier spectrum shows an almost pure sixth-order driving harmonic. For the *Metamorphose VI* the difference graph shows six positive peaks, with the height of the second peak clearly lower than the other ones (Figure 11.8, lower figure). Fourier analysis gives the fifth-order as the driving harmonic. Looking at the relationship between the original and fundamental curve of *Metamorphose VI*, the outside parts of the head and the four paws are longer than the outside parts of the tail. The harmonic driving this pattern is of fifth-order because of the prevalence, within the transformation, of the head and paws with respect of the tail. *Metamorphose VII* compared with *Metamorphose VI* displays a differential positive growth of the tail with respect to the head and the paws which is driven by the sixth-order harmonic. In this way the unique characteristics of each *Metamorphose* panel can be identified and characterized. The remaining sections of this chapter discuss the application of SAM to actual data.

11.5 Analysis of anthropological materials

11.5.1 The craniogram

Comparative results are reported for craniograms of Circeo 1, Cro-Magnon of STS 5, and of a modern dolichomorphic female cranium (Figure 11.14) (Pesce Delfino, et al., 1987a). The craniograms, standardized on the Frankfort plane, were positioned in the Cartesian plane by placing the sagittal axis parallel to the ordinate. The series of contour coordinates were then digitized starting at *glabella* and analyzed.

The total sum of the amplitudes constitutes an indicator of contour complexity. The highest value found was for STS 5 (14489), followed by Circeo 1 (11045),

Cro-Magnon (10926), and the modern cranium (7286). The spectra show very distinct characteristics. Of all *Homo* crania, for the driving second harmonic, Circeo 1 and modern were closest to each other, both in amplitude and phase (Figure 11.14 and Table 11.2) with the following amplitudes: modern skull (1337); Circeo 1 (1534); Cro-Magnon (1999); and STS 5 (2334). The second-order per-

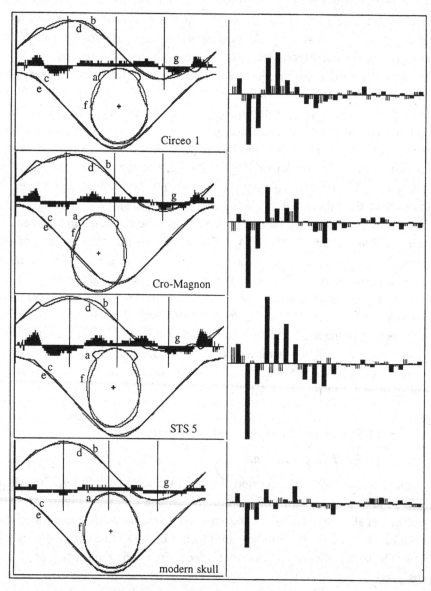

Fig. 11.14 Definition of fundamental shape and Fourier spectra obtained for the craniograms of Circeo 1, Cro-Magnon, STS 5, and a modern skull (refer to Figure 11.7 for details).

Table 11.2 Harmonic analysis applied to four hominid specimens. Fourier amplitude values and phase angles (upper row) for the first five harmonics obtained from the analysis of differences between the craniograms and their fundamental shape (curve). The amplitude values are reported as absolute values (upper row) and in percentages (lower row) as a function of the total consisting of 60 harmonics.

	1°	2°	3°	4°	5°
Circeo 1	495–246°	1534–83°	1020–83°	1102–265°	1298–260°
	4%	14%	9%	10%	12%
Cro-Magnon	345–323°	1999–92°	827–73°	1049–271°	410–257°
	3%	18%	8%	10%	4%
modern skull	273–266°	1337–87°	527–70°	367–269°	174–117°
	4%	18%	7%	5%	2%
Sts 5	733–232°	2334–96°	613–96°	2012–277°	909–287°
	5%	16%	4%	13%	6%

centage amplitude value for Cro-Magnon describes 18% of the total information, for Circeo 1 it describes 14%. The third-order harmonic for Circeo 1, which contributes to the description of the contour lateral expansion, had an amplitude of 1020 and a percentage of 9%.

Also for Circeo 1, the fourth- and fifth-order harmonics are larger than the third, both in absolute terms (1102 and 1298) and in percentage (10% and 12%). As these harmonics have similar phase values of 265° and 260°, they can be assumed to describe the morphology of the orbital constriction.

The values in Table 11.2 show that Neanderthal and Cro-Magnon crania are characterized by distinctive fourth- and fifth-order harmonics, although with similar phase values describing the shape of the post-orbital constriction. The Neanderthal cranium is also characterized by the third-order harmonic, which has the highest amplitude, describing the expansion in the lateral (horizontal) direction (Figure 11.14).

11.5.2 Profile comparisons

Shape differences in Hominid skulls have been classically considered to be a complex problem, both in terms of description and biological explanation. Together with the size increase from the infant to the adult form, allometries are able to provide a part of the biological explanation. In man, absolute facial skeletal growth is less than in other primates. In apes, facial growth vectors are mainly turned downwards and forwards, with an increasing trend in a supero-inferior direction but which reaches a minimum in man. Allometries are due to different combinations of bone deposition and resorption in craniofacial skeleton modeling, related to such diverse factors as brain development, chewing stresses, and braincase closure.

Table 11.3 Harmonic analysis applied to four hominid specimens. Amplitude (upper row) and phase values (lower row) listed separately for the first three harmonics. The amplitude values are reported as absolute values and in percentages as a function of the total consisting of the three harmonics shown here.

	1st harmonic ampl. & phase	2nd harmonic ampl. & phase	3rd harmonic ampl. & phase
STS 5	*21711/37%	21185/36%	15243/27%
	337°	11°	352°
Taung 1	8281/29%	*11832/41%	8657/30%
	349°	15°	358°
H. sapiens adult	*7792/45%	4723/28%	4483/27%
	97°	195°	345°
H. sapiens infant	5190/41%	*6345/51%	1016/8%
	172°	191°	9°

*maximum amplitude value

The goal of this section is to compare infant and adult frontofacial sagittal profiles in *Homo sapiens sapiens* and *Australopithecus africanus*, in order to provide morphological distance estimates, quantification of allometry, and topography, all in conjunction with a suitable mathematical model (Pesce Delfino et al., 1987b).

Plesianthropus transvaalensis (STS-5) displayed the largest amplitude values, followed by Taung 1, which seem to be related to facial protrusion and frontal inclination (Figure 11.11). Although the general trend of the Fourier spectrum for *P. transvaalensis* and Taung 1 was similar, there were differences in the magnitudes of the amplitudes. In contrast, the Fourier spectrum of the *H. sapiens* profile was different between the adult and infant specimens; that is, the ordering of the amplitude values.

Table 11.3 displays, for the first three harmonics, the maximum amplitude values (also their percentage of the total) and their phase values (in degrees). The first harmonic (i.e., the fundamental contributor) for *P. transvaalensis* and *H. sapiens* has the highest value in the adults, whereas with infants, Taung 1, and Modern man, the second harmonic has the largest value. The difference is more marked in *Australopithecus* than in Modern man. Phase differences for the first three harmonics were minimal between Taung 1 and *P. transvaalensis*. In modern *H. sapiens* only the third harmonic phase was comparable (Figure 11.9).

The result of summing harmonics, in terms of addition and subtraction, depends on the harmonic order and phase. The resulting sum is based on the addition of the second and third harmonics and subtraction of the first one for the frontal trait, addition of the first and third harmonics and subtraction of the second for the upper facial trait, and addition of all three contributors for the lower facial trait

(Figures 11.9 and 11.10). For Taung 1 the arrangement of the harmonic sum was the same as in *P. transvaalensis*. For the Modern adult the arrangement of the harmonic sum led to the addition of the first and third harmonics, subtraction of the second one for the frontal trait, addition of all three harmonics for the upper facial segment, and subtraction from the third harmonic of the first and second harmonic sum for the lower face. For the Modern infant the frontal trait and the lower face are described with the addition of the first and second harmonics, whereas the upper facial trait was defined with the subtraction of these terms from each other, whereas the contribution of the third harmonic was minimal. Distinct phase differences were found between adult and infant specimens of Modern man but not for the *Australopithecus* profiles.

Figure 11.6 shows the relationship between the four profiles and a straight line obtained from linear regression and used as a standard reference line in the Fourier detrended procedure. The continuous increase in angulation of this reference line (angular coefficient) from *P. transvaalensis* to Modern infant is apparent. Also shown are the superimpositions of the first, second, and third harmonic pairs for *Australopithecus* (on the left) and for Modern man (on the right). The dotted line refers to the adult specimens in each comparison. While the profile detrending procedure yields less pronounced differences, it strengthens the relationship with specimen shape. That is, a greater correspondence seems to be present between the *Australopithecus* adult and infant specimens than with the modern *H. sapiens* adult and infant — perhaps a reflection of delayed maturation in *H. sapiens*.[8]

11.5.3 Mid-sagittal shape analysis

This section is a description and comparison of three groups of midsagittal glabella-opisthocranion craniograms (based on Asiatic samples of H. erectus; Neanderthals and Upper Paleolithic samples from Europe; Vacca, et al., 1991). The examined craniograms were:

- Ten craniograms attributed to the Asiatic *erectus* form were: Pithecanthropus 1, Pithecanthropus 2, Pithecanthropus 4, Sinanthropus 10 (skull I Locus L), Sinanthropus 12 (skull III Locus L), Solo 1, Solo 5, Solo 6, Solo 11 (Weidenreich, 1943); Pithecanthropus 8 (Sartono, 1975).
- Eleven craniograms of Neanderthals: Circeo 1 and Saccopastore 1 (craniograms from Sergi, 1974); La Chapelle-aux-Saints, La Ferrassie 1(Heim, 1976), La Quina 5, Neanderthal, Le Moustier, Spy 1, and Spy 2 (Weidenreich, 1943); Amud I and Shanidar I (Suzuki and Takai, 1970);

---

[8] Editor's note: Although a provocative suggestion it must be cautioned that it is based on single specimens and not samples.

- Sixteen craniograms taken from samples from the European Upper Paleolithic: Predmosti 3, Predmosti 4, Predmosti 9, Predmosti 10, Predmosti 11 (Matiegka, 1934); Brno 1, Brno 2, Brno 3, Mladec I (Matiegka, 1929); Sungir 1, Pavlov 3, Markina Gora (Bunak and Gerasimova, 1984); Cro-Magnon, Combe-Capelle (Genet-Varcin, 1968); Obercassel (Weidenreich, 1943); and the skull from Riparo Villabruna (tomography).
- Also described and compared were: Petralona and Kabwe (Broken Hill; Murril, 1981); Steinheim (tomography; Dietrich-Adam, 1985).

The craniograms were normalized by adjusting the enlargement so that the distance between *glabella* and *opisthocranion* was the same for all the profiles. The 190 (x,y) points used to describe the craniogram were fitted with a Fourier series consisting of 94 harmonics (Vacca et al., 1991; Vacca, 1993). The craniogram was treated as an open curve. See Figures 1–8 in Vacca et al. (1991) for examples of individual traces of craniograms after normalization. The amplitude and phase values calculated from the first 15 sinusoidal components were used as variables for multivariate discriminant analysis. Each set of variables was tested in turn, to find the best combinations among the variables to produce the minimum percentage of error in classifying the groups, and the maximum distance between the centroids.

Subsequently, the relationship of Broken Hill, Petralona, and Steinheim to these groups was tested. Figure 11.12 illustrates the mean profiles and spectra obtained for each of the three groups. The acceptability of the fit of the Fourier function is determined by comparing the mean craniogram trace obtained from the total sum of all harmonics (Figure 11.12i) with the first 7 sinusoidal components (Figure 11.12h). In the same figure, the contributions of the first 7 terms are individually shown (Figure 11.12a–g). These 7 components, a reduction from 94 total contributors, contained over 75% of the total amplitude.

In order to limit the number of variables, only the amplitude and phase values of these 7 sinusoids were included in the multivariate discriminant analysis. A normality test (Kolmogorov-Smirnov two-tailed test) for the first amplitude of the Neanderthal group ($d = 0.30$, $p = 0.161$) did not reject the normality hypothesis.

Comparing the *erectus* group with the Neanderthal group (Figure 11.15a, b, c), we see that the amplitude values alone are not sufficient to separate the morphology of the two groups; the error of misclassification is 33%. However, using the phase values of only the first two harmonics they can be easily separated (0% error, distance between the centroids = 9). When all 7 phase values were used the separation was even better (0% error, distance between the centroids = 18.5). These results are explained as follows: for these two groups, *erectus* and Neanderthal, most of the differences seem to involve the sine coefficient of the second harmonic (Figure 11.12). For the Neanderthal group, this coefficient is

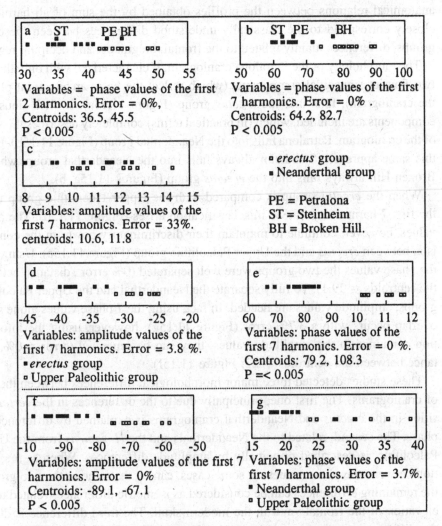

Fig. 11.15 Multivariate discriminant analysis of the craniogram traces of the erectus, Neanderthal, and upper Paleolithic groups. Also shown are the positions of single specimens (see text).

consistently smaller than for the *erectus* sample. With respect to phase (Figure 11.15), for the *erectus* group the phase value of the second harmonic is greater than for the first harmonic, whereas for the Neanderthal group this is reversed.

Thus, the major morphological differences between the groups seem to be explained by the peculiar relation between phase values, principally of the first and second harmonics. That is, these components in the Neanderthal group are opposite in phase, and describe the typical flattening of the upper part of the frontal bone and the posterior position of the vertex. In addition, we can observe that the anatomical relations between the profiles obtained by the sum of all harmonics closely correspond to the classically understood differences between these two groups, differences mainly related to the frontal, bregmatic, and occipital regions.

The morphology of the Petralona craniogram, if characterized with only the first two Fourier sinusoidal components (which reflect the basic architectural plan of the craniogram), falls into the *erectus* group (Figure 11.15a). If all 7 sinusoidal components are included, which (in practical terms) completely describes the shape of the craniogram, Petralona falls into the Neanderthal group (Figure 11.15b). Using this same approach, Steinheim always falls into the Neanderthal group, whereas Broken Hill always falls into the *erectus* group (Figures 11.15a, b).

When the *erectus* group is compared with the Upper Paleolithic group using the first 7 harmonics, a 3.8% misclassification arises (Figure 11.15d). The phase values, however, continue to maintain their discriminatory power in the comparison between *erectus* and the Upper Paleolithic group (Figure 11.15e). Using only the phase values the two groups were well separated (0% error , distance between the centroids = 29.1). To fully separate the Neanderthal and the Upper Paleolithic groups, amplitude values are needed. In fact, using the 7 phase values alone some overlap occurs with a 3.7% error (Figure 11.15g); however, using the information contained in the amplitude values, they can be separated (error = 0%, distance between the centroids = 22; Figure 11.15f).

These studies detected three major morphological differences between the sets of craniograms. The first one, principally due to the differences in the *vertex* position in the *erectus* and Neanderthal craniograms is explained by differences in phase. The second, related to the Neanderthals and the *H. sapiens* from the Upper Paleolithic, is described principally by amplitude differences. Whereas the relation between the phase values, in some cases, can be the same in the two groups, the remaining differences can be considered as isometric differences related to the elevation of the frontal vault in the modern skull. The third difference is due to the differences in *vertex* position and differences in vault height in the *erectus* and *H. sapiens* from the Upper Paleolithic specimens and can be described with both amplitude and phase differences, even though phase values alone were able to distinguish the two groups.

The Petralona morphology seems to fill the discontinuity between the *erectus* and Neanderthal craniograms. Moreover, this morphology shows hints at features that are only fully expressed as typical features in the Neanderthal group of the Würmian pleniglacial.

11.5.4 Sexual differentiation of the human pelvic bone

The single pelvic bone (os coxae) can be divided into two evolutionary, functionally, and causally related independent subsystems: the ischiopubic and sacroiliac.

This "*incisura ischiadica major*" (sciatic notch) can be considered as one of the most sexually differentiated characters within the sacroiliac segment and is widely used for sexing unknown material. Nevertheless, the use of discriminant functions based on sciatic notch dimensions display overlap of the discriminant scores for both sexes (Novotny and Vancata, 1985; Novotny et al., 1993). Male (n = 97) and female (n = 98) os coxae of known sex (Czech and German origin), from the collections of Anatomical Institutes in Prague and Brno (Czechoslovakia) and dating from the last century (~1880–1890s) were examined. The silhouettes of the *incisura ischiadica major* were obtained using an optical projection technique with the illumination provided by a distant light source. The outline of the segment between *spina ischiadica* and *spina iliaca posterior inferior* was examined up to where the *latero-superior* segment of the *incisura* meets the extremity of the *facies auricularis*. The profiles were then normalized with a depth measurement using an optical scaling value. Each profile was scaled to the same depth. Also, the

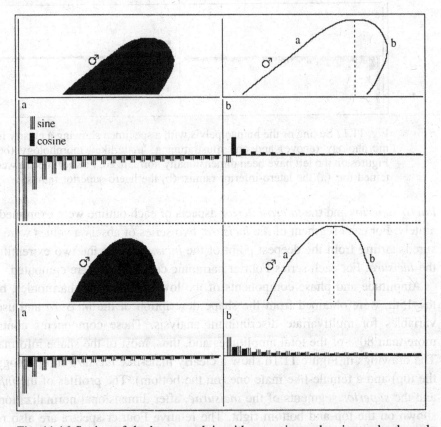

Fig. 11.16 Sexing of the human pelvis with a specimen showing a clearly male morphology (above) and one illustrating a "female-like" morphology (below). Figures on the left are have been dimensionally normalized. Fourier spectra were obtained for: (a) the latero-inferior ramus; (b) the latero-superior ramus.

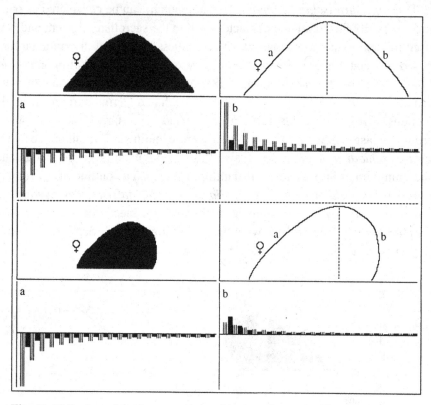

Fig. 11.17 Sexing of the human pelvis with a specimen showing a clearly female morphology (above) and one illustrating a "male-like" morphology (below). Figures on the left have been dimensionally normalized. Fourier spectra were obtained for: (a) the latero-inferior ramus; (b) the latero-superior ramus.

latero-superior and the *latero-inferior* aspects of each outline were examined separately. For each segment of the *incisura*, two series of abscissa values were measured starting from the deepest point of the *incisura* up to the two extremities of the *incisura*. For each series Fourier harmonic coefficients were computed.

Amplitude and phase components of the low-order Fourier harmonics, below the 15th, were obtained from the shape description of the *incisura* and used as variables for multivariate discriminant analysis. These components contained more than 80% of the total amplitudes and, thus, most of the shape information. The drawings in Figure 11.16 show a clearly male-like *incisura* morphology (on the top) and a female-like male one (on the bottom). The profiles of the *inferior* and the *superior* segments of the *incisurae,* after dimensional normalization, are shown on the top and bottom right. The relative Fourier spectra are also represented. In contrast, Figure 11.17 illustrates an *incisura* that is clearly female (top figure) with a male-like female one (bottom figure). Again, the profile of the *in-*

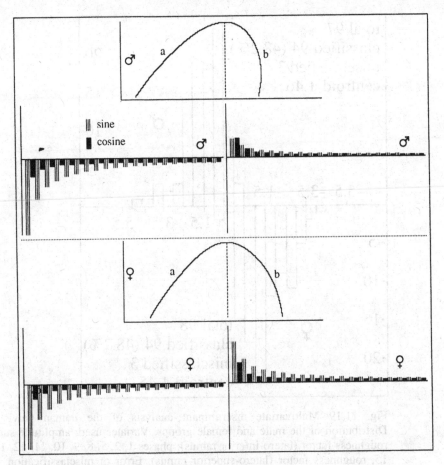

Fig. 11.18 Mean outlines of the human pelvis for male (N = 97) and female (N = 98) groups with their associated Fourier spectra for: (a) the latero-inferior ramus; (b) the latero-superior ramus.

ferior and the *superior* segments of the *incisura*, after dimensional normalization, are depicted on the right (top and bottom). The relative Fourier spectra are also illustrated as before. In Figure 11.18, the mean Fourier spectra coefficients for the two *incisura* segments, inferior and superior, are shown for both sexes.

The best multivariate discriminant was found using the Fourier amplitude and phase parameters of both the *latero-superior ramus* and the *latero-inferior ramus* of the *incisura* (Figure 11.19). By utilizing 14 parameters (phase angles 1, 2, 5, 8, 9, 10, 11, 12, 13, 14, 15; together with the roughness factor for both the *latero-superior ramus* and the *latero-inferior ramus;* and the amplitude sum). Utilizing these variables, a correct classification was obtained for 96% of examined cases. These results are markedly superior compared to the best discriminant

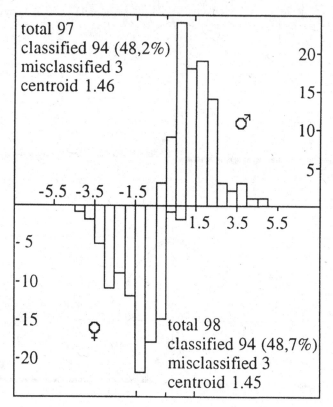

Fig. 11.19 Multivariate discriminant analysis of the human pelvis data. Distribution of the male and female groups. Variates used: amplitude sum and roughness factor (latero-inferior ramus); phases 1, 2, 5, 8, 9, 10, 11, 12, 13, 14, 15, roughness factor (latero-superior ramus). Error of misclassification = 4%. (F = 25.5; df = 14, 180; p < 0.0005).

function obtained with conventional parameters of the *incisura*. In that situation, using the same sample, the discriminant function generated a correct classification for only 65% of the cases.

11.6 Data from pathology

Using SAM, applications for morphological diagnosis in pathology have been performed for the following: (1) node shape and metastatic spreading in breast carcinoma (Giardina et al., 1989a; Giardina et al., 1994); (2) large bowel dysplasia (Bufo et al., 1990); (3) follicular carcinoma and follicular adenoma of the thyroid (Giardina et al., 1989a; 1989b; Giardina et al., 1994); (4) cell populations in non-Hodgkin lymphomas (Ricco et al., 1985; 1986; 1989); (5) liver dysplasia and carcinoma (Pesce Delfino et al., 1986b; Pollice et al., 1988); (6) gastric carcinoma

(Bufo et al., 1987); (7) serous borderline tumors of the ovary (Resta et al., 1992); (8) hand mirror cell leukemia (Liso et al., 1989) and (9) malignant glial tumors (Ricco et al., 1994). Some of these are discussed below.

11.6.1 Differential diagnosis of non-Hodgkin lymphoma subtypes

A number of cellular types (centrocytes, centroblasts, immunoblasts, lymphoblasts, lymphocytes) are involved in non-Hodgkin lymphoma allowing the classification of prevalent types based on the morphology of the nuclei.

Lymphocytes showed the lowest computed values for the parameters (from Boxes 7 and 9); whereas, for centrocytes, very high values were demonstrated for the same parameters. The Fourier amplitudes displayed intermediate values for the centroblasts, immunoblasts, and lymphoblasts. These amplitudes increased from centroblasts to immunoblasts. Multivariate discriminant analysis provided an error-free classification when comparing centrocytes to all the other cellular types.

In the attempt to differentiate centroblasts from immunoblasts (difficult by simple inspection) a 17% misclassification error initially occurred using a set of 12 nuclear parameters (four from Box 5, four from Box 7, and four from Box 9). Of the set, the Fourier parameters included the mean of the amplitudes, the first harmonic amplitude, the maximum harmonic amplitude, and its percent value. With the addition of the sum of Fourier amplitudes and the percent of the first harmonic amplitude, the discriminant analysis reduced the misclassification rate to 3%.

11.6.2 Biological significance of tumor shape in breast cancer

The shape of tumor nodes was matched with the occurrence of axillary metastases, which have been, up to now, the most meaningful prognostic factors for this disease. One parameter from Box 5, together with sum of Fourier amplitudes and mean amplitude value were found to be highly significant (Giardina, et al., 1989a; 1989b; Giardina et al., 1994; Figure 11.3). Thus, local irregularities of the neoplastic node outline can be related to the aggressiveness of metastatic spreading.

11.6.3 Epithelial tumors of the ovary

In recent years, borderline epithelial tumors of the ovary have been investigated with morphometric techniques to allow for a differential diagnosis of benign from malignant neoplasms. In order to enhance the discriminatory power, nuclei of benign ovarian serous neoplasms (n = 60), nuclei of serous borderline tumors (n = 60), and nuclei of serous carcinomas (n = 60) were examined using the SAM.

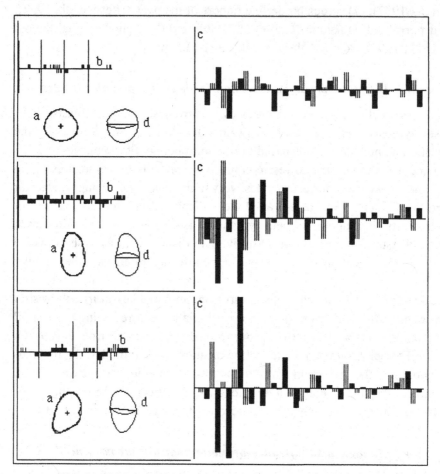

Fig. 11.20 Neoplastic nuclei of the ovary identified as benign, borderline and malignant cells: (a) original cell boundary superimposed on the centroid with the fundamental shape; (b) graph of the differences between the original outline and the fundamental shape; (c) the Fourier spectrum; and (d) the asymmetry evaluation.

An increase was observed in the parameters of symmetry from benign to borderline and malignant tumors. Further, the Fourier analysis parameters showed the highest magnitude for the carcinomatous nuclei, which may be due to the irregularity of their profiles (Figure 11.20). It was concluded that borderline serous tumors of the ovary were different from their benign counterparts in terms of the degree of nuclear asymmetry. The carcinomatous nuclei were significantly more asymmetric and show a major degree of fine contour irregularity when compared with benign and borderline serous tumors.

A nosological continuum was not found between benign, borderline and malignant ovarian tumors, beyond the aspect of nuclear morphology. Borderline tumors should not be considered as a transitional stage between two extremes, but represent an autonomous nosologic entity with a distinctive morphology.

11.7 Forensic anthropology

The identification of unknown skeletal remains by comparing the skull with a photograph taken while the subject was alive is a common procedure in forensic anthropology. Using SAM, an approach has been developed (Pesce Delfino, et al., 1990b) which decreases the subjective evaluation and provides numerical values for systematic comparisons. The soft-tissue photograph must be clear and suitable for appropriate alignment with the corresponding cranial segments. A convenient and easy-to-use procedure was developed, using a video camera and prism to check the degree of skull-photo superimposition (Pesce Delfino, et al., 1986a).

In skull/photo superimpositions, identification is based on a number of parameters including Fourier spectra for each skull and photo. The result of such a photo/skull matching procedure is presented in Table 11.4. Since the Fourier spectrum is composed of harmonics ordered in an identical way for all cases and the terms are independent of each other (orthogonality), the correlation between the Fourier spectrum of the photo profile and the corresponding skull profile can be computed. This correlation is based on the amplitude and phase computed from the Fourier coefficients. The correlation coefficient is calculated from the linear regression between the Fourier amplitudes from the photo profile and the corresponding skull profile. In contrast to the use of linear regression to derive the correlation coefficient for the amplitudes, the correlation coefficient was directly computed for the phase values. The reciprocal of these correlation coefficients was then multiplied by the other two parameters to generate a final estimate (final evaluator) used for the identification procedure (Table 11.4). Also computed is the RMS and the allometry and isometry differences.

In addition to the final estimate (above), an approach was developed to evaluate the contribution of the Fourier parameters that make up the final estimate. A first partial evaluator was computed by multiplying the RMS with the allometry/isometry percent difference. A second partial evaluator was obtained by multiplying the first partial evaluator by the reciprocal of the amplitude linear correlation. A third partial evaluator was obtained by multiplying the first partial evaluator by the correlation coefficient for the phase angle (Table 11.5). Comparison of the third evaluator with the final estimate showed a higher weighting of phase over amplitude, but both parameters are needed for a correct final evaluator.

Table 11.4 Skull/face superimposition: for true and false comparisons in oblique and lateral view. The values of the parameters are derived as follows: A from Box 5, Figure 11.2; B from Box 9, Figure 11.2; and C from Box 7, Figure 11.2. These parameters were then multiplied by each other to obtain the final evaluator. The lowest values are for the correct (true) comparisons (Case E refers to the oblique view and Cases A, M, I, and L to the lateral view).

Superimposition match	-A- Minimized square root of mean square error (L.A.—5)	-B- Amplitude linear correlation reciprocal (L.A.—9)	-C- Phase pearson correlation reciprocal (L.A.—9)	-D- Allometry and isometry % difference (L.A.—7)	Final evaluator (×1000)
Oblique view					
True					
Case E	1.16	1.002	1.01	0.72	845
False					
Face E/skull A	2.73	1.02	1.03	1.92	5507
Face E/skull F	3.7	1.03	1.57	3.3	19745
Face E/skull G	3.34	1.04	1.43	2	9934
Lateral view					
True					
Case A	1.36	1.01	1.02	1.55	2171
Case H	0.967	1.058	2.74	0.66	1850
Case I	1.56	1.07	1.98	2.73	9022
Case L	0.849	1.003	1.097	4.352	4065
False					
Face A/skull B	4.62	1.14	2.46	3.14	40683
Face A/skull C	3.29	1.11	3	2.21	24212
Face A/skull D	3.59	1.14	2.32	3.78	35890
Face L/skull M	3.386	1.029	1.305	6.948	31592
Face L/skull N	1.581	1.004	1.072	9.194	15645

Table 11.5 Three partial evaluators were obtained by multiplying parameters A, B, C, and D computed in Table 11.5, together with the final evaluator. The third partial evaluator, using the Fourier phase values, is closer to final evaluator than the second partial evaluator which only uses the harmonic amplitude values.

Superimposition match	First partial evaluator FP = A × D	Second partial evaluator (×1000) SP = FP × B	Third partial evaluator (×1000) TP = FP × C	Final evaluator (×1000)
Oblique view				
True				
Case E	0.835	837	843	845
False				
Face E/skull A	5.241	5346	5398	5507
Face E/skull F	12.21	12576	19170	19745
Face E/skull G	6.68	6947	9552	9934
Lateral view				
True				
Case A	2.108	2129	2150	2171
Case H	0.638	675	1748	1850
Case I	4.259	4557	8433	9022
Case L	3.695	3706	4053	4065
False				
Face A/skull B	14.506	16537	35685	40683
Face A/skull C	7.27	8070	21810	24212
Face A/skull D	13.57	15470	31482	35890
Face L/skull M	23.526	24208	30701	31592
Face L/skull N	14.536	14594	15583	15645

Twelve superimpositions were investigated from the forensic cases before Italian Courts. In each instance, the disappeared person was positively identified.

11.8 Summary and conclusions

We began by noting Thompson's suggestion: The study of natural language can eventually lead to restatement in the language of mathematics. A careful analysis of Sergi's work, and others, led us to conclude that shape has a variety of mathematically describable properties of which two are particularly significant; these are symmetry and boundary perturbations of the outline. We have expressed these two properties through a fundamental curve (a polynomial smoothing fit) and various descriptors derived from a Fourier analysis of the difference between the fundamental curve and the original (digitized) outline. Subsequent sections have shown the applicability of the approach to a variety of datasets drawn from anthropology, pathology, orthodontics, and forensics.

Acknowledgments

The authors wish to thank Dr. T. Cipriani, Dr. C. Ferri, Mr. G. Colacicco, Mr. G. Inchingolo, Mr. C. Lattanzio, and Mr. D. Vinci for their valuable assistance in the SAM project and their help in the preparation of this manuscript.

References

Aherne, W. A. & Dunnill, M. S. (1982). *Morphometry.* Edward Arnold.

Bookstein, F. L. (1978). *The Measurement of Biological Shape and Shape Change. Lecture Notes in Biomathematics.* Berlin & Heidelberg: Springer-Verlag.

Bufo, P., Ricco, R., Potente, F. & Troia, M. (1987). Displasia gastrica e carcinoma: morfometria analitica della forma nucleare. *Boll. Soc. It. Biol. Sper.* **63,** 341–7.

Bufo, P., Ricco, R., Potente, F., Troia, M., Serio, G. & Pesce Delfino, V. (1990). Using analytical morphometry to distinguish severe dysplasia and large bowel carcinoma. *Boll. Soc. It. Biol. Sper.* **66,** 143–50.

Bunak, V. V. & Gerasimova, M. M. (1984). Verchnepaleoliticeskij cerep Sungir 1 i ego mesto V Rjadu drugich verchnepaleoliticeskich cerepov. In *Sungir, Antropologiceskoe Issledovanie.* Moscow: Academy of Sciences of USSR, NAUKA.

Diaz, G., Quacci, D. & Dell'Orbo, C. (1990). Recognition of surface modulation by elliptic Fourier analysis. *Comp. Meth. Prog. Biomed.* **31,** 57–62.

Dietrich-Adam, K. (1985). The chronological and systematic position of the Steinheim skull. In *Ancestors, the Hard Evidence*, ed. E. Delson. New York: Alan R. Liss Inc.

Dullemeijer, P. (1974). *Concepts and Approaches in Animal Morphology.* Assen: Van Gorcum.

Genet-Varcin, E. (1968). Considerations morphologiques sur l'homme de Cro-Magnon. In *L'Homme de Cro-Magnon.* Paris: Centre de Recherches Anthrop. Preist. et Etn., Arts et Metiers Graphiques.

Giardina, C., Lettini, T., Gentile, A., Serio, G., De Benedictis, G., Ricco, R. & Pesce Delfino, V. (1989a). Relationship between primary tumor shape and biological behaviour in breast cancer. *Tumori* **75**, 117–22.

Giardina, C., Pollice, L., Serio, G., Pennella, A., Vacca, E., Mastrogiulio, S., Potente, F., Ricco, R. & Pesce Delfino, V. (1989b). Differential diagnosis between thyroid follicular adenoma and carcinoma: analytic morphometric approach. *Path. Res. Pract.* **185**, 726–8.

Giardina, C., Ricco, R., Serio, G., Vacca, E., Pennella, A., Renzulli, G., Punzo, C., Borzacchini, A. M. & Pesce Delfino, V. (1994). Nuclear shape and axillary metastases in breast cancer. *Acta Cytologica.* **38**, 341–6.

Heim, J. L. (1976). *Les Hommes Fossiles de La Ferrassie*. Paris: Archives de Institut de Paleontologie Humain, Masson.

Holloway, R. L. (1981). Exploring the dorsal surface of Hominoid brain endocasts by stereoplotter and discriminant analysis. *Phil. Trans. R. Soc. Lond.*, B. **292**, 155.

Kuhl, F.P. & Giardina, C. (1982). Elliptic Fourier features of a closed contour. *Comp. Graph. Image Proc.* **18**, 236–58.

Jacobshagen, B. (1982). Variations in size and shape of the orbital contour. A comparison between man and the greater apes using Fourier analysis. *Anthropos* **21**, 113–30.

Johnson, D. R., O'Higgins, P., McAndrew, T. J., Adams, L. M. & Flinn, R. M. (1986). Measurement of Biological shape: a general method applied to mouse vertebrae. *J. Embryol. Exp. Morph.* **90**, 363–77.

Lazotti, L. & Tarantino, V. (1984). *Linguaggio Visuale e Beni Culturali*. Firenze: Bulgarini.

Lestrel, P. E. (1974). Some problems in the assessment of morphological size and shape differences. *Yearbk Phys. Anthrop.* **18**, 140–62.

Lestrel, P. E. (1982). A Fourier analytic procedure to describe complex morphological shapes. In *Factors and Mechanisms Influencing Bone Growth*. ed. A. D. Dixon & B. G. Sarnat. New York: Alan R. Liss, Inc.

Lestrel, P. E. (1989). Some approaches toward the mathematical modeling of the craniofacial complex. *J. Craniofac. Genet. Develop. Biol.* **9**, 77–91.

Liso, V., Lettini, T., Specchia, G., Pavone, V., Pennella, A., Marzano, N., Ricco, R. & Pesce Delfino, V. (1989). Hand-mirror cells in acute lymphoblastic leukemia (A.L.L.): an attempt at cellular classification by morphometric analytical parameters using the S.A.M. work-station. *Path. Res. Pract.* **185**, 715–8.

Locher, J. L. (1971). *The World of M. C. Escher*. Amsterdam: Meulenhoff International.

Masashi, I., Terutaka, I., Yoshitaka, F. & Kichiro, O. (1992). Sex determination by discriminant function analysis of lateral cranial form. *Forensic Sci. Int.* **57**, 109–17.

Matiegka, J. (1929). The skull of fossil man Brno III, and the cast of its interior. *Anthropologie* **7**, 79.

Matiegka, J. (1934). *Homo predmostensis, fosilni clovek z Predmosti na Morave*. Prague: Nacladem Ceske Akademie ved Umeni.

Moellering, H. & Raynor, J. N. (1982). The dual axis Fourier shape analysis of closed cartographic forms. *The Cartographic J.* **19**, 53–9.

Murril, R. I. (1981). *Petralona Man*. Springfield: C. Thomas Publisher.

Novotny, V. & Vancata, V. (1985). Systems aspect of the sexual dimorphism in Human Lower Limb. In *Evolution and Morphogenesis*, ed. J. Mlikovsky & V. J. A. Novak. Praha: Academia.

Novotny, V., Vacca, E., Vancata, V. & Pesce Delfino, V. (1993). Differenze szssuali rilevabili sulla incisura ischiadica major del bacino dell'uomo: confronto tra analisi metrica e analisi della forma. *Antrop. Contemp.* **16**, 229–37.

O'Higgins, P. & Williams, W. (1987). An investigation into the use of Fourier coefficients in characterizing cranial shape in primates. *J. Zool. Lond.* **211**, 409–30.

Pesce Delfino, V. & Ricco, R. (1983). Remarks on analytic morphometry in biology: procedure and software illustration. *Acta Stereol.* **2**, 458–68.

Pesce Delfino, V., Lettini, T., Potente, F., Vacca, E. & Ricco, R. (1986a). Descrizione e classificazione delle strutture biologiche. Illustrazione della procedura e di software dedicato. In *Atti del Convegno I.C.O. Graphics*. Bergamo: ETAS.

Pesce Delfino, V., Pollice, L., Russo, S., Pagniello, G., Potente, F. & Vacca, E. (1986b). Analytical morphometry of normal, displastic and neoplastic hepatic cells. *Quant. Im. Anal. Cancer Citol. Histol.* **7**, 318–20.

Pesce Delfino, V., Potente, F., Vacca, E., Lettini, T. & Lenoçi, R. (1987a). Analytical morphometry in fronto-facial profile comparison of Taung 1, P. transvaalensis, H. sapiens infant and H. sapiens adult. In *Int. Symp. Biological Evolution Proceedings.* ed. V. Pesce Delfino. Bari: Adriatica.

Pesce Delfino, V., Vacca, E., Lettini, T. & Potente, F. (1987b). Analytical description of cranial profiles by means of Kth order polynomial equations. Procedure and application on P. transvaalensis (STS 5). *Anthropologie* **26**, 47–55.

Pesce Delfino, V., Potente, F., Vacca, E., Lettini, T., Ragone, P. & Ricco, R. (1990a). Shape evaluation in medical image analysis. *Europ. Microscopy Anal.* **7**, 21.

Pesce Delfino, V., Ragone, P., Vacca, E., Potente, F. & Lettini, T. (1990b). Valutazione quantitativa delle trasformazioni allometriche delle strutture biologiche. *Boll. Soc. It. Biol. Sper.* **66**, 263–9.

Pesce Delfino, V., Vacca, E., Potente, F., Lettini, T. & Ragone, P. (1991). Morfometria analitica del cranio neandertaliano del monte Circeo (Circeo 1). In *Il Cranio Neandertaliano Circeo 1. Studi e Documenti. Ministero per i Beni Culturali e Ambientali*—Museo Naz. Preist., ed. Etn. L. Pigorini. Roma: Istituto Poligrafico e Zecca dello Stato.

Pollice, L., Russo, S., Maiorano, E., Pagniello, G., Ricco, R. & Pesce Delfino, V. (1988). Hepatocellular dysplasia: Immunoistochemical and morphometrical evaluation. *Appl. Pathol.* **6**, 73–81.

Resta, L., Ricco, R., Colucci, G. A., Troia, M., Russo, S., Vacca, E. & Pesce Delfino, V. (1992). A new approach to the histologic study of ovarian tumors by analytical morphometry. *Int. J. Gynecol. Cancer.* **2**, 307–13.

Ricco, R., De Benedictis, G., Giardina, C., Bufo, P., Resta, L. & Pesce Delfino, V. (1985). Suggestions for morphometrical analytical evaluators of lymphoid population in non neoplastic lymphnodes. *Quant. Citol. Histol.* **7**, 288–93.

Ricco, R., De Benedictis, G., Giardina, C., Bufo, P., Gentile, A., Lettini, T. & Pesce Delfino, V. (1986). Performing analytical morphometry on nuclear shapes in non-H lymphomas: an attempt of classification. *Quant. Im. Anal. in Cancer Citol. and Histol.* **7**, 297–9.

Ricco, R., De Benedictis, G., Lettini, T., Pesce Delfino, V., Gentile, A., Pennella, A. & Serio, G. (1989). Non-Hodgkin's lymphoma diagnosis aided by the SAM system. *Path. Res. Pract.* **185**, 719–21.

Ricco, R., Serio, G., Caniglia, D. M., Cimmino, A., Lettini, T., Lozupone, A. & Pesce Delfino, V. (1994). Size and shape evaluation of astrocytoma nuclei with the shape analytical morphometry software system. *Anal. Quant. Cytol. Histol.* **16**, 345–9.

Sartono, S. (1975). Implication arising from Pithecanthropus VIII. In: *Paleoanthropology. Morphology and Paleoecology.* ed. R. H. Tuttle. Paris: Mouton.

Sergi, G. (1900). *Specie e Varietà Umane.* Torino: F.lli Bocca.

Sergi, G. (1911). *L'uomo.* Torino: F.lli Bocca.

Sergi, S. (1974). *Il Cranio Neandertaliano del Monte Circeo (Circeo I)*, a cura di Ascenzi. Roma: Acc. Naz. Linc.

Suzuki, H. & Takai, F. (1970). *The Amud Man and his Cave Site*. Tokyo: The University of Tokyo.

Ungerer, E. (1966). *Die Wissenschaft von Leben der Problemlage der Biologie in der Letzen Jahrzehnten*. Friburg-Munchen: Verlag K. A.

Thompson, W. (1942). *On Growth and Form*. (2nd ed.) Cambridge: Cambridge University Press.

Vacca, E. (1993). Morfometria analitica del craniogramma sagittale mediano glabella-opisthocranion in Homo. *Antrop. Contemp.* **16,** 335–43.

Vacca, E., Potente, F. & Pesce Delfino, V. (1991). I craniotipi secondo G. Sergi: un approccio analitico. *Antrop. Contemp.* **14,** 209–20.

Weibel, E. R. (1994). Design of biological organisms and fractal geometry. In *Fractals in Biology and Medicine*. ed. T. F. Nonnenmacher, G. A. Losa, & E. R. Weibel.

Weidenreich, F. (1943). The skull of Sinanthropus pekinensis, a comparative study on a primitive hominid skull. *Paleont. Sinica.*, Ser. D, **10,** 1.

12

Analyzing Human Gait with Fourier Descriptors

TERUO UETAKE

Tokyo University of Agriculture and Technology

12.1 Walking and Fourier analysis

One of the distinct human characteristics is bipedalism. This means of locomotion has existed throughout human history for more than three and a half million years ago judging from the fossilized footprints at Laetoli in northern Tanzania (Leakey, 1987). Scholars throughout the world have long been interested in human bipedal locomotion and have attempted to analyze it by various means.

12.1.1 Dimensional measurements

To understand the nature of human locomotion, we must recognize that the walking cycle which is composed of dimensional and temporal measurements. Figure 12.1 illustrates the components making up the walking cycle. These dimensional measurements are composed of both right and left step lengths, foot angles, and step widths. A step length is defined as the distance from one heel point to the other; for example, the right step length is expressed as the distance from the left heel point to the right heel point. Foot angle is defined as the angle of a foot from a center line drawn between both feet. Step width is the distance between right and left heel points normal to the center line.

12.1.2 Temporal measurements

Walking is the repeated alternation of one leg followed by the other, with one limb remaining in contact with the ground. Double support time occurs when both legs are in contact with the ground. Single support time occurs when only one leg is in contact with the ground. For a given leg, the stance phase takes place while the leg is in contact with the ground, and the swing phase transpires when the leg is off the ground. Walking primarily involves the motion of the lower limbs, although the upper limbs are partially involved. A detailed description of human locomotion requires knowledge of the bones and muscles of the lower limbs.

294

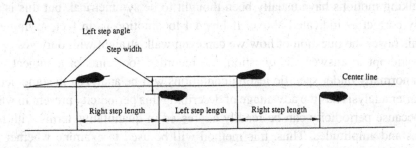

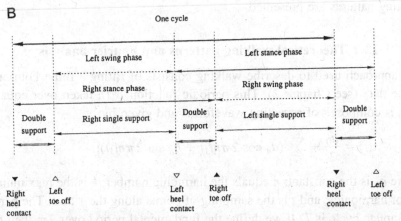

Fig. 12.1 Dimensional measurements (a) and temporal measurements (b) used to describe the walking cycle.

Given that the walking utilizes many bones and muscles, the question arises whether the proper functioning of this musculo-skeletal system is in some way related to the visual system. If so, one test would be to establish whether one can really walk straight with the eyes closed. Prior to testing this assertion, a few preliminary issues need to be discussed. Schaeffer (1928) concluded that a man with closed eyes walks, runs, swims, rows a boat, and drives, more or less, in a curved or "clock spring" spiral path. More recently, Nigorikawa (1988) reported that an individual usually veers to the right or left when walking with closed eyes. Thus, it is not readily apparent just what the primary reason might be for this deviation from a straight path. Some aspects that are thought to be related to this deviation from a straight path can be mentioned. Normally walking individuals have a lateral difference in step length (Hills and Parker, 1991), as seen from kinematic investigations (Hannah et al., 1984; Gundersen et al., 1989; Laassel et al., 1992), and from plantar pressure measurements (Herzog et al., 1989; Laassel et al., 1992).

Walking motions have usually been thought to be symmetrical, but this is not strictly correct as indicated above. If bipedal locomotion is, in fact, asymmetrical, this raises the question of how we can even walk straight with our eyes open. In an attempt to answer this question, the recorded footprints of a subject who walks normally under specific testing conditions will be analyzed in some detail.

Fourier analysis has the advantage of describing this periodicity present in walking, because periodicity can be readily expressed in quantitative terms with harmonics and amplitudes. Thus, this method will be used to examine whether the path of footprints is really straight or not. The next section describes how this method is applied to analyze human bipedal locomotion, and some theoretical walking patterns are presented.

12.2 Theoretical walking patterns and Fourier analysis

The approach used to describe walking consists of fitting a finite Fourier series to the data (see Chapter 2). This periodic function, $x(t)$, taken over equal intervals, is composed of repeating waveforms and given by:

$$x(t) = \frac{a_0}{2} + \sum_{n=1}^{k} (a_n \cos 2\pi n f_1 t + b_n \sin 2\pi n f_1 t), \tag{12.1}$$

where a_0 is the constant; n equals the harmonic number; k is the maximum number of harmonics; and t is the sampling of points along the x-axis. The period for a complete cycle is T. If we define the fundamental period over 2π, then the frequency is equal to the inverse of the period or:

$$f_1 = \frac{1}{T}. \tag{12.2}$$

If the sampling of t-points or observations are chosen with equal divisions over the interval (generally with a period of 2π) then the coefficients a_0, a_n, and b_n can be computed from:

$$a_0 = \frac{1}{T} \int_0^T x(t)dt, \tag{12.3}$$

$$a_n = \frac{2}{T} \int_0^T x(t)\cos 2\pi n f_1 t \, dt, \qquad (n = 0, 1, 2, \ldots k) \tag{12.4}$$

$$b_n = \frac{2}{T} \int_0^T x(t) \sin 2\pi n f_1 t \, dt. \qquad (n = 1, 2, \ldots k) \tag{12.5}$$

This Fourier series can also be defined as:

$$x(t) = \frac{a_0}{2} + \sum_{n=1}^{k} X_n \cos(2\pi n f_1 t - \theta_n), \tag{12.6}$$

where X_n refers to the nth amplitude and θ_n refers to the nth phase angle, which are given as:

$$X_n = \sqrt{a_n^2 + b_n^2} \qquad\qquad (n = 1, 2, 3, \ldots k) \qquad\qquad (12.7)$$

$$\theta_n = \tan^{-1}\left[\frac{b_n}{a_n}\right]. \qquad\qquad (n = 1, 2, 3, \ldots k) \qquad\qquad (12.8)$$

Suppose that an artificial (or theoretical) subject left a trail of 64 (32 right and 32 left) footprints with step widths as shown in Figure 12.2. Here, the measurements, in the direction of walking, are indicated by a plus symbol if to the left of the center line and by a minus sign if to the right of the center line. This series of 64 step widths can be also viewed as data along the ordinate (y-axis) with the corresponding step number along the abscissa (x-axis) (see Table 3 in Uetake, 1992 for the data that correspond with Figure 12.3). Fourier analysis was then fitted to initially analyze some *theoretical* patterns of step width changes (Figure 12.3).

These data were then processed by a microcomputer program developed by Hino (1977), which computed cosine (a_n) and sine (b_n) coefficients, as well as their amplitudes (X_n) for each harmonic (see Table 4 in Uetake, 1992).

The fit with the Fourier series was determined from the sum of the residuals between the observed data and the reconstructed or predicted data using 35 harmonics. These values, in mm, for each case shown in Figure 12.3 (Case A = 0.000000, Case B = 0.185195, Case C = 0.002872) agreed well with the original data, and substantiate the use of the Fourier series for the analysis of walking patterns.

If this theoretical subject could walk in an absolutely straight line, as in case A (Figure 12.3a), then the cosine or sine coefficients (and the corresponding amplitudes) with appreciable magnitude would only appear at the 32nd harmonic, and values for all other harmonics would be zero. The large magnitude at the 32nd harmonic is a consequence of the data sampling scheme being used. Namely, there are 64 steps, which alternate forming a "sawtooth" data stream if the points are connected with straight lines. This configuration yields exactly 32 peaks. The 32nd harmonic is, thus, an exact fit to the data, when all the other harmonics are zero.

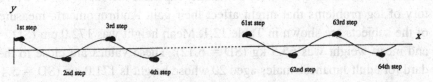

Fig. 12.2 Application of Fourier analysis to the walking track composed of theoretical footprints. Compare with Figure 12.3a.

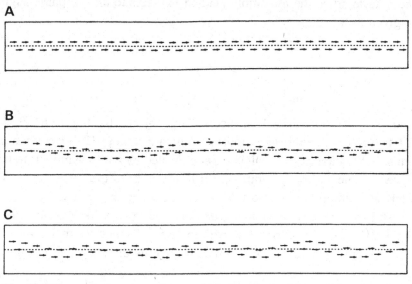

Fig. 12.3 Theoretical pattern of step-width change. Straight walking (a), and meandering walking with periodicity (b and c). Refer to Table 12.1 for the data used to prepare this figure. From Uetake (1992). Can we really walk straight? *Am J. Phys. Anthrop.* 89:19–27. Reprinted by permission of John Wiley and Sons, Inc.

If this theoretical subject walks in a meandering manner (but in a regular or periodic fashion) as in case B or C, peaks would not only appear at the 32nd harmonic but at the 2nd harmonic for case B, and at the 4th harmonic for case C (see Figure 12.3b and c) . Using Fourier analysis in this manner, one can discern the pattern of the step width change. The appearance of harmonics with significant amplitudes in any other than the 32nd harmonic would demonstrate that the individual walked meanderingly rather than in a straight line. Using Fourier analysis, an investigation was then conducted to see if walking in a straight line is possible with actual human subjects.

12.3 Experimental walking

12.3.1 Subjects

The experimental subjects were 20 healthy Japanese men aged 18–45 with no history of leg problems that might affect their gait. Anthropometric measurements of the subjects are shown in Table 12.1. Mean height was 172.0 cm (SD = 5.55) and mean weight was 65.5 kg (SD = 6.12). These values are close to the standard for adult Japanese males aged 20 whose height is 171.0 cm (SD = 5.41) and weight is 63.1 kg (SD = 7.52) (Nakanishi, 1989). Thus, the subjects in this study can be regarded as approximately average. A lateral difference between left and right sides was demonstrated in the forearm length but not in the other measure-

Table 12.1 Means for anthropometric measurements of the
experimental subjects (mm).

	Left		Right	
	$\overline{X}$	SD	$\overline{X}$	SD
Arm length	31.8	1.43	30.9	1.87
Forearm length[a]	23.0	1.32	24.2	1.14
Hand length	19.2	0.80	19.3	0.81
Bi-epicondyle of arm	6.7	0.36	6.7	0.29
Bi-styloid proc. of forearm	5.6	0.29	5.7	0.27
Hand breadth	8.3	0.38	8.4	0.40
Arm girth	27.5	1.54	27.9	1.49
Forearm girth	25.6	1.34	26.2	1.18
Thigh length[b]	46.8	1.83	46.8	1.73
Leg length	36.2	2.55	36.0	2.57
Foot length	25.2	1.04	25.2	1.00
Foot height	7.5	0.50	7.5	0.50
Bi-epicondyle of thigh	9.4	0.65	9.5	0.71
Bi-epicondyle of leg	7.4	0.38	7.4	0.37
Foot breadth	10.3	0.60	10.2	0.58
Thigh girth	52.7	3.37	52.9	3.48
Leg girth	36.8	1.94	36.9	1.55

(a): The lateral difference is statistically significant ($p < 0.01$).
(b): Thigh length = (iliac height − tibial height) × 0.93
Anthropometrical measurements were taken following the procedures of Martin and Seller
(1957). From Uetake (1992) Can we really walk straight? *Am. J. Phys. Anthrop.* 89:19–27.
Reprinted by permission of John Wiley and Sons, Inc.

ments of the upper or lower limbs. Use of a leg to kick a ball was examined to
ascertain functional leg dominance. Seventeen of the subjects had right leg dominance, the remaining three had left leg dominance.

12.3.2 Experimental walking

A flat white paper sheet 0.8 × 50 m in size was placed on the floor, and the subject was asked to walk at his normal speed as straight as possible toward a target
60 m away. The subject donned socks soaked with red ink before each walking
trial. These socks had pedal switches attached beneath the heel and the great toe
of both feet. These switches were made by the author using two fine wires and a
rubber mat containing fine-grained iron, which was compressed with each step
allowing current flow from one wire to the other. The switch was sealed with
vinyl tape to prevent electrical shorts. A circuit, consisting of the pedal switch, a
small battery, and a data recorder, was placed to the side of the experimental walk-
way. The circuit was closed whenever the pedal switch contacted the floor, and

Table 12.2 Cadence, mean step width (cm), and the leg, right or left, used for kicking a ball for the 20 experimental subjects.

Subj.	BFS*	Cadence	$\overline{X}$	Left SD	CV	$\overline{X}$	Right SD	CV
1	R	0.95	5.6	2.91	52.0	−3.8	2.61	68.7
2	R	0.85	0.2	4.19	2095.0	−3.5	3.95	112.9
3	R	0.92	6.3	3.28	52.1	0.7	3.82	545.7
4	L	0.95	4.5	2.25	50.0	−3.2	2.01	62.8
5	R	0.93	0.6	3.43	571.7	−2.8	3.17	113.2
6	R	0.86	−5.2	6.76	130.0	−2.8	2.24	80.0
7	R	1.00	1.5	2.24	149.3	−4.3	2.29	53.3
8	R	0.92	1.9	1.95	102.6	−5.2	1.29	24.8
9	R	0.85	−2.0	1.78	89.0	−6.0	2.05	34.2
10	R	0.81	3.2	2.52	78.8	−0.5	3.09	618.0
11	R	0.78	0.3	2.30	766.7	−4.7	1.91	40.6
12	R	0.93	6.4	2.49	38.9	−1.1	2.59	235.5
13	R	0.93	11.5	2.87	25.0	7.1	2.64	37.2
14	R	1.05	4.5	1.32	29.3	−2.5	1.24	49.6
15	L	0.93	1.9	1.91	100.5	−4.2	1.99	47.4
16	R	1.06	1.5	2.23	148.7	−3.4	2.10	61.8
17	L	1.06	2.2	1.86	84.5	−4.9	2.17	44.3
18	R	0.96	2.1	1.52	72.4	0.8	1.89	236.3
19	R	0.81	12.4	1.58	12.7	5.9	1.72	29.2
20	R	0.96	−1.3	2.63	202.3	−9.9	2.29	23.1

*: Better foot side for kicking a ball.
From Uetake (1992) Can we really walk straight? *Am. J. Phys. Anthrop.* 89:19–27. Reprinted by permission of John Wiley and Sons, Inc.

the time of contact was recorded. The pedal switches used in this experiment were designed to be as small as possible so that they would not affect the experimental walking. The subject was instructed to stand in the middle of one end of the paper, to start with his left foot, and to walk at normal speed to the other end of the paper. Subjects left an average of 72.8 (SD = 6.47) footprints in this experiment. The cadence (cycle/sec) of each subject was calculated as the number of strides divided by the time.

12.3.3 Cadence in experimental walking

The cadences are shown in Table 12.2. They ranged from 0.78 (No.11) to 1.06 (Nos.16 and 17), which are comparable to earlier studies of cadence at a normal walking speed: 0.93 (SD = 0.084; Finley and Cody, 1970) and 0.83–1.00 (Molen et al., 1972). Therefore, the values in this study can be presumed to lie within normal limits.

12.3.4 Fourier analysis of the pattern of change in walking

Because the walking pattern during the first four steps of the cycle was often erratic, only steps 5–68 were usually analyzed in this study. However, if a subject walked less than 68 steps, his steps from the fifth to the last were analyzed. The distance from the most posterior point of the heel impression to the center line of the paper was measured with a digitizer. This electronic digitizer was used to input the rectangular coordinates into a microcomputer, which immediately processed the data. As described in Section 12.2, the 64 sequential step width distances represented the ordinate, and step number the abscissa. These values were then submitted as data to the Fourier analysis program.

12.4 Results

12.4.1 Fourier analysis of step widths

Table 12.3 shows the results of the Fourier analysis, in which the subjects are ranked by the magnitude of the amplitude associated with the harmonic number. The number in parentheses indicates the percentage of magnitude of the amplitude at each harmonic, but relative to the value of the 32nd harmonic (or as a proportion of the 32nd). Fifteen subjects among the 20 had the highest amplitude at the 32nd harmonic (column one). The next-highest amplitude fell into four groups: eleven at the first harmonic, four at the second, one at the fourth, and one at the 24th (column two). The next three highest amplitudes (columns three, four, and five) show considerable variability suggesting increasing randomness in contrast to periodicity. Nevertheless, as discussed in Section 12.2, since two sequential steps make up a cycle, which is 1/32 of the total period (or 2π), and this is closely represented by the 32nd harmonic, it is reasonable to expect a high magnitude in that amplitude for a majority of the cases.

12.4.2 Typical pattern of step width change

Figure 12.4 shows four typical tracks chosen from the twenty experimental subjects. The actual measurements and the values for the cosine and sine coefficients with their associated amplitudes for the step widths of these four subjects are available from the author. The sum of the residual values (mm) as the difference between the actual walking pattern and the computed one are: 0.000216 for Case A, 0.000162 for Case B, 0.242216 for Case C, and 1.973979 for Case D.

Case A, subject No.14, illustrates a subject who walked relatively straight. While Case B (subject No. 6), Case C (subject No. 17), and Case D (subject No. 5), show tracks that contain not only high values for the 32nd harmonic, but also high values of the 1st, 2nd, and 4th harmonics, respectively.

Table 12.3 Ranking order (columns one through five) based on highest magnitude of the amplitude associated with the harmonics. Number in parentheses is the proportion of the magnitude with reference to the 32nd harmonic (see text).

Subj. no.*	1st	2nd	3rd	4th	5th
1	32 (100)	1 (49.1)	2 (35.6)	4 (25.5)	26 (18.0)
2	1 (100)	2 (58.8)	32 (52.7)	3 (47.0)	20 (30.8)
3	1 (100)	32 (79.9)	2 (48.4)	7 (38.8)	3 (36.6)
4	32 (100)	1 (35.3)	9 (24.4)	5 (19.4)	10 (18.1)
5	32 (100)	4 (95.5)	1 (90.1)	6 (75.4)	2 (72.1)
6	1 (100)	32 (81.6)	10 (33.5)	8 (29.7)	12 (28.9)
7	32 (100)	2 (71.6)	3 (34.6)	11 (29.6)	23 (28.2)
8	32 (100)	1 (39.0)	15 (27.8)	31 (25.4)	6 (22.0)
9	32 (100)	1 (78.7)	3 (40.2)	4 (33.8)	17 (33.5)
10	1 (100)	32 (64.2)	3 (41.2)	2 (28.1)	20 (20.2)
11	32 (100)	1 (68.5)	2 (45.1)	7 (26.1)	22 (24.6)
12	32 (100)	1 (65.8)	30 (42.3)	2 (37.9)	4 (35.5)
13	32 (100)	1 (95.1)	3 (80.9)	26 (55.3)	2 (53.5)
14	32 (100)	1 (18.9)	21 (15.0)	24 (14.6)	23 (13.3)
15	32 (100)	1 (47.0)	6 (29.8)	12 (27.5)	4 (24.3)
16	32 (100)	1 (53.9)	3 (45.3)	5 (37.8)	13 (32.4)
17	32 (100)	2 (46.8)	1 (28.7)	19 (20.1)	3 (20.0)
18	1 (100)	2 (78.5)	23 (62.3)	32 (60.1)	11 (48.1)
19	32 (100)	1 (47.5)	2 (24.7)	3 (20.2)	5 (18.2)
20	32 (100)	24 (32.9)	23 (25.0)	14 (20.1)	4 (20.0)

*: Subject number
From Uetake (1992) Can we really walk straight? *Am. J. Phys. Anthrop.* 89:19–27. Reprinted by permission of John Wiley and Sons, Inc.

The results of the Fourier analysis of Case A shown in Figure 12.4 indicated that the highest amplitude was associated with the 32nd harmonic, and the next highest one with the 1st harmonic. The magnitude of the amplitude of the 1st harmonic, however, was moderate as were the remaining other harmonics. This suggests that although subject No. 14 walked relatively straight compared to Nos. 5, 6, and 17, the gait was not completely straight, a contention that is reinforced by Figure 12.4. In contrast, subject No. 6 had never walked straight. The Fourier results displayed a high 1st harmonic with a low 32nd. The amplitudes of other harmonics were also appreciable in magnitude. Again, the lack of a straight gait is clearly discernible in Figure 12.4b.

These results suggest that, strictly speaking, no one can walk straight. The results of the Fourier analysis for the subjects shown in Figures 12.4b (Cases C and D) hint at the fact that some individuals who walk in a meandering fashion substitute periodicity as a mechanism for straight walking. These findings raised another interesting question; namely, what is the reason that we cannot walk straight? Possible explanations for this phenomenon are explored in the next section.

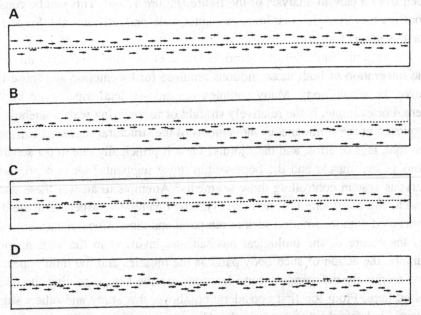

Fig. 12.4 Typical pattern of step width change for four experimental subjects. Relatively straight walking (a), long periodic meandering (b), moderate periodic meandering (c), and relatively short periodic meandering (d). From Uetake (1992). Can we really walk straight? *Am J. Phys. Anthrop.* 89:19–27. Reprinted by permission of John Wiley and Sons, Inc.

12.5 Discussion

12.5.1 Some interpretations of the results of experimental walking

Looking at walking from the viewpoint of kinesiology, the partial motion of the upper or lower limbs is first integrated with the motion of the foot, and then our walking motions are, in turn, aided by resistance of the ground with the plantar surface. This action and reaction takes place alternately with continuous locomotion. Consequently, one can derive considerable information from the footprints made by the plantar surface during walking. For example, the fossilized track of hominid footprints discovered by Leakey (1987) in Laetoli in northern Tanzania, revealed a number of characteristics of early hominid locomotion. From those footprints one can approximate the stature (White, 1980; Robbins, 1987; Tuttle, 1987), the cadence of the walk (Charteris et al., 1981,1982; Alexander, 1984; Tuttle, 1987) as well as the contour pattern of the footprints (Day and Wickens, 1980; White and Suwa, 1987). Footprints have also been used to diagnose abnormal walking (Dougan, 1924). Thus, they are a valuable source of information about locomotion.

From the experimental walking data, one can understand that even a subject (No.14) who appeared to walk nearly straight was not actually able to do so, as seen from a careful analysis of the figure (Figure 12.4a). This can be considered confirmatory evidence that humans cannot walk perfectly straight. Some reasons for this are as follows.

The experimental trail of footprints left after walking represents the result of the integration of both those motions required for locomotion and those that involve the whole body. Many complex musculo-skeletal aspects are implicated here. For example, is the relatively straight or meandering walk somehow a consequence of the simultaneous movements of the structural segments consisting of the foot, legs or arms, and their joints? Or is it principally due to the separate actions of the muscle and the bone within those segments? Or is it primarily the nervous system controlling those segments? Attempts to answer these questions lead to two considerations; namely: (1) those quantitative aspects that are directly concerned with locomotion such as temporal and dimensional measurements; and (2) the nature of the biological mechanisms involved in the walking process, namely, the action of such body parts as the muscles and the limb bones. These two seemingly independent considerations are the key to solving the questions raised here. From the first consideration above, this study and others suggest a lateral, or left/right difference in the measurement of the walking motion. Many cinematic, or dynamic, studies of the lower limbs during walking have also demonstrated a lateral difference or asymmetry in the measurements. These include lateral differences in the motion of the knee and ankle (Hannah et al., 1984; Gundersen et al., 1989; Laassel et al., 1992) and in the plantar ground reaction force (Hamill et al., 1984; Herzog et al., 1989; Laassel et al., 1992). Consequently, the dimensional and temporal measurements during walking differed for the right and left sides (Chatinier and Rozendal, 1970; Hills and Parker, 1991). The experimental subjects here also displayed these lateral differences (Uetake, 1992).

The second consideration above, the determination of functional dominance and structural asymmetry of limbs, which seem to induce this lateral difference in walking motion, is now taken up. Generally, the length of the lower limbs is different from one side to the other, the left limbs being slightly longer than the right (Wolanski, 1957; Friberg, 1983; Friberg and Kvist, 1988). Also, the circumference, as a size measure, may differ but the dominant side seems to be more related to sports or lifestyle (Wolanski, 1957; Kimura and Hattori, 1967). Moreover, many of us seem to have an asymmetry in tibial torsion with more torsion on the right side than the left (Hutter and Scott, 1949; Staheli and Engel, 1972; Malekafzali and Wood, 1979; Clementz, 1988).

Numerous studies of functional asymmetry in our lower limbs have been reported (Nachshon et al., 1983; Plato et al., 1985; Chapman et al.,1987), and the subjects in the current study also had a dominant side as seen from kicking a ball. Many of the subjects had right side dominance. This suggests that our legs and feet, like our hands, have an independent role in behaviors such as kicking a ball, supporting a standing posture, and in the taking-off prior to a jump.

12.5.2 Concluding remarks

From the experimental walking data one can recognize at least two factors: first, individuals are unable to walk straight even if they want to, and second, the pattern of meandering differs from person to person, with some displaying a distinct periodicity in this meandering walk. In sum, numerous factors have been considered as possible causes for the meandering walk, but a primary one, if it exists, has not been identified. It is suggested, however, that these factors are acting in some interdependent and simultaneous way to produce the meandering form of walking. One indisputable fact is that all of the subjects reached their destination in whatever way they walked. This can be considered as the result of feedback from the visual system. Thus, the deviation and subsequent track correction, in some cases with a distinct periodicity, can be considered as a consequence of this feedback mechanism.

References

Alexander, R. M. (1984). Stride length and speed for adults, children, and fossil hominids. *Am. J. Phys. Anthrop.* **63**, 23–7.

Chapman, J. P., Chapman, L. J. & Allen, J. J. (1987). The measurement of foot preference. *Neuropsychologia* **25**, 579–84.

Charteris, J., Wall, J. C. & Nottrodt, J. W. (1981). Functional reconstruction of gait from the Pliocene hominid footprints at Laetoli, northern Tanzania. *Nature* **290**, 496–8.

Charteris, J., Wall, J. C. & Nottrodt, J. W. (1982). Pliocene hominid gait: New interpretations based on available footprint data from Laetoli. *Am. J. Phys. Anthrop.* **58**, 133–44.

Chatinier, K. D. & Rozendal, R. H. (1970). Temporal symmetry of gait of selected normal human subjects. *Proc. Koniklijke Nedarlandse Akad. Weternschappen Seriese C* **74**, 353–61.

Clementz, B. G. (1988). Tibial torsion measured in normal adults. *Acta Orthop. Scand.* **59**, 441–2.

Day, M. H. & Wickens, E. H. (1980). Laetoli Pliocene hominid footprints and bipedalism. *Nature* **286**, 385–7.

Dougan, S. (1924). The angle of gait. *Am. J. Phys. Anthrop.* **7**, 275–9.

Finley, F. R. & Cody, K. A. (1970). Locomotive characteristics of urban pedestrians. *Arch. Phys. Med. Rehabil.* **51**, 423–6.

Friberg, O. (1983). Clinical symptoms and biomechanics of lumbar spine and hip joint in leg length inequality. *Spine* **8**, 643–51.

Friberg, O. & Kvist, M. (1988). Factors determining the preference of takeoff leg in jumping. *Int. J. Sports Med.* **9**, 349–52.

Gundersen, L. A., Valle, D. R., Barr, A. E., Danoff, J. V., Stanhope, S. J. & Mackler, L. S. (1989). Bilateral analysis of the knee and ankle during gait: An examination of the relationship between lateral dominance and symmetry. *Physical Therapy* **69**, 640–50.

Hamill, J., Bates B. T. & Knutzen, K. M. (1984). Ground reaction force symmetry during walking and running. *Res. Quart. Exer. Sport* **55**, 289–93.

Hannah, R. E., Morrison, J. B. & Chapman, A. E.(1984). Kinematic symmetry of the lower limbs. *Arch. Phys. Med. Rehabil.* **65**, 155–8.

Herzog, W., Nigg, B. M., Read, L. J. & Olsson, E. (1989). Asymmetries in ground reaction force patterns in normal human gait. *Med. Sci. Sports Exerc.* **21**, 110–14.

Hills, A. P. & Parker, A. W. (1991). Gait asymmetry in obese children. *Neuro Orthopedics* **12**, 29–33.

Hino, M. (1977). *Spectrum Analysis*. Tokyo: Asakura. (in Japanese)

Hutter, C. B. & Scott, W. (1949). Tibial torsion. *J. Bone and Joint Surg.* **31A**, 511–18.

Kimura, K. & Hattori, O. (1967). On the asymmetries in limb-circumferences in Japanese fencing (Kendo) players. *Bulletin of the Faculty of Physical Education, Tokyo University of Education* **6**, 35–9. (in Japanese)

Laassel, E. M., Voisin, P. H., Loslever, P. & Herlant, M. (1992). Analyse de la dissymetrie des deux membres inferieurs au cours de la marche normale. *Ann. Readaptation Med. Phys.* **35**, 159–73.

Leakey, M. D. (1987). The hominid footprints. In *Laetoli: A Pliocene Site in Northern Tanzania*, ed. M. D. Leakey & J. M. Harris. Oxford: Clarendon Press.

Malekafzali, S. & Wood, M. B. (1979). Tibial torsion—A simple clinical apparatus for its measurement and its application to a normal adult population. *Clin. Orthop. Rel. Res.* **145**, 154–7.

Molen, N. H., Rozendal, R. H. & Boon, W. (1972). Fundamental characteristics of human gait in relation to sex and location. *Proc. Kon. Ned. Acad. Wet. Ser. C.* **75**, 215–33.

Nachshon, I., Denno, D. & Aurand, S. (1983). Lateral preferences of hand, eye, and foot: Relation to cerebral dominance. *Int. J. Neurosci.* **18**, 1–10.

Nakanishi, M. (1989). *Physical Fitness Standards of Japanese People*. Laboratory of Physical Education, Tokyo Metropolitan University. Tokyo: Fumaido. (in Japanese)

Nigorikawa, T. (1988). A preliminary study of walking bias phenomenon which is related to ring-wandering. *Ann. Physiol. Anthropol.* **7**, 99–106.

Plato, C. C., Fox, K. M. & Garruto, R. M. (1985). Measures of lateral functional dominance: Foot preference, eye preference, digital interlocking, arm folding, and foot overlapping. *Hum. Biol.* **57**, 327–34.

Robbins, L. M. (1987). Hominid footprints from site G. In *Laetoli: A Pliocene Site in Northern Tanzania*, ed. M. D. Leakey & J. M. Harris. Oxford: Clarendon Press.

Schaeffer, A. A. (1928). Spiral movement in man. *J. Morphol. Physiol.* **45**, 293–398.

Staheli, L. T. & Engel, G. M. (1972). Tibial torsion. *Clin. Orthop. Rel. Res.* **86**, 183–6.

Tuttle, R. H. (1987). Kinesiological inferences and evolutionary implications from Laetoli bipedal trails G-1,G-2/3, and A. In *Laetoli: A Pliocene Site in Northern Tanzania*, ed. M. D. Leakey & J. M. Harris. Oxford: Clarendon Press.

Uetake, T. (1992). Can we really walk straight? *Am J. Phys. Anthrop.* **89**, 19–27.

White, T. D. (1980). Evolutionary implications of Pliocene hominid footprints. *Science* **208**, 175–6.

White, T. D. & Suwa, G. (1987). Hominid footprints at Laetoli: Facts and interpretations. *Am. J. Phys. Anthrop.* **72**, 485–514.

Wolanski, N. (1957). Vwagi na temat asymetrii budowy ciala czlowieka w zwiazku z asymetria funkcji konczyn. *Kultura Fizyczna* **1**, 59–69.

13

Elliptic Fourier Descriptors of Cell and Nuclear Shapes

GIACOMO DIAZ, CORRADO CAPPAI, MARIA DOLORES SETZU,
SILVIA SIRIGU AND ANDREA DIANA
Cagliari University

13.1 Cell shapes

A few general issues concerning the shape of cells, their visualization, spatial considerations, and resolution, need to be examined prior to discussing the use of Fourier analysis. This may somewhat complicate matters, but will allow a better focus on the quantitative evaluation of cell shapes, as well as the suitability and applicability of the proposed methods.

13.1.1 Basic cell shapes

Cell shapes may be divided into four main categories. The first applies to the vast majority of cells, and includes forms with convex or concave surfaces as well as short cytoplasmic extensions of the type found in epithelial, connective tissue, and smooth and cardiac muscle cells (Figures 13.1a–b, 13.1f–k). The second category includes giant structures, known as syncytia, produced by the fusion of several cells. An example of the latter is the long cylinders of skeletal muscle fibers (Figure 13.1c). The third category includes cells with branching processes, which form extended networks. The shape of these latter cells corresponds more to a logic graph (dendrogram or tree structure) than to a simple surface. Nerve cells belong to this group (Figure 13.1d). The fourth category is made up of those forms that differ topologically from the previous ones. Examples are endothelial cells, which flatten and curve to form the void cylinders of blood capillaries (Figure 13.1e) and Schwann cells, which form the myelin tubes around the axons of the peripheral nervous system.

13.1.2 Shape determinants

The variety of cell shapes is a result of tissue differentiation. However, shape changes may also arise between cells of the same tissue type as well as within

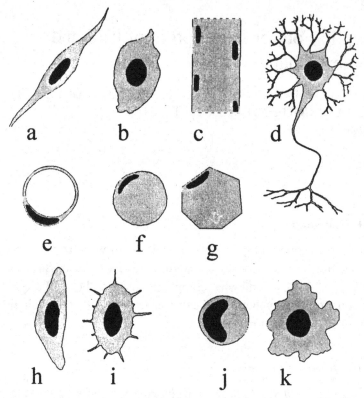

Fig. 13.1 Examples of cell shapes. Drawings follow the conventional iconography and do not correspond strictly to sections: (a) fibrocyte; (b) endocrine epithelial cell; (c) skeletal muscle fiber; (d) multipolar neuron; (e) capillary endothelial cell; (f) isolated fat cell; (g) clustered fat cell; (h,i) chondroblast and chondrocyte; (j,k) peripheral blood monocyte and macrophage. Sizes are arbitrary.

the same cell observed at different times. These morphological domains, which extend across tissues, within tissues, and within cells, suggest the presence of a hierarchy of factors which influences the expression of cell shape. It is evident that cell shape is the result of interactions between the cell and environment. At times the extracellular environment prevails, as in the case of the polyhedral shape of adjacent fat cells, a consequence of compression (Figure 13.1f–g). At other times an active role of the cell is observed, such as the growth of pseudopods in chondrocytes (Figure 13.1h–i) and macrophages (Figure 13.1j–k). Among cell organelles, the plasmalemma (membrane plasma) and the cytoskeleton appear to be most important for the control of the cell shape (Kirschner and Mitchinson, 1986). However, it is not yet clear how the cell compartments interact to give rise to the wide variety of shapes which characterizes different tissues.

13.1.3 Shape visualization

Cells can be visualized in 3-D by confocal microscopy techniques, provided their boundaries show sufficient contrast. This is easy to obtain with isolated cells from cultures, tissue imprints, and so on. In contrast, the *continuum* between cells within tissue, in both horizontal and vertical directions, renders the segmentation of cell outlines and the subsequent 3-D reconstruction problematic. This may be mediated with microinjection of vital stains to fill the cytoplasm. Although this method allows for the examination of a limited number of cells, there is no assurance against possible artifacts. On the other hand, the examination of single sections raises serious problems of image interpretation. One example is the striking contrast between the small size and the smooth outline of peripheral blood lymphocytes observed in sections using transmission electron microscopy, and the large size and folded surface of the *same* cells when observed as intact cells with scanning electron microscopy (Figure 13.2). Moreover, section profiles may appear discontinuous, with highly convoluted shapes like those of mitochondria, neutrophil nuclei, and so forth. These facts, well known in the field of quantitative stereology (Weibel, 1979), require special consideration in the evaluation of shapes.

13.1.4 Spatially referenced and unreferenced shapes

Nearly all forms of animal and plant organisms are characterized by points of spatial reference, or landmarks, whose location and point homology determine anatomical relationships (Bookstein, 1984). It is likely that cells, too, have a spa-

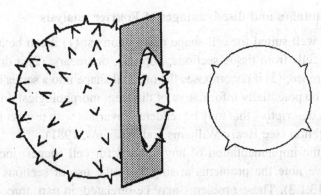

Fig. 13.2 Problems of section interpretation. Visualization of the morphological aspect a sphere, characterized by its true size and the presence of several spines, contrasted with the appearance of sections, which are smaller and surrounded by a lesser number of spines.

tially referenced organization, possibly controlled by the cell center, which contains the pair of centrioles and the microtubule organizing center (McIntosh, 1983). Various examples of linear, radial, and other arrangements of cell organelles confirm such a hypothesis (Diaz et al., 1987). However, the lack of unique, permanent, and recognizable structures in the cytoplasm makes it impossible to consider the cell shape on the basis of conventional anatomical criteria.

13.1.5 Shape and resolution

The definition of the shape of any object (e.g., a cell, a leaf, or a whale) is implicitly associated with size (cells, 3–300 micrometers; leaves, 0.5–50 centimeters; and whales, 3–30 meters, respectively). However, in an attempt to distinguish between objects of the *same* class we need to focus on microcharacteristics within a more restricted range but also with a higher degree of precision. As far as cells are concerned, the increase in resolution may also produce unexpected changes in cell surface morphology, which expands in a nonlinear fashion, well documented by fractal theory (Mandelbrot, 1983). Cells that appear to be perfectly smooth through an optical microscope (average resolution: 0.3 μm) may display increased complexity at the electron microscopic level (average resolution: 0.001 μm) due to the presence of cilia, microvilli, microlamines, deep invaginations, and so on. Thus, the more we attempt to distinguish similar cells on the basis of shape differences, the more we have to rely on descriptors that can probe and explore various dimensional domains. The so-called *form factor*, rapidly obtained as the area/perimeter length ratio, which is not related to the spatial resolution, does not meet the requirements of an adequate descriptor.

13.2 Advantages and disadvantages of Fourier analysis

Fourier analysis is well suited for cell shape description as: (1) it can be applied to 2-D outlines of cells from tissue sections, imprints, or smears; (2) it does not require spatial reference; (3) it decomposes the original shape into a series of subshapes which may be potentially informative of different morphological domains, and; (4) it provides descriptors that may be rendered invariant with respect to size, translation, and rotation (see Healy-Williams and Williams, 1981).

With regard to the implementation of any method for cell shapes, including Fourier analysis, we note the problems arising from the use of sections (mentioned in Section 13.1.3). These problems may be obviated, in part, through the optimization of section sampling and the assumption of some *a priori* hypotheses regarding cell shape. A second, more specific problem with Fourier analysis consists in the difficulty of accounting for singular, nonperiodic contour features.

This can arise with the presence of a single flagellum, pseudopod, invagination, or other single feature. Although with a sufficiently large number of harmonics the fit can be as precise as is desired, the inclusion of such a singular feature may result in a substantial alteration of the whole Fourier spectrum. This can obstruct the capability of revealing the true periodic components of the contour. This point has been considered the most defective aspect of Fourier analysis (Bookstein et al., 1982).[1] In this regard, new alternative methods, such as Wavelet analysis (Argoul et al., 1989), are being currently explored.

13.3 Different Fourier methods for closed contours

Various methods exist for the harmonic analysis of closed contours; these include: (1) the "raw polar" method (Ehrlich and Weinberg,1970); (2) the "tangent angle" function (Zahn and Roskies,1972); (3) the "dual-axis" approach (Moellering and Rayner, 1983); and (4) the "elliptic" formulation (Kuhl and Giardina, 1982). Although it is difficult to precisely determine the accuracy and efficiency of each of these methods, a comparison of some of these methods was provided by Rohlf and Archie (1984).

With respect to cell shapes, methods based on polar coordinates may need to be excluded, as they are unable to process heavily convoluted outlines with respect to an arbitrarily chosen internal starting point, as radii drawn from that starting point to the outline may result in multiple values.[2]

13.4 Elliptic Fourier analysis and applications in cytology

In comparative studies, elliptic Fourier analysis (EFA) has proven to be very efficient with respect to flexibility and consistency of results (Rohlf and Archie, 1984). Since the equations are readily available they will not be given here (Kuhl and Giardina, 1982; Diaz et al., 1989; Chapter 2, this volume). Each EFA harmonic is described by four coefficients, which account for the size, angulation, and displacement of the starting point of an ellipse. For brevity, the term ellipse will implicitly refer to the elliptic shape associated with each harmonic. Likewise, the terms semi-major axis and semi-minor axis refer to the amplitudes of the harmonic ellipses (a schematic illustration is provided in Figure 13.3a). The harmonic ellipse's angle of rotation and starting point, in turn, depend on the angle of rotation and starting point of the contour outline (see Kuhl and Giardina (1982) for details).

[1] Editor's note: This, and other issues regarding the supposed defficiencies of Fourier analysis as boundary-outline functions, have been answered elsewhere (Ehrlich et al., 1983; Read and Lestrel, 1986).

[2] Editor's note: Conventional Fourier analysis is constrained to forms that can be represented by single-valued functions (see Chapter 2).

Suggestions have been made concerning the reorientation of the outline by rotating the first harmonic's major axis to coincide with the abscissa (*x*-axis) and subsequently repositioning the outline's starting point on the same axis. In this way, a normalization is obtained which is "self-consistent, based only on the intrinsic, shape properties of the contour" (Kuhl and Giardina, 1982). Such an option (Rohlf, 1993) is appropriate and advisable for sections of spatially referenced objects (i.e., identifiable landmarks). However, such a reorientation may be ambiguous and misleading with randomly oriented cell sections. In those cases it may be more useful to consider the major and minor axes alone, disregarding the rotation angle and the starting point (Diaz et al., 1989; 1990).

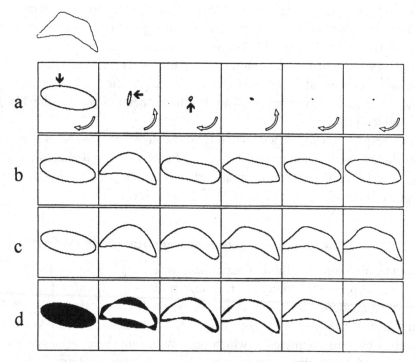

Fig. 13.3 Geometric features of EFA. The series is truncated to the first six harmonics. The original outline is shown on the upper left corner. (a) Individual traces of the harmonic ellipses, showing the elliptic shape, magnitude, rotation angle, rotation direction (white arrows), and starting point (black arrows). The starting point is indicated only for the first three harmonics, as the subsequent ones are too small to be visible. (b) Each harmonic is separately summed to the first harmonic to illustrate the individual contributions (not to scale). (c) Synthesis, approximating to the original contour by summing the contributions of all six harmonics. (d) The contribution of each increasing harmonic is displayed in black. That is, the difference (in black) between the 2nd harmonic superimposed on the 1st harmonic; then the 3rd harmonic superimposed on the 1st plus the 2nd harmonic; and so on.

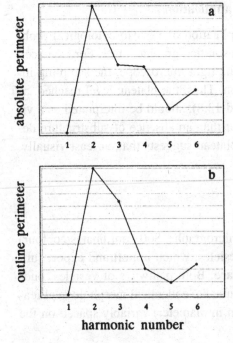

Fig. 13.4 Single-parameter evaluation of the elliptic harmonics. Data refer to the outline of Figure 13.3. (a) Absolute perimeter length of each harmonic, given by the elliptic perimeter multiplied by the frequency. (b) Contribution of each harmonic to the outline perimeter through the inverse transform.

The first harmonic represents the ellipse that most closely approaches the contour. Therefore, this component is most affected by the actual size of the cell and, consequently, is less informative of its shape. The size effect can be removed by normalizing the magnitude of remaining ellipses with respect to the first one. In this way, the harmonics are made invariant to translation, rotation, and size.

It should be noted that the major and minor axes cannot be considered, strictly speaking, as statistical variables. Indeed, there is no formal difference between the two axes, except size; that is, the data being assigned to either one or the other axis solely on the basis of values as seen "posteriori." Generally, principles of statistics require an independent definition and evaluation of variables. However, it is possible to circumscribe this problem by estimating a single parameter for each harmonic in terms of its absolute perimeter. That is, the perimeter of the ellipse is multiplied by its frequency (Figures 13.3a and 13.4a), or the perimeter is synthesized (i.e., a summation of harmonics) up to a particular harmonic number (Figures 13.3c and 13.4b). The latter is of greater interest as it accounts for the actual contribution of each harmonic within the harmonic series. Moreover, the same parameter appears to be suitable for the evaluation of the fractal dimension of the contour, based on the relationship between the synthesized outline perimeter and the harmonic wavelength (Mandelbrot, 1983).

13.5 Some examples of the EFA application

This section illustrates examples of the application of EFA to cytological problems. All data were obtained using a high-resolution graphic tablet. For some cell outlines, particularly those observed with the light microscope, the sampling interval was below the actual image resolution. Thus, the plateau level reached, as shown in some curves (see Figures 13.5 and 13.8), should be interpreted as a visual limit to the available information rather than an absence of substructure detail. At the same time, the linearity in the plateau suggests that, at least visually, there is a virtual absence of noise.

13.5.1 Endothelial cells

Capillary endothelial cells (from electron micrographs, 800 coordinates/contour) are characterized by an intense traffic of vesicles, which import and export substances into and out of the extracellular space. Both pathways of vesicles, budding off from, or fusing with the plasmalemma, have identical morphological features in the form of small pits (about 70 nm in diameter) variably spaced on the cell surface.

If we process the same endothelial cell section twice, once tracing the concavity of pits carefully (a normal cell profile, Figure 13.5a) and once passing straight over them, in effect a smoothing process (the rectified cell profile, Figure 13.5b), an interesting relationship between the morphological and harmonic components is found. As shown in Figure 13.5, for both normal and rectified outlines, the harmonic perimeter length increases in a nearly identical way up to the 16th harmonic. After this harmonic the curves diverge. The rectified profile perimeter soon reaches a plateau, whereas the normal profile perimeter, after a short bending, continues to increase up to the 100th harmonic.

Curves (with the plateaus excluded) can be linearized by a log-log transformation in agreement with the properties of fractal objects. The above correlation is confirmed by inspecting the inverse transform of the normal profile (Figure 13.6). Small invaginations, corresponding to pits, begin to appear with the 17th harmonic. Prior to this harmonic the profile is quite similar to the rectified form.

13.5.2 Cartilage cells

Differentiation of cartilage cells or chondrocytes (from electron micrographs, 1200 coordinates/contour) involves an enlargement of the cell body and the development of several thin cytoplasmic spines, coinciding with the secretion of proteoglycans and collagen (Figure 13.7). From the examination of a sample of chondrocytes, it would appear that cytoplasmic spines start to be represented from the

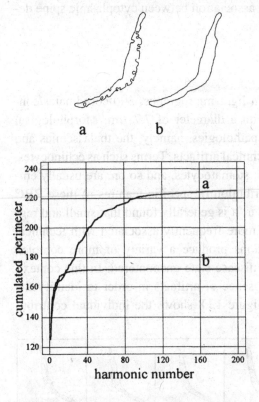

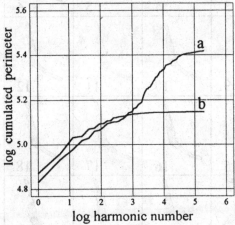

Fig. 13.5 EFA applied to an endothelial cell outline (a) containing all original details and (b) after straightening (smoothing) of the membrane pits. The upper plot shows two initially coincident curves which then begin to diverge. The rectified outline curve reaches a plateau after the 16th harmonic, whereas the original outline curve continues to increase up to the 100th harmonic, after which it plateaus. The lower plot displays the log-log transformation of the same data. A linear relationship is shown (plateau excluded), which reveals the fractal character of the cell outline.

14th harmonic onwards. This allows for the removal of the effect of gross cell shape by excluding the first 13 harmonics from analysis. Graphically, the procedure consists of aligning the seven log-log plots on the 14th harmonic, with the result that three of the curves could be closely superimposed. The remaining four curves were distributed above and below this central group. The three groups of curves, classified on the basis of low, mid, or high slope, correspond to different cell clusters. This finding confirms the association between cytoplasmic spine development and cell stage differentiation.

13.5.3 Red blood cells

Normal red blood cells, or RBC (from light micrographs, 300 coordinates/contour), are biconcave, rounded cells with a diameter of 7.7 μm. Morphological changes in RBC are typical of blood pathologies, namely, the thalassemias and anemias, but they can also appear as technical artifacts. Terms such as echinocytes, acanthocytes, spherocytes, elliptocytes, stomatocytes, and so on, are usually employed to describe the extent and distribution of specific features of these RBC shapes (Bessis et al., 1973). In particular, it is generally found that small and regular crenations of the cell surface are more frequently associated with technical artifacts, whereas pathological conditions produce a variety of more complex forms and patterns, depending on the disease state encountered. In this context, EFA may be useful to remove the presence of artifacts in order to identify the more specific pathological changes. Figure 13.8 shows the individual contribu-

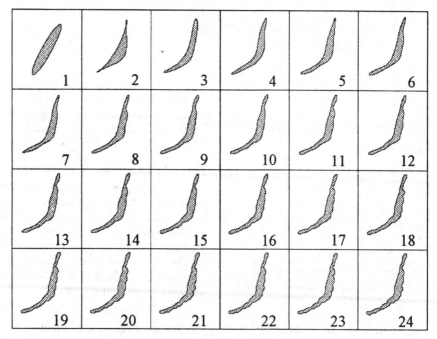

Fig. 13.6 Inverse elliptic Fourier transform of the endothelial cell of Figure 13.5a. No signs of the membrane pits are visible throughout the first 16 harmonics. Prior to the 16th harmonic, the synthesized outline is similar to the rectified outline of Figure 13.5b. Small membrane crenations begin to appear from the 17th harmonic on.

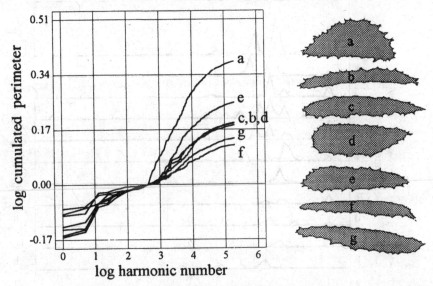

Fig. 13.7 EFA of seven chondrocyte outlines characterized by a series of spikes, which vary in extension, thickness, and number. The seven plots are shifted, or normalized to the 14th harmonic to equalize for differences in the gross cell shape (see text). The graph shows three closely coincident curves (c, b, d) two higher ones (a, e) and two lower (g, f) traces.

tions of the first 60 harmonics obtained from a series of 12 RBC profiles with contour irregularities, sampled from peripheral blood films of healthy subjects. The comparison of the numerical data with the cell outlines in the right margin allowed the detection of at least three components: (1) the 2nd harmonic peak, accounting for the cell ovalization; (2) major peaks between harmonics 10 and 20, accounting for typical echinocytic artifacts; and (3) sporadic peaks between harmonics 20 and 35 accounting for minor perturbations of the RBC surface. Moreover, it can be observed that the three components may be singularly or simultaneously present in a given cell, suggesting the existence of independent causes responsible for the different phenomena.

13.5.4 Nucleoplasmic shape ratio

The nucleoplasmic ratio (from light micrographs, 300 coordinates/contour) is a classic morphometric parameter which expresses the relative amount of cell volume occupied by the nucleus. In normal conditions, the nucleoplasmic ratio characterizes cells of different species and different tissues. However, changes in the nucleoplasmic ratio may also result from pathologic conditions involving cell hypertrophy, nuclear pyknosis, and so on (Diana et al., 1993). In an analogy to the

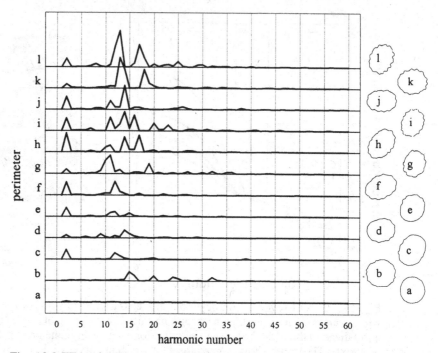

Fig. 13.8 EFA of twelve normal red blood cell outlines with morphological alterations due to artifacts (echinocytes). Only cell (a) shows a normal, smooth surface. The eleven irregular configurations, (from b to l), show a relationship between the occurrence of major crenations and the concentration of peaks between the 13th and 18th harmonics. The degree of cell ovalization is accounted for by the peak at the 2nd harmonic.

volumetric ratio above, it is possible to conceive of a shape ratio between cell and nucleus. Indeed, there are cases of perfectly spherical cells containing very irregular nuclei (i.e., blood neutrophils) as well as irregular cells containing perfectly spherical nuclei (i.e., neurons). The nucleoplasmic shape ratio, of course, cannot be evaluated as a single estimate, but may be conveniently assessed with EFA when applied in parallel to both the cell and its nuclear outlines. Figure 13.9 shows the *difference* between the perimeter outline of the cell and its nucleus, calculated for the first 60 harmonics for ten normal peripheral blood lymphocytes. Positive and negative values indicate a predominant complexity of either the cell or nuclear outline, respectively. That is, perimeter increases as the convolutions or invaginations increase, so it is possible to get a larger perimeter for the nucleus compared to the cell outline. In general, the nucleus appears to be more convoluted than the cell. The majority of changes were found in the first 15 harmonics. A steady-state plateau condition, probably due to the limit of digital resolution, is reached after the 30th harmonic. It is interesting to observe that some of

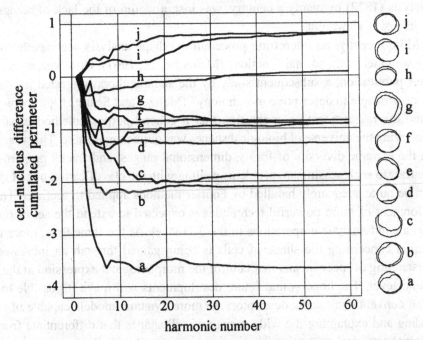

Fig. 13.9 EFA applied to the cell and nuclear outlines of ten normal peripheral blood lymphocytes. The graph shows curves above (+) and below (−) the zero-difference line, which indicates the presence of dominating complexity of either the cell outline (+) or the nuclear outline (−). Major changes occur in the first ten harmonics. It is interesting to observe that the apparently "strange" nuclei (c and g), are, in fact, better adapted to the cell shape than are the other nuclei with more "regular" shapes (a and b) (see text).

the nuclei that could be classified as morphologically "irregular" (Figure 13.9c, g), in fact, are better adapted to the cell shape than others which are apparently more "regular" (Figure 13.9a, b). This fact suggests a reconsideration of the morphological criteria used for the assessment of cytopathological changes. This is in agreement with evidence that shows that the majority of cell alterations, including tumors, involve a disarrangement of the cytoskeleton (Ben-Ze'ev, 1985) with the consequent disruption of the structural link between nucleus and cell membrane.

13.6 Closing remarks

The idea of a numerical description of biological forms has inspired the development of different approaches, which have yielded alternate results. An early approach, dating to 1753, according to Lartigue (1918), consisted of a curve which approximated the outline of the human face. That method, antedating Fourier analysis (1822) by nearly a century, was lost, a victim of the lack of recognition of its novel viewpoint.

More recently, an interesting procedure of shape analysis was developed, using weighted orthonormal functions (Meltzer et al., 1967). However, a year after their publication, a subsequent study by the same authors suggested the "impotence principle in descriptive morphology" (Meltzer and Searle, 1968), which implied the rather negative conclusion that a general-purpose method capable of specifying the universe of biological shapes was quite unrealizable. This was based on the extreme diversity of forms, dimensional ranges, and fractal properties. In contrast to such pessimism, problems dealing with the classification of cell shapes can be quite adequately handled by Fourier methods applied to sections. The development of more powerful techniques is expected to extend the set of optical, digital, and analytical operations to the 3-D level. At the same time, more information concerning the shape of cells is being gained through an increased understanding of mechanisms that control the morphological expression at the molecular level. The hope is that future developments will make it possible to pass from conventional, static descriptors, to more dynamic models capable of reproducing and explaining the wide variety of cell shapes that differentiate from the original spherical zygote.

Acknowledgments

This work was supported by the CNR Target Project on Biotechnology and Bioinstrumentation. The authors wish to thank Carlo Dell'Orbo and Daniela Quacci for the micrographs of cartilage tissue. The linguistic assistance of Anne C. Farmer is gratefully acknowledged.

References

Argoul, F., Arnéodo, A., Grasseau, G., Gagne, Y., Hopfinger, E. J. & Frish, U. (1989). Wavelet analysis of turbulence reveals the multifractal nature of the Richardson cascade. *Nature* **338,** 51–3.

Ben-Ze'ev, A. (1985). The cytoskeleton in cancer cells. *Biochim. Biophys. Acta* **780,** 197–212.

Bessis, M., Weed, R. & Leblond, P. (1973). *Red Cell Shape: Physiology, Pathology, Ultrastructure*. New York: Springer.

Bookstein, F. L., Strauss, R. E., Humphries, J. M., Chernoff, B., Elder, R. L. & Smith, G. R. (1982). A comment on the use of Fourier methods in systematics. *Syst. Zool.* **31,** 85–92.

Bookstein, F. L. (1984). A statistical method for biological shape comparisons. *J. Theor. Biol.* **107,** 475–520.

Diana, A., Setzu, M., Sirigu, S. & Diaz, G. (1993). Nuclear patterns of apoptotic and developing neurons of superior cervical ganglion of newborn rat. *Int. J. Devl. Neuroscience* **11**, 773–80.

Diaz, G., Cossu, M. & Riva, A. (1987). Quantitative ultrastructural approach to the study of the spatial relationships among cell organelles. I. Cytological organization of human exocrine epithelia. *J. Electr. Micr. Tech.* **7**, 167–75.

Diaz, G., Quacci, D. & Dell'Orbo, C. (1990). Recognition of cell surface modulation by elliptic Fourier analysis. *Comp. Meth. Progr. Biomed.* **31**, 57–62.

Diaz, G., Zuccarelli, A., Pelligra, I. & Ghiani, A. (1989). Elliptic Fourier analysis of cell & nuclear shapes. *Comput. Biomed. Res.* **22**, 405–14.

Ehrlich, R. & Weinberg, B. (1970). An exact method for characterization of grain shape. *J. Sedim. Petrol.* **40**, 205–12.

Ehrlich R, Baxter Pharr Jr., R. & Healy-Williams, N. (1983). Comments on the validity of Fourier descriptors in systematics: A reply to Bookstein et al., *Sys. Zool.* **32**, 202–6.

Kirschner, M. & Mitchinson, T. (1986). Beyond self-assembly: From microtubules to morphogenesis. *Cell* **45**, 329–42.

Kuhl, F. P. & Giardina, C. R. (1982). Elliptic Fourier features of a closed contour. *Comput. Graph. Image Proc.* **18**, 236–58.

Healy-Williams, N. & Williams, D. F. (1981). Fourier analysis of test shape of Planktonic foraminifera. *Nature* **289**, 485–87.

Lartigue, A. (1918). Lettres à l'Académie des Sciences. Paris: Sur l'unification des forces et des phénomènes de la Nature.

Mandelbrot, B. B. (1983). *The Fractal Geometry of Nature*. New York: Freeman and Co.

McIntosh, J. R. (1983). The centrosome as an organizer of the cytoskeleton. *Mod. Cell Biol.* **2**, 115–42.

Meltzer, B., Searle, N. H. & Brown, R. (1967). Numerical specification of biological form. *Nature* **216**, 32–6.

Meltzer, B. & Searle, N. H. (1968). Impotence Principle in descriptive morphology. *Nature* **217**, 1289–90.

Moellering, H. & Rayner, J. N. (1983). The harmonic analysis of spatial shapes using dual axis Fourier shape analysis (DAFSA). *Geograph. Analysis* **13**, 64–77.

Read, D. W. & Lestrel, P. E. (1986). A comment upon the uses of homologous-point measures in systematics: A reply to Bookstein et al., *Syst Zool* **33**, 241–53.

Rohlf, F. J. (1993). *NTSYS-pc. Numerical Taxonomy and Multivariate Analysis System*. New York: Exeter Software.

Rohlf, F. J. & Archie, J. W. (1984). A comparison of Fourier methods for the description of wing shape in mosquitoes (Diptera: Culicidae). *Syst. Zool* **33**, 302–17.

Weibel, E. R. (1979). *Stereological Methods*. New York: Academic Press.

Zahn, C. T. & Roskies, R. Z. (1972). Fourier descriptors for plane closed curves. *IEEE Trans. Comp.* **21**, 269–81.

14

Cranial Base Changes in Shunt-Treated Hydrocephalics: Fourier Descriptors

PETE E. LESTREL AND JAN Å. HUGGARE

UCLA School of Dentistry
Karolinska Institute

If you want to understand function, study structure.
Francis H. C. Crick (1988)

14.1 Introduction

It is now well established that the growth of the brain is a major factor in shaping the craniofacial complex, particularly the cranial vault (Moss and Young, 1960). Numerous studies have also suggested that the cranial base (CB) plays a role in determining the shape of the craniofacial complex (Bjork, 1955; Enlow, 1976; Kerr and Adams, 1988). Intracranial pressure may also be implicated in the postnatal development of the CB. For example, the prolonged increased intracranial pressure in uncontrolled hydrocephalus (HC) results in well-defined cranial vault as well as CB changes (Kantomaa et al., 1987), which have been confirmed by experimentally induced HC in rats (Hoyte, 1991).

The most important factor in controlling HC is to maintain an adequate subdural pressure (Hakim and Hakim, 1984). A valve-regulated shunt device which insures unidirectional cerebrospinal fluid (CSF) circulation, has markedly improved prognosis for HC children (Fernell et al., 1988). However, this treatment modality is still prone to various complications such as shunt obstructions, mechanical failures, infections, and overdrainage of CSF (Serlo, 1985).

Earlier studies have shown definitive abnormalities in CB morphology. For example, a smaller *sella turcica* and the *planum sphenoidale* thickened and displaced superiorly (Kaufman et al., 1970). A cephalometric evaluation of five adults with untreated HC revealed that the CB was flat and anteriorly elongated (Kantomaa et al., 1987). A cross sectional study of 37 shunt-treated HCs, 7 to 8 years of age, compared with age- and sex-matched controls, found a thickened calvarium and increased CB flexure among subjects with prolonged treatment (Huggare et al., 1986; 1989; 1991). Two- and four-year follow-up studies showed that these morphological changes were progressive (Huggare et al., 1988; 1992).

Although these studies have documented changes in the hydrocephalic CB with respect to other craniofacial structures, no studies are available that have treated the shunt-treated CB as a single morphological unit. Furthermore, those studies were all based on conventional cephalometric approach (CMA) consisting of distances and angles, which are difficult to effectively apply because of the irregular configuration of the CB. CMA was originally developed for the measurement of *regular* geometric objects, and not for complex *irregular* forms of the type encountered in the craniofacial complex. Moreover, considerable evidence is now available indicating that much of the morphological information of interest (especially with respect to shape) is perceived as outline phenomena. This boundary information plays a fundamental role in pattern recognition analyses (Spoehr and Lehmkuhle, 1982; van Otterloo, 1991; Nadler and Smith, 1993). Thus, CMA is ill-suited for the purposes required here.

Clearly, an alternative method is needed if a more complete analysis of changes in the form, especially the boundary-outline, is desired. This method should be able to describe subtle differences in shape and allow for the precise numerical description of the contour of the CB as seen on the lateral radiograph. The use of Fourier descriptors facilitates the fitting of curves to any complex 2-D form (see Chapter 2). Once the Fourier equation has been determined empirically, precise comparisons can be readily visualized and statistically analyzed to characterize differences in size and shape between the shunt-treated HCs and their age- and sex-matched controls. Earlier studies of the cranial base using Fourier descriptors include Lestrel and Roche, 1986; Lestrel, 1989a; 1989b; and Lestrel et al., 1993.

14.2 Some theoretical considerations

14.2.1 Elliptical Fourier functions

The procedure of elliptical Fourier functions (EFFs) represents a parametric solution (Kuhl and Giardina, 1982). That is, the set of closely located points (x,y) on the contour is represented with a pair of equations written as functions of a third variable (t). Any irregular 2-D bounded outline can be approximated with a polygon by connecting the observed data points with straight lines. Note that the distances between these data points need not be equidistant, a usual requirement with conventional Fourier descriptors.

The Kuhl and Giardina formulation is set up in Cartesian coordinates as the derivation of a pair of equations (x and y) as functions of a third variable (t). These parametric functions are defined in $x(t)$ as

$$x(t) = A_0 + \sum_{n=1}^{N} a_n \cos nt + \sum_{n=1}^{N} b_n \sin nt, \qquad (14.1)$$

and in y(*t*) as

$$y(t) = C_0 + \sum_{n=1}^{N} c_n \cos nt + \sum_{n=1}^{N} d_n \sin nt, \qquad (14.2)$$

where n equals the harmonic number, N equals the maximum harmonic number and the interval is over 2π. See Kuhl and Giardina (1982) and Chapter 2 for details.

14.2.2 Goodness-of-fit considerations

Once the EFFs have been generated, there is the question of the goodness-of-fit. Since the EFF is a convergent series, it is simply a matter of summing enough terms in the series to insure a satisfactory fit, subject to the Nyquist frequency criterion which requires that the number terms in the series (harmonics) not exceed 1/2 the number of observed data points. This goodness-of-fit can be tested by computing the residual or difference between the observed data points and their expected or predicted values derived from the EFF. The truncation of the series solution is based on a residual that is less than the errors associated with tracing and digitizing. Values less than one- to two-tenths of a millimeter, averaged over the whole outline, are generally obtainable with 20–30 harmonics.

14.2.3 Morphometrics and point-homology

Morphometrics, as a generalized procedure, can be viewed as a mapping of the visual information into a numerical (symbolic) representation (Read, 1990; Lestrel, in press). Two distinct representations of form have been developed in the last three decades. The first one is dependent on *homologous points* or landmarks and excludes the curvature between points. The second one focuses on the curvature or the *boundary outline* of the form. These two representations, independently applied, have generated useful information while focusing on different aspects of the form. Both representations, homologous point and boundary outline, have advantages as well as constraints.

Homologous-point representations include CMA as well as deformation methods such as biorthogonal grids (Bookstein, 1978), finite elements (Cheverud and Richtsmeier, 1986) and thin plate splines (Bookstein, 1991). All these methods require that the points used to depict the morphology of interest stand in a one-to-one relationship, or mapping, between specimens. In other words, anatomical points, to be considered as homologous, must be unambiguously located across specimens.

However, there is considerable controversy with respect to such issues as: (1) what is meant by homology, initially defined by Owen (1848) to refer to whole biological structures such as organs and not to points; and (2) which of the so-called homologous points actually satisfy the mapping criterion.

One way to alleviate the scarcity of legitimate homologous landmarks is to accept a less stringent class of points, called here *pseudo-homologous* points. These are often geometrically constructed, bisections and so forth, which although not strictly homologous in the 1:1 mapping sense, can be located with sufficient precision to make them operationally useful.[1]

The second representation, boundary outline, consists of techniques such as eigenshape analysis (Lohmann, 1983), medial axis analysis (Blum, 1973) and Fourier descriptors (Ehrlich and Weinberg, 1970; Anstey and Delmet, 1972; Lestrel, 1974; O'Higgins, 1989). These methods focus on the outline and generally exclude homologous points, except for a center and a starting point.

Two drawbacks with Fourier descriptors have been that: (1) they are generally interpreted using harmonics, phase angles, amplitudes, and so on, notions more familiar in an engineering or pattern recognition context than in the biological setting that is being addressed here; and (2) once the outline has been fitted with a Fourier descriptor such as the EFF, the carefully gathered homologous-point information is largely lost.

To clarify what is meant by this loss, consider linear regression, the fitting of a straight line through an observed set of data points. Assuming that the data are linear and a good fit is obtained (Pearson's $r \approx 0.95$), one can then use the fitted line as a predictor. At this point the observed dataset is of less concern since the major emphasis has shifted to the regression line and its parameters. That is, one could view the locus of the predicted points (of the regression line) as beads *on the line*, which can slide up and down (their coordinates continuously changing) while the line (the equation) remains unchanged. That is, the predicted points on the line are now independent from the observed point data used to derive the regression line (using least squares). In an analogous way, the boundary information modeled with the closely fitting EFF is also independent of the original or pseudo-homologous-point data. Two considerations flow from the above issues: (1) is it possible to combine the two above representations, landmark and boundary; and (2) can one develop a procedure that lends itself to a more direct interpretation than spectral analysis techniques consisting of harmonics, phase angles, and so on. The approach developed below requires that: (1) the boundary curvature is accurately replicated with a set of closely located points; and (2) the original point homology, or pseudo-homology, is maintained.

Since the EFF is an efficient convergent series solution,[2] one can maintain a high degree of correspondence between the observed polygon describing the form and the EFF representation, insuring a close analog of the observed form.

---

[1] The issue of operationally useful points and homology is taken up in Chapter 4.
[2] Efficient in two senses which are: (1) that it is computationally efficient using algebraic computations instead of integral calculus which is required with conventional Fourier analysis; and (2) does not require an unduly high number of series terms to insure a satisfactory fit to the observed outline.

Differences between the observed pseudo-homologous points and the actual curve are quite small, usually fractions of a millimeter. Thus, it is possible to use the function to compute a set of predicted x and y coordinate estimates which emulate the observed values. This technique can be visualized, more or less heuristically, as if the observed pseudo-homologous points have been "moved" onto the function. Although some loss is inevitable in regions of sharp curvature, this is usually minor and quite localized, affecting only small regions of curvature. In this fashion the homologous landmarks as well as other observed points are now positionally placed on the boundary outline of the form. In other words, point homology is being maintained.

14.2.4 Size, shape, and orientation

Three further considerations need to be addressed before meaningful comparisons between samples can be made. These are: (1) positional-orientation; (2) size-standardization; and (3) superimposition (Parnell and Lestrel, 1977).

Analysis of morphological forms using EFFs, although dependent on size as well as shape, is also dependent on the orientation of the form in 2-D space. Kuhl and Giardina (1982) proposed a useful way to orient the form. Their approach consists of rotating the form until the major axis of the first harmonic ellipse is parallel to an axis, in this case the x or horizontal axis. This common orientation has been applied to all morphological forms (compare Figure 14.1 with Figure 14.3).

Besides the problem of orientation in 2-D, there remains the need to normalize for differences in size. This was accomplished by scaling the polygon up or down, so that the bounded area for every specimen is equal to a constant value of 10,000 mm^2. The ratio of 100 divided by the square root of the original or actual area is used as a scaling factor by which all the EFF coefficients are multiplied.

Once a suitable positional-orientation and size-standardization have been computed, forms need to be superimposed on the centroid (an unbiased center) so that size and shape changes can be documented.

14.3 Materials and methods

14.3.1 Subjects

Since 1984 the craniofacial development of 94 shunt-treated HCs has been studied at the Institute of Dentistry, University of Oulu, Finland. Several patients have been followed up two, four, and eight years after the initial radiographs were taken. In the present study, lateral cephalograms from the first examination of 32 shunt-treated patients, 6 to 19 years of age, were used. The sample was evenly divided into 16 males and 16 females.

Most of the subjects had developed HC during the first or second postnatal year and had their first shunt operation immediately after the diagnosis. Failure of shunt function required replacement with a new shunt device. Such repeated operations were common, with an average of 3.9 operations per patient. The subjects generally cooperated, and many of them managed to maintain normal lives. Their motor skills, however, were somewhat disturbed (Serlo, 1985).

Each of the treated subjects was matched with a control based on age, sex, and ethnic background. Age-matching, based on chronological age, was kept within a period of 3 months. HC females (mean age 12.23 ± 3.74) were matched with controls (mean age 12.25 ± 3.69), and HC males (mean age 12.25 ± 3.90) were matched with controls (mean age 12.43 ± 3.96).

The control subjects were drawn from the same geographical region as the treated ones. Some of the control children were part of a nontreated orthodontic group who had participated in a survey on craniofacial morphology in northern Finland (Huggare, 1986; 1987), whereas others were patients coming in for orthodontic care at the Institute of Dentistry, University of Oulu. In the latter cases the data were limited to pretreatment X-rays. The data were then partitioned into treatment and sex and groups. Each of these groups was further subdivided into young and old subgroups, equally divided ($n = 8$).

14.3.2 Tracing procedures

The CB outlines were traced on 0.003 matte acetate sheets with a 0.3 mm lead pencil. The outline was drawn from *basion* (BA) proceeding anterosuperiorly along the dorsal clivus to the posterior aspect of the hypophyseal fossa, *dorsum sellae*, and continuing along the inferior border of the fossa into the anterior CB terminating at the sphenoid registration point (SE). From SE, the outline was continued along the cross section of the greater wings of the sphenoid (where the anterior border of the middle cranial fossa is defined). This outline was then extended along the inferior border of the middle cranial fossa and continued along the inferior border of the CB, ending at *basion*. This generated the bounded figure (Figure 14.1).

14.3.3 Point locations

The point locations were precisely defined with the following procedure. Each CB outline tracing was taped onto a light table for visibility and the following method utilized to locate points 1 through 54. Initially, a line was drawn anteriorly from BA through the most inferior aspect of the hypophyseal fossa and ex-

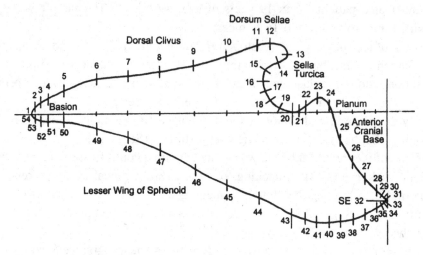

Fig. 14.1 The fifty-four points used to characterize the CB. The tracing was initially oriented with a horizontal line drawn from *basion* through the most inferior aspect of the hypophyseal fossa. This line was then divided (vertical line) into two segments at the most inferior aspect (Point 20). Refer to Table 14.1 for a description of the locations of these points.

tended past the SE point. A perpendicular was then constructed to this tangent point at the hypophyseal fossa (point 20) and extended inferiorly until it intersected (point 43) with the inferior border of the CB (Figure 14.1). The distance from point 1 (BA) to point 20 was identified as S1. This distance was bisected repeatedly to produce 8 equal divisions. These 8 divisions were then extended vertically to the superior outline of the CB defining points 5 to 11, as well as inferiorly to define points 44 through 50 (Table 14.1).

The distance from point 20 to a vertical line extended from point 32 (SE) was identified as S2. This line was similarly bisected repeatedly to produce another 8 divisions. These divisions were also extended superiorly to define points 22–28 and inferiorly to define points 36–42. Points 4 and 51, as well as 29 and 35, are additional bisections. However, the distances on the outline between 1–4, 29–32, and 32–35 represented trisections (these are identified with asterisks in Table 14.1). Point 12 was the most superior aspect of the fossa. Points 13 and 16 were the most anterior and posterior points within the curvature of the fossa. The outline between points 13 and 16 was also trisected, generating points 14 and 15. Points 18 and 21 are the locations of bisections, as are points 52, 53, and 54 (Table 14.1). Once the 54 points were located on the CB outline, the data was submitted to a program written in Basic[3] that controlled the digitizing, plotting, com-

---

[3] Power Basic, Version 3.0c, Power Basic, Inc., 316 Mid Valley Center, Carmel, CA 93923.

Table 14.1. The two columns on the left describe
the line segments that were bisected to generate the
points shown in the two columns on the right. Line
segments identified with asterisks contain trisections, hence the two points in the columns on the
right. Points 1, 12, 13, 16, 20, 32, and 43 represent
anatomic or constructed landmarks.

Cranial Base Point System			
Points defining line segments		Generated points	
S1	S2	S1	S2
1–20	20–32	8,47	25,39
1–8	20–25	6,49	23,41
1–6	20–23	5,50	22,42
6–8	23–25	7,48	24,40
8–20	25–32	10,45	27,37
8–10	25–27	9,46	26,38
10–20	27–32	11,44	28,36
Points located on CB outline		Generated points	
Above S1	Above S2	S1	S2
1–5	28–32	4	29
16–20	20–22	18	21
16–18	—	17	—
18–20	—	19	—
1–4*	29–32*	2,3	30,31
13–16	—	14,15	—
Below S1	Below S2	S1	S2
1–50	32–35*	51	33,34
1–51	—	53	—
1–53	—	54	—
51–53	—	52	—

putation of EFF coefficients, and the computation of expected distances for statistical processing.

14.3.4 Digitization

Each CB tracing was oriented on a Houston Instruments HIPAD digitizer as follows. First a center was established directly on the digitizer surface by equally dividing the digitizing area and extending two lines, vertically and horizontally to

the borders of the digitizing pad, to create a set of cross-hairs. The tracing was then placed on the digitizer with point 20 on the intersection of the cross-hairs. The CB tracing was oriented so that the line from point 1 (BA) to point 32 (SE) was coincident with the horizontal axis of the digitizer cross-hairs. The SE point was aligned to the right of the origin. All points, 1 to 54, were then digitized in succession with a dedicated 286 IBM clone used only for data entry. Points were subsequently plotted using a plotter[4] and superimposed on the originally traced points to check for digitizing errors.

14.3.5 Computation of EFFs

The digitized CB points were then transferred to a 16MHZ 386DX PC with an 80387 coprocessor (required by the EFF program) and submitted to a specially written EFF program which computed two Fourier functions for each specimen, one uncorrected for size, yielding *size and shape* data, and the other standardized on the area, representing *shape only* data. Both of these EFFs were rotated by the software so that the major axis of the first harmonic ellipse was parallel to the *x*-axis according to the Kuhl and Giardina (1982) procedure. The EFF was truncated at 25 harmonics. The goodness-of-fit was checked by calculating the residual or difference between the expected fit from the EFF and the observed data. The curve-fit was considered acceptable, allowing for a slight loss at *basion* and in the vicinity of SE. The fit for an individual case is shown as a stepwise procedure in Figure 14.2.

With 10 harmonics the residual fit averaged over the whole outline was 0.542 mm. With 25 harmonics the residual value dropped to 0.120 mm. The mean residual based on the total database was 0.130 ± 0.017 mm ($n = 275$). This value was less than the errors associated with both the tracing and digitizing procedures.

14.3.6 Computation of distances

Once the EFFs had been computed for each specimen, the functions were used to generate a new set of closely spaced expected points on the CB outline, maintaining homology. These points, consisting of the original 54 observed points and an additional set of 54 points (bisections between the original 54 points) yielded 108 expected points. These expected points were then used to generate plots for each individual specimen as well as mean plots for each subgroup to assess the changes in CB form between the shunt-treated HCs and their age- and sex-matched controls. These mean values were then plotted and superimposed on the centroid to provide a visual assessment of the CB changes with treatment, sex, and age.

---

[4] Hewlett-Packard HP 7040A 2-pen plotter, 16399 W. Bernardo Drive, San Diego, CA 92127.

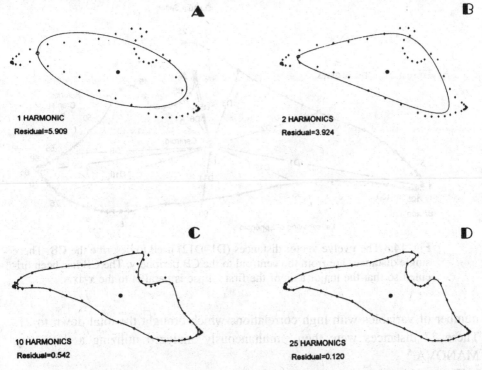

Fig. 14.2 Computer-generated illustration of the elliptical Fourier function (EFF) fitted to an individual CB. Observed data points are shown as crosses and the EFF curve fit is displayed as a solid line.

Besides visualization, these expected points were also used to compute distances from the centroid to the CB outline. These distances are illustrated in Figure 14.3.

The means for the 12 distances, broken down by treatment and sex, and their associated standard deviations are available from the senior author. Besides these 12 distances from the centroid to the CB outline, their components in the x- and y-directions were also computed although not listed for brevity. The justification for computing these components was predicated on the grounds that change in form, although moderate in the combined vector direction, may be significant in the x-component (horizontal) direction or y-component (vertical) direction.

14.4 Results

The 12 distances chosen to depict the major regions of the CB outline (Figure 14.3) with their x- and y-component directions, resulted in a total of 36 variables. Before these distances were submitted to a MANOVA routine for analysis, a correlation matrix was generated to test the relationship between variables. On the whole, correlations were low; nevertheless, it was deemed desirable to remove a

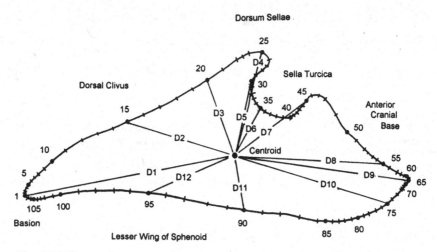

Fig. 14.3 The twelve vector distances (D1–D12) used to describe the CB. These single distances are from the centroid to the CB periphery. The CB has been oriented so that the major axis of the first ellipse is parallel to the *x*-axis.

number of variables with high correlations which brought the total down to 27. These 27 distances were then simultaneously analyzed utilizing a three-way MANOVA.[5]

The results can be broken down into two parts: one based on *size and shape* data and the other on *shape only* data. In each case the 27 variables (distances and their *x*- and *y*-components) were used. The three contrasts tested with the MANOVA were the differences between treatment and controls (of primary interest), sexual dimorphism, and age.

14.4.1 Size and shape data

Focusing on the size and shape data first, Figure 14.4 displays the CB sex contrast with mean female controls superimposed on mean male controls.

The superimposition is on the centroid. As discussed earlier, these data have been oriented so that the first, or principal ellipse, has been rotated so that it is parallel to the *x*-axis for all specimens. This orientation applies to all figures. Figure 14.5 depicts the mean shunt-treated female HCs superimposed on their mean male counterparts.

The size difference seen in both the controls and the shunt-treated hydrocephalic comparisons was also reflected in the 3-way MANOVA table where the sex effect was significant (Wilk's Lambda = 0.304, p < 0.01). While not illustrated

---

[5] CSS: Statistica, Statsoft, Inc., 2325 E. 13th Street, Tulsa, OK 74104.

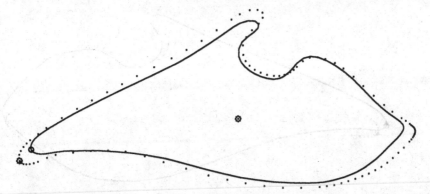

Fig. 14.4 Computer-generated plot of the mean outline of the female control group (solid line) versus the male control group (dotted line). The data have been normalized for orientation only, so both size and shape differences are involved. Superimposition is on the centroid.

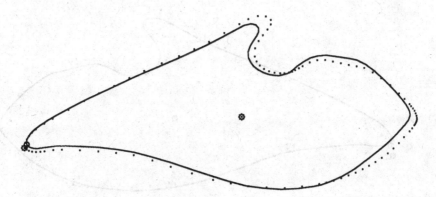

Fig. 14.5 Computer-generated plot of the mean outline of the female HC group (solid line) versus the male HC group (dotted line). The data have been normalized for orientation only, so both size and shape differences are involved. Superimposition is on the centroid.

here, the age contrast, mean young versus mean old, was also statistically significant (Wilk's Lambda = 0.304, p < 0.01) for the size and shape data.

Of greater interest is the comparison between the shunt-treated HCs and the controls. Figure 14.6 depicts the female hydrocephalic means superimposed on female controls. Both size and shape differences can be discerned. Figure 14.7 shows the same superimposition but for males. These differences are also statistically significant (Wilk's Lambda = 0.231, p < 0.001).

At this juncture the question that arises is where within the CB are these differences localized? This can be discerned from the univariate F-tests. For the female versus male comparison, statistically significant differences were found for

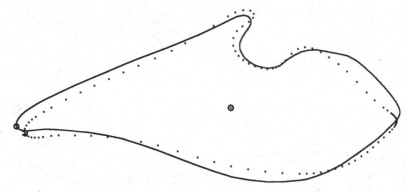

Fig. 14.6 Computer-generated plot of the mean outline of the female HC group (solid line) versus the female control group (dotted line). The data have been normalized for orientation only, so both size and shape differences are involved. Superimposition is on the centroid.

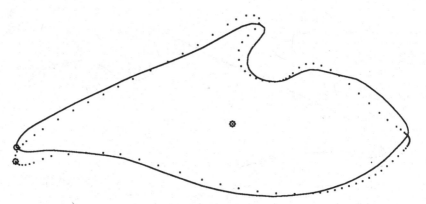

Fig. 14.7 Computer-generated plot of the mean outline of the male HC group (solid line) versus the male control group (dotted line). The data have been normalized for orientation only, so both size and shape differences are involved. Superimposition is on the centroid.

8 distances and 2 components: **D1** ($p < 0.05$), **D3** ($p < 0.005$), **D4** ($p < 0.001$), **D5** ($p < 0.05$), **D7** ($p < 0.05$), **D8** ($p < 0.005$), **D9** ($p < 0.001$), **D10** ($p < 0.001$), **Y1** ($p < 0.01$) and **Y4** ($p < 0.001$). For the young versus old contrast significant differences were computed for 10 distances and 4 components: **D1** ($p < 0.001$), **D2** ($p < 0.05$), **D3** ($p < 0.01$), **D5** ($p < 0.001$), **D6** ($p < 0.005$), **D7** ($p < 0.001$), **D8** ($p < 0.01$), **D9** ($p < 0.01$), **D10** ($p < 0.005$), **D11** ($p < 0.005$), **Y2** ($p < 0.005$), **Y5** ($p < 0.001$), **Y6** ($p < 0.005$), and **Y7**($p < 0.05$).

Focusing on the treatment differences between the shunt-treated HCs and controls, although only one distance is significant, 8 components are significant. The *x*-component distances: **X5** ($p < 0.01$) and **X11** ($p < 0.05$) and the *y*-component directions: **Y1** ($p < 0.001$), **Y2** ($p < 0.001$), **Y4** ($p < 0.005$), **Y5** ($p < 0.005$), **Y8** (p < 0.001) and **Y10** (p < 0.05) are sensitive, in contrast to the actual vector distances from the centroid to the CB outline. In fact, only one such distance, **D11** (p < 0.001), was found to be statistically significant. Figures 14.6 and 14.7 generally reinforce this interpretation.

14.4.2 Shape only data

A different picture emerges when the size-standardized data were analyzed. Figure 14.8 displays the superimposition of mean female controls on mean male controls, whereas Figure 14.9 depicts the HC mean sex comparison.

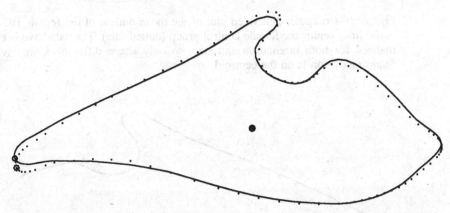

Fig. 14.8 Computer-generated plot of the mean outline of the female control group (solid line) versus the male control group (dotted line). The data have been normalized for both orientation and size, so only shape differences are involved. Superimposition is on the centroid.

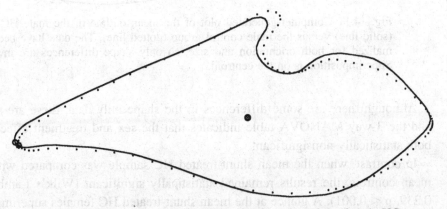

Fig. 14.9 Computer-generated plot of the mean outline of the female HC group (solid line) versus the male HC group (dotted line). The data have been normalized for both orientation and size, so only shape differences are involved. Superimposition is on the centroid.

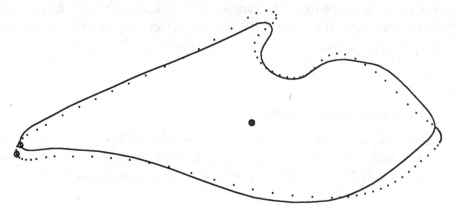

Fig. 14.10 Computer-generated plot of the mean outline of the female HC group (solid line) versus the female control group (dotted line). The data have been normalized for both orientation and size, so only shape differences are involved. Superimposition is on the centroid.

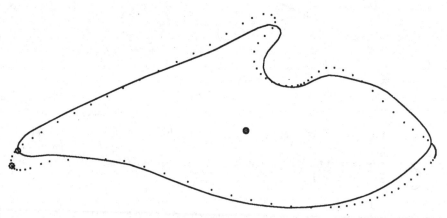

Fig. 14.11 Computer-generated plot of the mean outline of the male HC group (solid line) versus the male control group (dotted line). The data have been normalized for both orientation and size, so only shape differences are involved. Superimposition is on the centroid.

Although there are some differences in the shape only data, these are small, and the 3-way MANOVA table indicates that the sex and treatment effects are both statistically nonsignificant.

In contrast, when the mean shunt-treated HC sample was compared with the mean controls, the results remained statistically significant (Wilk's Lambda = 0.239, $p < 0.001$). A glance at the mean shunt-treated HC females superimposed on the mean female controls (Figure 14.10) discloses these shape differences. They are similarly present in the superimposition of the mean shunt-treated HC males on the mean male controls (Figure 14.11).

Finally, the univariate F-tests indicate that, in contrast to the size and shape results discussed above, 6 vector distances as well as 8 x- and y-components are now statistically significant. These are: **D3** ($p < 0.001$), **D4** ($p < 0.005$), **D5** ($p < 0.001$), **D8** ($p < 0.05$), **D10** ($p < 0.005$), **D11** ($p < 0.001$), **X5** ($p < 0.01$), **Y1** ($p < 0.001$), **Y2** ($p < 0.005$), **Y4** ($p < 0.001$), **Y5** ($p < 0.001$), **Y7** ($p < 0.05$), **Y8** ($p < 0.001$) and **Y10** ($p < 0.005$).

14.5 Discussion and conclusions

With respect to the size and shape data, statistically significant changes using the 3-way MANOVA, were found for all three contrasts: treatment, sex, and age. With respect to both sex and age differences, a majority of the distances were found to be statistically significant. Not surprisingly, size is a pronounced factor between the sexes (sexual dimorphism) as well as for age.

Although there are few reports specifically dealing with sexual dimorphism in the CB, the results of two studies (Knott, 1969; 1971) suggest that both the pre- and post-sphenoid aspects of the CB are longer in boys than girls over the ages of six to fifteen years. Although not comparable to the results here, the overall size differences in the male and female controls are, however, quite evident (Figure 14.4). In contrast, the sexual dimorphism in the HC group was much less evident (Figure 14.5). Reasons for this lack of sexual dimorphism in size are not readily apparent, but must be associated in some unspecified way with the presence of HC.

Once size has been corrected for, differences between the sexes, as well as with age, became statistically insignificant. Since size is a major factor in both growth and sexual dimorphism, these results are not unexpected. However, the difference between the shunt-treated HCs and the controls remained statistically significant after the correction for size. This indicated that after size correction, significant differences in shape remain.

Reasons for this difference in shape between the HCs and controls suggests a complex shift in the developmental pathway leading to the HCs. Anatomically, a possible explanation may be the following. The dural connections from the coronal suture are firmly attached to the CB along the lesser wings of the sphenoid bone, whereas the lambdoid suture is attached to the petrous ridges of the CB. Increased "stretching" in these dural folds could be responsible for an upward bending of the sphenoidal and clival planes, leading to the kind of changes observed in this study (Figures 14.6 and 14.7 and Figures 14.10 and 14.11).

It is emphasized that the twelve distance vectors utilized in this study were not always the most appropriate measure of shape change. That is, in a number of cases, rather than the actual distance, its component, either in the x or y direction, was more sensitive to changes.

The basic theme here has dealt with the need for numerical methods that can accurately characterize the shape of complex morphological shapes and how, once determined, such forms can be meaningfully compared. EFFs represent one such method that facilitates the simultaneous analysis of a number of variables designed to measure subtle shape differences of complex irregular outlines such as those found in the craniofacial complex.

References

Anstey, R. L. & Delmet, D. A. (1972). Genetic meaning of zooecial chamber shapes in fossil bryozoans: Fourier analysis. *Science* **177**, 1000–2.

Bjork, A. (1955). Cranial base development. *Am. J. Orthod.* **41**, 198–225.

Bookstein, F. L. (1978). *The Measurement of Biological Shape and Shape Change.* Lecture notes in biomathematics. New York: Springer-Verlag. **24**, 1–191.

Bookstein, F. L. (1991). *Morphometric Tools for Landmark Data.* Cambridge: Cambridge University Press.

Blum, H. (1973). Biological shape and visual science. *J. Theoret. Biol.* **38**, 205–87.

Cheverud, J. M. & Richtsmeier, J. T. (1986). Finite-element scaling applied to sexual dimorphism in rhesus macaque (*Macaca mulatta*) facial growth. *Syst. Zool.* **35**, 381–99.

Ehrlich, R. & Weinberg, B. (1970). An exact method for the characterization of grain shape. *J. Sed. Petrol.* **40**, 205–12.

Enlow, F. H. (1976). The prenatal and postnatal growth of the human basicranium. In *Symposium on Development of the Basicranium.* ed. J. F. Bosma, Jr. Bethesda: NIH.

Fernell, E., Hagberg, B. & Hagberg, G. (1988). Epidemiology of infantile hydrocephalus in Sweden. A clinical follow-up study in children born at term. *Neuropediat.* **19**, 135–42.

Hakim, S. & Hakim, C. (1984). A biochemical model of hydrocephalus and its relationship to treatment. In *Hydrocephalus.* ed. K. Shapir. New York: Raven Press.

Hoyte, D. A. N. (1991). Hydrocephalus in the infant rat: A further look at the basicranial synchondroses. In *Fundamentals of Bone Growth: Methodology and Applications.* ed. D. Dixon, B. Sarnat & D. A. N. Hoyte. Boca Raton, Florida: CRC Press.

Huggare, J. Å. (1986). Head posture and craniofacial morphology in adults from northern Finland. *Proc. Finn. Dent. Soc.* **82**, 199–208.

Huggare, J. Å. (1987). A cross sectional study of head posture and craniofacial growth in children from the north of Finland. *Proc. Finn. Dent. Soc.* **83**, 5–15.

Huggare, J. Å., Kantomaa, T. J., Rönning, O. V. & Serlo, W. S. (1986). Craniofacial morphology in shunt-treated hydrocephalic children. *Cleft Palate J.* **23**, 261–9.

Huggare, J. Å., Kantomaa, T. J., Rönning, O. V.& Serlo, W. S. (1988). Basicranial changes in shunt-treated hydrocephalic children: A two-year report. *Cleft Palate J.* **25**, 308–12.

Huggare, J. Å., Kantomaa, T. J., Serlo, W. S. & Rönning, O. V. (1989). Craniofacial morphology in untreated and shunt-treated hydrocephalic children. *Acta Neurochir (Wien)* **97**, 107–110.

Huggare, J. Å., Kantomaa, T. J., Serlo, W. S. & Rönning, O. V. (1991). CB and Facial Prognathism: A Radiocephalometric study on shunt-treated hydrocephalics. In *Fundamentals of Bone Growth: Methodology and Applications.* ed. D. Dixon, B.

Sarnat & D. A. N. Hoyte. Boca Raton, Florida: CRC Press.

Huggare, J. Å., Kantomaa, T. J., Rönning, O. V. & Serlo, W. S. (1992). Craniofacial growth in shunt-treated hydrocephalics: a four-year roentgenocephalometric follow-up study. *Child's Nerv. Syst.* **8,** 67–9.

Kantomaa, T. J., Huggare, J. Å., Rönning, O.V. & Wendt, L. (1987). Cranial base morphology in untreated hydrocephalics. *Child's Nerv. Syst.* **3,** 222–4.

Kaufman, B., Sandström, P. H. & Young, H. F. (1970). Effects of prolonged cerebrospinal fluid shunting on the skull and brain. *Radiol.* **97,** 537–42.

Kerr, W. J. S. & Adams, C. P. (1988). Cranial base and jaw relationship. *Am. J. Phys. Anthrop.* **77,** 213–20.

Knott, V. B. (1969). Ontogenetic change of four cranial base segments in girls. *Growth* **33,** 123–42.

Knott, V. B. (1971). Changes in cranial base measurements of human males and females from 6 years to early adulthood. *Growth* **35,** 145–58.

Kuhl, F. P. & Giardina, C. R. (1982). Elliptic Fourier features of a closed contour. *Comp. Graph. Imag. Proc.* **18,** 236–58.

Lestrel, P. E. (1974). Some problems in the assessment of morphological size and shape differences. *Yearbk. Phys. Anthropol.* **18,** 140–62.

Lestrel, P. E. & Roche, A. F. (1986). Cranial base shape variation with age: A longitudinal study of shape using Fourier analysis. *Hum. Biol.* **58,** 527–40.

Lestrel, P. E. (1989a). Some approaches toward the mathematical modeling of the craniofacial complex. *J. Cran. Genet. Develop. Biol.* **9,** 77–91.

Lestrel, P. E. (1989b). A new method for analyzing complex two-dimensional forms: Elliptical Fourier functions. *Am. J. Hum. Biol.* **1,** 149–64.

Lestrel, P. E. Morphometrics of craniofacial form. In *Fundamentals of Craniofacial Growth.* ed. D. Dixon, D. A. N. Hoyte & O. Ronning. Boca Raton, Florida: CRC Press. (in press)

Lestrel, P. E., Bodt, A. & Swindler, D. R. (1993). Longitudinal study of cranial base shape changes in *Macaca nemestrina. Am. J. Phys. Anthrop.* **91,** 117–29.

Lohmann, G. P. (1983). Eigenshape analysis of microfossils: A general morphometric procedure for describing changes in shape. *Math. Geol.* **15,** 659–72.

Moss, M. L. & Young, R. W. (1960). A functional approach to craniology. *Am J. Phys Anthrop.* **18,** 281–92.

Nadler, M. & Smith, E. P. (1993). *Pattern Recognition Engineering.* New York: John Wiley & Sons.

O'Higgins, P. (1989). Developments in cranial morphometrics. *Folia Primatol.* **53,** 101–24.

Owen, R. (1848). Report on the archetype and homologies of the vertebrae skeleton. *Rep. 16th Meeting Brit. Assoc. Adv. Sci.* 169–340.

Parnell, J. N. & Lestrel, P. E. (1977). A computer program for fitting irregular two-dimensional forms. *Comp. Prog. Biomed.* **7,** 145–61.

Read, D. W. (1990). From multivariate to qualitative measurement: Representation of shape. *Hum. Evol.* **5,** 417–29.

Serlo, W. S. (1985). Shunt treatment of hydrocephalus in children. *Acta Universitasis Ouluensis,* Series D Medica No 130. Oulu: University of Oulu.

Spoehr, K. T. & Lehmkuhle, S. W. (1982). *Visual Information Processing.* San Francisco: W. H. Freeman.

van Otterloo, P. J. (1991). *A Contour-Oriented Approach to Shape Analysis.* New York: Prentice-Hall.

15

A Numerical and Visual Approach for Measuring the Effects of Functional Appliance Therapy: Fourier Descriptors

WON MOON

Private Dental Practice, Fullerton, CA

15.1 Introduction

Functional appliances (FAs) are orthodontic/orthopedic appliances designed to treat patients with dental malocclusions and skeletal imbalances (e.g., retrognathia and prognathia). Despite the relatively long history of FA treatment, there continues to be controversy as to their mode of action, due in part to conflicting findings between investigators.

These appliances have been more commonly used in Europe than in the United States. American orthodontists were more likely to treat with fixed appliances or braces, whereas Europeans have been more interested in altering masticatory function to correct malocclusions, usually with the use of removable appliances. There are numerous designs all generally known as "functional appliances." Some of these include specialized features for dental arch expansion, altering the facial vertical dimension, myofunctional therapy, and so on. The theoretical basis of FA treatment is in general the principle that a "new pattern of function," dictated by the appliance, leads to the development of a "new morphological pattern" (Ricci, 1983). This new functional pattern is, presumably, generated by the sagittal and vertical alteration of the mandibular position in young patients (usually in early mixed dentition stages), resulting in orthodontic (primarily affecting dental units) and orthopedic (structural alterations affecting bone) adaptations to a new jaw position (Figure 15.1).

These adaptations affect the arrangement of the teeth within the jaws, and are intended to produce: (1) an improvement in the occlusion; (2) an altered relationship of the jaws; and (3) changes in the amount and direction of growth of the jaws. Although FAs have been designed to treat all types of malocclusions, they seem to be most effective in treating dental and skeletal Class II malocclusions,[1] particularly cases with mandibular deficiency (Bishara et al., 1981). Thus

---

[1] Editor's note: The Class divisions of I, II, and III were originally derived by Angle, around 1900, who attempted to provide a three-fold typology of increasing malocclusion as encountered in clinical dentistry. This typology involves both dental and skeletal elements. Class I is more- or-less normal; Class II is protrusive with a retrognathic profile; and Class III has a marked protrusion of the mandible with a prognathic profile (see Graber, 1966; Enlow, 1975 for details).

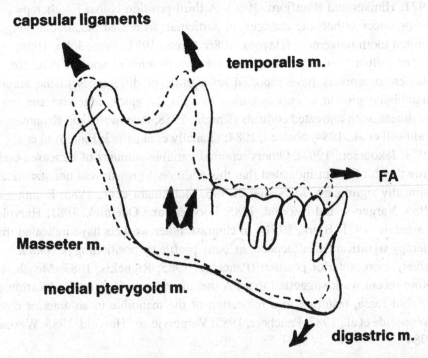

Fig. 15.1 Proposed muscular adaptations in response to FA therapy. Tensions created by the stretching of masseter, medial pterygoid, temporalis, and digastric muscles, may play a significant role in the remodeling process of the mandible undergoing FA therapy (dotted to solid line).

for Class IIs, either as removable or fixed intra-oral FAs, they work by forcing the patient's mandibular posture forward.

There are several possible ways through which the treatment with FAs can lead to a satisfactory correction. The attainment of an acceptable treatment outcome is probably due to a combination of orthodontic and orthopedic adaptations to the new pattern of function in conjunction with the FA (Lestrel and Kerr, 1992; Carels and van der Linden, 1987; Frankel, 1984; Cohen, 1983; Bjork, 1951).

15.1.1 Orthodontic and orthopedic processes in FA therapy

The controversial nature of FA therapy has led to various schools of thought regarding the exact mode of FA action. One group has suggested that changes were mainly dentoalveolar (Hamilton et al., 1987; Caldwell et al., 1984; Gianelly et al., 1984; Robertson, 1983; Graber, 1975). Another group, although accepting that some changes were primarily dentoalveolar, suggested that there were also orthopedic effects involved (McNamara, 1985; McNamara et al., 1985; Vargervic and Harvold, 1985; Pancherz 1984; Gianelly, et al., 1983; Harvold and Vargervic,

1971; Hiniker and Ramfjord, 1966). A third position is that FA therapy primarily produces orthopedic changes, in particular, increased mandibular length and limited tooth movement (Haynes, 1986; Eirew, 1981; 1976; Joho, 1968).

The influence of FAs on mandibular growth remains another debatable issue. Numerous workers have reported no significant differences in the amount of mandibular growth when comparing two FA therapies, edgewise and headgear treatment, with untreated controls (Knight, 1988; Adenwella and Kronman, 1985; Caldwell et al., 1984; Nielsen, 1984; Gianelly et al., 1983; Baumrind et al., 1981; 1978; Jakobsson, 1967). Others reported variable amounts of increased condylar growth with FA but indicated that this additional growth was not statistically or clinically significant (McNamara, 1986; McNamara et al., 1985; Remmer et al., 1985; Vargervic and Harvold, 1985; Forsberg and Odenrick, 1981; Harvold and Vargervic, 1971; Bjork, 1951). In contrast, other workers have indicated that FA therapy significantly influenced the bony profile by positioning *pogonion* in a relatively more anterior position (Pancherz, 1985; Righellis, 1983; Meach, 1966). More recent work suggested that FA therapy has an effect on the location of the glenoid fossa, resulting in a relocation of the mandible in an anterior direction (Woodside et al., 1987; Pancherz, 1985; Vargervic and Harvold, 1985; Wieslander, 1984).

15.1.2 Measurement of FA therapy effects

Although FA therapy is generally accepted as an effective method for treatment of Class II malocclusions, there is a lack of consensus regarding the relative amounts of orthodontic and orthopedic corrections involved (DeVincenzo and Winn, 1989). Much of this controversy is due to the choice of variables used to describe the morphological changes presumed to occur with treatment. This subjectivity in the variable selection process may hinder the objectivity required in the analysis and evaluation of data. Thus, conclusions drawn by various authors have differed significantly, while the data and methods used are very similar, varying only in the lines or angles they have chosen as measures. Figure 15.2 illustrates an example which could lead to different conclusions simply by choosing different variables to be measured. Variables A, B, and C were all used to represent the length of mandible. They measure from *pogonion* to the: (1) top of the condyle; (2) center of the condyle; and (3) articulare point.

With an identical amount of increase in the length of the mandible, the changes observed would be *different* with each variable. For example, variable C does not really represent the length of mandible, since it is dependent on the cranial base and introduces an additional element of variability. Furthermore, any condylar length changes due to remodeling within the glenoid fossa aspect would not be

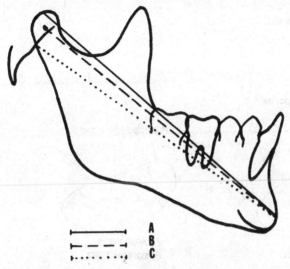

Fig. 15.2 Three different distances, A, B, and C, of mandibular length as measured with CMA (see text).

measurable with variables B or C. Moreover, all of these measurements are unable to describe the subtle mandibular shape changes arising with the treatment. Some of the study limitations regarding FA therapy, which included inadequate methods for describing the various changes thought to be occurring, were discussed by Bishara et al., (1981). Thus, the use of the conventional metrical approach (CMA), consisting of linear distances, angles, and ratios, may not always be the best choice for describing irregular morphologies found in the craniofacial complex, especially morphologies that are undergoing subtle, complex changes in response to FA therapy.

The above discussion suggests that more sophisticated and sensitive methods are required. Certainly, a better method that describes size and shape changes occurring with FA therapy would lead us to a more complete understanding of this widely used but still debated treatment.

15.2 Fourier descriptors

Elliptical Fourier functions (EFF) utilized here have a number of attractive properties (Lestrel, 1989a; 1989b; Kuhl and Giardina, 1982). This technique fits a function to a set of closely spaced observed points on the outline or boundary of any shape (Kuhl and Giardina, 1982). A unique feature of the EFF is that it is largely independent of anatomical landmarks, which can introduce a potential for a personal bias, as discussed earlier, using CMA (Figure 15.2). EFFs provide a

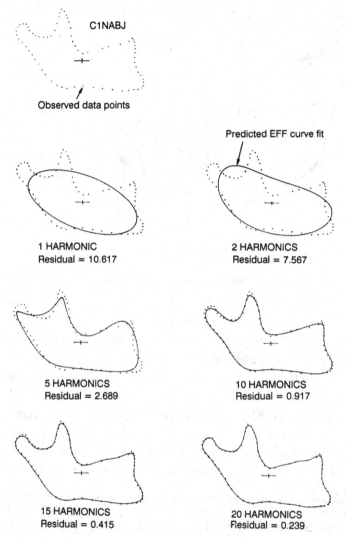

Fig. 15.3 A computer-generated graphic of the mandible described with 72 points. The solid line represents the predicted EFF curve as a stepwise fit with increasing harmonics and the dots represent the observed data being fitted. The computed mean residual is the difference between the predicted points (calculated from the EFF) and the observed points, averaged over all 72 points. With 10 harmonics the residual was less than a millimeter, with 20 harmonics the value dropped to 0.24 mm.

precise analog of the observed morphology and facilitate the extraction of the boundary information that is largely missing from landmark-based techniques. Figure 15.3 illustrates the EFF as a step-wise procedure being fitted to the mandible described with 72 points. With 20 harmonics the mean residual dropped to less than 0.24 mm which is well below the tracing and digitizing errors. All mandibles in this study were standardized for size by scaling the bounded *area* of all specimens, up or down, to a standard area (see Chapter 2).

15.3 Samples used in the FA study

Lateral cephalometric radiographs of patients treated with FA formed the basis of the present study. Patients were selected for two studies: (1) cross-sectional, composed of treatment cases ($n = 22$) and controls ($n = 22$); and (2) longitudinal ($n = 13$). In order to collect homogeneous samples, the following selection criteria were used:

- All subjects were treated by the same clinician using a Dawn Sells (DS) FA (DeVincenzo et al., 1987);
- The subjects were in skeletal Class II relationship (ANB, an angle formed by A-point, *nasion* and B-point, $\geq 3°$), as well as displaying a dental Class II malocclusion (i.e., the mandibular dentition is positioned posteriorly relative to the maxillary dentition);
- The mandibular plane angle[2] (SN-Go-Gn) of each subject was limited to $32° \pm 3°$;
- FA therapy was performed during the adolescent growth spurt as determined by clinical judgment (e.g., height changes, secondary sex characteristics, and hand/wrist radiographs); and
- All subjects were Caucasian females.

All head films were shot with an anode-to-midsaggittal plane distance of 152.4 cm and a midsaggittal plane-to-film distance of 14.5 cm. This yielded a radiographic enlargement factor of 9.5%. Patients were selected for two separate studies, one cross-sectional ($n = 22$) and the other longitudinal ($n = 13$). Tables 15.1 and 15.2 describe the characteristics of the cross-sectional and longitudinal samples utilized in this study.

15.3.1 Study I: cross-sectional data

Two sets of lateral cephalometric radiographs (each with $n = 22$) from patients who fulfilled the above selection criteria were obtained. These patients were treated with FA therapy (DS). Lateral cephalometric radiographs were taken *before* (t_1) and *after* (t_2) treatment (right before the delivery of the FA and immediately after the removal of the appliance). The difference ($t_2 - t_1$) is a measure of shape changes due to both growth and treatment. This group served as a treatment group (Table 15.1).

---

[2] Editor's note: The mandibular plane angle is commonly used by orthodontists as a measurement of facial pattern in the vertical direction, 32° being the norm. It is an angle between the two straight lines drawn from *sella* to *nasion* and from *gonion* to *gnathion*.

Table 15.1. Description of the cross-sectional samples used in the study.

Control Group	
Sample:	$n = 22$ Caucasoid females
Mean age:	11 years 6 months $\pm$ 14 months
MPA:	32.3 $\pm$ 1.5 degrees
ANB:	5.9 $\pm$ 1.5 degrees
Duration length:	9.1 months $\pm$ 6.6 months
Treatment Group	
Sample:	$n = 22$ Caucasoid females
Mean age:	11 years 5 months $\pm$ 14 months
MPA:	31.8 $\pm$ 1.8 degrees
ANB:	6.0 $\pm$ 1.4 degrees
Duration length:	14.1 months $\pm$ 4.6 months

Table 15.2. Description of the longitudinal samples used in the study.

Sample:	$n = 13$ Caucasoid females
Mean age:	11 years 5 months $\pm$ 17 months
MPA:	31.8 $\pm$ 2.3 degrees
ANB:	6.4 $\pm$ 1.8 degrees
Control period length (CP):	8.6 months $\pm$ 6.0 months
Treatment period length (TP):	9.1 months $\pm$ 4.8 months
Difference (TP-CP):	0.5 months $\pm$ 3.7 months

The control group was composed of patients who also fulfilled the above selection criteria but did *not* receive any type of treatment (Table 15.1). Various orthodontic treatment procedures were performed on these patients, but only *after* this period. Two consecutive sets of lateral cephalometric radiographs (each with $n = 22$) were taken on these patients as well, for observation and monitoring of craniofacial developmental growth. Therefore, any changes that occurred between these two sets of radiographs (that is, again measured as differences: $c_2 - c_1$) can be presumed to be solely due to growth.

While in FA therapy, treatment consisted of advancing the mandible in 3 mm increments at 2 month intervals (DeVincenzo and Winn, 1989). During FA therapy some patients had a palatal intrusion appliance placed to deliver an intrusive force to the maxillary incisors in order to reduce a deep over-bite and allow the mandible to be advanced (DeVincenzo and Winn, 1987). No patient had any kind of additional intraoral or extraoral force applied to correct the Class II anomaly, except what was derived from the FA itself. All patients wore the FA full-time, even while eating. The treatment was deemed completed when a Class I molar relationship was obtained.

15.3.2 Study II: longitudinal data

For the longitudinal study, cephalometric radiographs of patients ($n = 13$) were chosen, who satisfied the selection criteria described earlier but who also went through the observational period *without any type of treatment* before the beginning of FA therapy. Lateral cephalometric radiographs were taken at the *beginning* (t_1) and *end* (t_2) of this observational period. The difference ($t_2 - t_1$) can be ascribed to growth. Subsequently, FA therapy was performed on these patients, and a third (t_3) radiograph, taken at the end of the treatment. Thus, the longitudinal study was designed to compare the effect of FA therapy during the subsequent treatment period ($t_3 - t_2$) with the effect of natural growth during the initial control period ($t_2 - t_1$). Thus, three sets of lateral cephalometric radiographs (initial, progress, and final) from the 13 patients were used. These patients were also treated by the same clinical protocols and by the same clinician using the DS functional appliance (Table 15.2). Between the initial and progress radiographs (control period), no treatment was performed. The FA therapy was initiated immediately after the progress radiograph, and the final radiograph was taken at the time of removal of the appliance (treatment period). That is, the patient served as her own control. The length of time of the control period was approximately equal to the length of the treatment period for each patient. This was done to insure that the changes thought to be occurring during the control and treatment periods were comparable in length.

15.4 Methods

Mandibles were traced using the lateral cephalograms of all subjects, both control and treatment groups. The tracings included outlines of mandible and labial surface of mandibular incisors and the occlusal surfaces of teeth. On each tracing, some of the commonly used anatomical landmarks such as B-point, PM, *pogonion, gnathion, menton, gonion*, and so on, as well as the cusp tips of the first molar and incisor, were marked. Using these anatomical landmarks as references, a further set of closely located points along the outline of the mandible was created by bisections and/or trisections of the reference points. Each tracing consisted of a set of 72 closely spaced points imbedded on the boundary of the mandible to closely describe its shape (Figure 15.3). The definitions of these 72 points are available from the author. The points marked on each tracing were then digitized using a HIPAD digitizer.[3]

The digitized points were subsequently plotted[4] using an HP 7470A and the 72 points superimposed on the original tracing to check for errors. After the accuracy of digitizing was confirmed, the (x,y) coordinates of each digitized point were corrected for radiographic enlargement (0.950). EFFs for each mandible were then computed from these digitized points, and the goodness-of-fit, or residual (the difference between the expected values calculated from EFF and the digitized observed data) calculated. Values larger than 0.4 mm, suggesting the presence of errors, were redigitized. The goodness-of-fit must be below the errors of tracing and digitizing.

All EFFs were subsequently standardized for size using the area bounded by each EFF to control for inter-subject size variability, as well as the elimination of any size changes due to growth as well as treatment. Further, each mandible, was oriented in such a way that the long axis of the EFF was made parallel to the x-axis (Kuhl and Giardina, 1982). Because all EFFs were *size-standardized* and *oriented* with respect to the x-axis, the EFF representations approximated each other closely when they were superimposed on their centroids. A mean EFF was then calculated for each sample, generating an average shape for the mandible. Utilizing the mean EFF, a set of 150 equally spaced *predicted* points on the mandibular boundary outline was generated (Figure 15.4).

When the mean EFFs were superimposed on their respective centroids, the difference in shape between the two EFFs can be visualized (e.g., Figure 15.5).

The greater the difference between the mean EFF traces at any aspect of the bounded mandibular contour, the greater the amount of mean shape change. In order to localize and quantify these changes, predicted distances using the EFF were computed from the centroid to the 150 equally spaced points on the bounded figure (Figure 15.4). These 150 distances were averaged for ten regions on the mandibular margin.[5] Differences between samples were then subjected to statistical analyses which are described in Section 15.5.3 below.

15.5 Results

Two independent analyses were carried out, one based on cross-sectional data and the other on the longitudinal series. In the cross-sectional study, changes in mandibular shape of the 22 subjects who have undergone FA therapy (treatment

---

[3] Houston Instruments, 8500 Cameron Road, Austin, Texas 78753.

[4] Hewlett-Packard, 16399 W. Bernardo Drive, San Diego, CA 92127.

[5] Editor's note: With these 150 equally spaced points, any semblance of homology is technically lost. This was, in part, compensated for by averaging along localized regions of the mandibular contour (see Chapters 14 and 16 where homology is maintained).

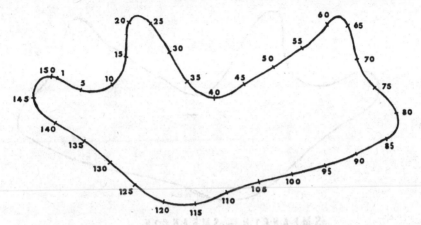

Fig. 15.4 Utilizing the EFF, a set of 150 equally spaced points was generated. This figure has been rotated so that the major axis of the first harmonic ellipse was made parallel to the *x*-axis. This orientation was used to provide a common basis for superimposition.

group) were compared to the changes that occurred during growth of the 22 subjects who were never treated with any type of appliance (control group). In the longitudinal study, the mandibular shape changes observed in the 13 subjects while no treatment was rendered (control period) were compared to the changes observed in the *same* subjects after the FA therapy (treatment period).

15.5.1 Study I: cross-sectional results

The shape change in the treatment group was calculated by comparing the predicted distances from the two mean EFFs derived from the two sets of lateral cephalometric radiographs taken before and after the FA treatment (see Section 15.5.3). The shape change in the control group was calculated similarly by comparing distances from the two EFFs derived from the two sets of radiographs taken at "time-one" and "time-two" (note that no treatment was performed between "time-one" and "time-two" here).

Figure 15.5 illustrates the means (mean EFFs) of the four mandible samples (each with $n = 22$) that were utilized in the cross-sectional study. Two of the mandibular outlines are superimposed on the centroid as the *control* groups (SMEANC1N and SMEANC2N; Figure 15.5, top) and the other two outlines are also superimposed, on the centroid, as the *treatment* groups (SMEANS2N and SMEANS4N; Figure 15.5, bottom). When the mean EFFs, serving as analogs of the mean control groups (SMEANC1N and SMEANC2N), were superimposed on the centroid, no appreciable amount of shape change was present with growth

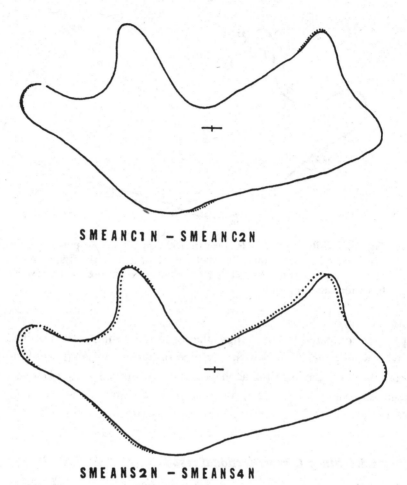

SMEANC1N – SMEANC2N

SMEANS2N – SMEANS4N

Fig. 15.5 Cross-sectional study. SMEANC1N (dotted) — SMEANC2N (solid) represents the superimposition of the mean EFFs generated from the *control* group. SMEANS2N (dotted) — SMEANS4N (solid) represents the superimposition of the mean EFFs generated from the *treatment* group. All outlines have been superimposed on their respective centroids and oriented so that the major axis of the first harmonic ellipse is parallel to the *x*-axis.

alone, as shown here. However, when the treatment groups (SMEANS2N and SMEANS4N) were superimposed on the centroid, a considerable amount of shape change was discerned, especially at the incisor, condyle, and gonial areas. In order to further localize these areas of morphometric change arising from FA therapy, the mean EFFs (SMEANS2N and SMEANS4N) were superimposed on different regions such as the condyles. This illustrates a clockwise rotation of the body and coronoid process after the treatment, implying that the condyle has tipped posteriorly relative to the rest of mandible. Moreover, the magnitude of the rotation at the body was much greater than that at the coronoid process, indicating that there may have been changes in other areas as well. When the EFFs were superimposed at the lower border of mandibles, the whole ramus was tipped posteriorly relative to the body of mandible. There was also a considerable amount of the incisor movement in an anterior direction.

15.5.2 Study II: longitudinal results

In the longitudinal study, the shape changes during the control period were compared to the changes within the *same* individuals during the subsequent treatment period. The shape change during the control period was calculated by comparing the EFFs derived from the two sets of radiographs taken at "time-one" and "time-two." The changes during the treatment period were calculated by comparing the EFFs derived from the radiographs taken at "time-two" and "time-three." The FA therapy began immediately after the "time-two" and terminated right before the "time-three" period. Thus, means (each with $n = 13$) of the three sets of mandibles were obtained at "time-one", "time-two", and "time-three."

Figure 15.6, top, displays the mean EFFs of "time-one" and "time-two" (LMEANC1N and LMEANC2N) superimposed on centroid.

Again, no appreciable amount of shape change has occurred with growth alone during this control period. However, when the mean EFFs of the "time-two" and "time-three" groups (LMEANC2N and LMEANS4N) were superimposed on the centroid (Figure 15.6, bottom), considerable change was discerned at the incisor, condyle, and gonial areas. When the LMEANC2N and LMEANS4N groups were superimposed at the condyles a posterior tipping of the condylar neck was apparent. When the EFFs were superimposed at the lower border of the mandible the ramus was tipped posteriorly relative to the body of mandible, and a considerable amount of the incisor movement in an anterior direction was noted.

There were no appreciable shape changes in the mandible during the control period (between LMEANC1N and LMEANC2N), but a considerable amount of change was present during FA therapy (i.e., between LMEANC2N and LMEANS4N), results identical to those in the cross-sectional study. Thus, the longitudinal study results were consistent with those obtained from the cross-sectional study. The major changes were a forward movement of incisor, posterior tipping of condylar neck and posterior tipping of ramus away from the body of mandible (or opening of gonial angle).

15.5.3 Localization of the areas with shape change

While the shape changes due to FA treatment are readily visualized for both the cross-sectional and the longitudinal studies (Figures 15.5 and 15.6), there remains the issue of whether these visual differences are statistically significant. To ad-

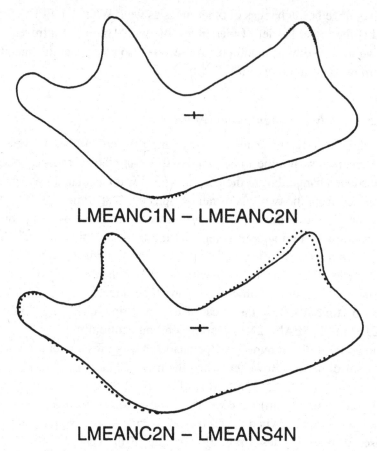

LMEANC1N – LMEANC2N

LMEANC2N – LMEANS4N

Fig. 15.6 Longitudinal study. LMEANC1N (dotted) — LMEANC2N (solid) represents the superimposition of the mean EFFs generated from the initial and progress radiographs (control period). LMEANC2N (dotted) — LMEANS4N (solid) represents the superimposition of the mean EFFs generated from the progress and final radiographs (treatment period). All outlines have been superimposed on their respective centroids and oriented so that the major axis of the first harmonic ellipse is parallel to the x-axis.

dress this question, the distances from the centroid to the outline of the mean mandibular border between superimposed pairs of EFFs were averaged for ten localized areas that described the major structural components of the mandible (Figure 15.4). These were: (1) the incisal aspect, points 59–77; (2) the alveolar border, 41–59; (3) the anterior ramus, 32–41; (4) the coronoid process, 17–32; (5) the sigmoid notch, 5–17; (6) the condyle head, 141–5; (7) the posterior ramus, 131–141; (8) gonial aspect, 114–131; (9) the inferior mandibular border, 90–114; and (10) the chin, 77–90. Although point homology is only approximate, the set of 150 points are sufficiently closely spaced, and as there is minimum visual difference between samples, one can consider that structural homology as being largely maintained.

These ten sets of distances from the centroid were then averaged over the samples for each group in the two studies. One-way MANOVAS were then computed to compare the distances between the treatment and control groups for both the cross-sectional and longitudinal studies.

Turning to the cross-sectional results first, statistically significant shape differences were found (Wilk's Lambda = 0.256, p < 0.001). Univariate F-tests were significant for six of the ten localized regions. These were: (1) incisal aspect, p < 0.001; (2) the sygmoid notch, p < 0.005; (3) condyle head, p < 0.01; (4) gonial process, p < 0.001; (5) inferior mandibular border, p < 0.001 and (6) the chin, p < 0.01. For the longitudinal study, using a repeated-measures design, statistically significant shape differences were also found (Wilk's Lambda = 0.264, p < 0.01). Univariate F-tests were significant for three of the ten localized areas: (1) the incisal aspect, p < 0.05; (2) the condyle head, p < 0.05; and (3) the inferior mandibular border, p < 0.05. For both the cross-sectional as well as the longitudinal data, the treatment sample values were larger than the controls. These statistical results reenforce the visual display of the mandibular differences (Figures 15.5 and 15.6).

These results indicate that there were appreciable shape changes in the mandible with FA therapy. Moreover, these shape changes were primarily localized to the incisor, condyle, and mandibular inferior border margins. These changes resulted in an anterior movement of the incisor, a posterior tipping of condyle, and opening of the gonial angle. These changes were statistically significant in all three regions for both the cross-sectional and longitudinal studies.

15.6 Discussion and conclusions

Mandibular growth is clearly sex- and age-dependent (Carter, 1987; Buschang et al., 1988; 1986; Bishara et al., 1981) as well as related to the facial patterning involving both the sagittal and vertical planes (Buschang et al., 1986; Harris, 1962). Thus, treatment samples and controls must be matched as closely as possible for these variables in order to reduce variation. Since the response to FA therapy is presumed to be dependent on mandibular growth, the above variables need to be carefully controlled. The relative magnitude of the observation period in the treatment and control groups may also be critical for an understanding of the changes occurring with FA treatment. The treatment and control periods were comparable in the longitudinal study (Table 15.2); however, the observation period for the control group was shorter than that for the treatment group in the cross-sectional study (Table 15.1). Since all EFFs were standardized for size in the present studies, any differences in size between treatment and controls were largely eliminated. Thus, only shape changes were present for comparison. There was only a small amount of shape change observable in the cross-sectional control group, whereas the changes observed in the corresponding treatment group were significant. These results were independently supported by the longitudinal study of comparable duration.

The control subjects in the cross-sectional study, and the subjects during the control period in the longitudinal study, maintained their original position and function of the jaws. Thus, any changes observed in these two groups could be attributed to the natural remodeling processes occurring with growth. In the present study, it was demonstrated that the control group mandibles grew largely *without* changes in shape, maintaining the original jaw function. The stability of the original shape during the growth interval was quite striking.

It is logical to assume that with FA therapy there should be a considerable amount of dentoskeletal remodeling and subsequent reshaping of the form of the jaws. As a patient functions with the mandible postured forward, a tension in the posterior direction would be applied to the head of condyle by the stretched capsular ligaments, which may explain the posterior tipping of condyle (or remodeling of condylar head) with FA therapy as demonstrated in the present studies. In this new environment, the masseter, medial pterygoid, temporalis and digastric muscles are also stretched (Figure 15.1). A recent photoelastic study (Ishioka, 1991), using the bionator and activator, also indicated that the stress on the mandible from the muscle attachments was most pronounced at the condylar neck and gonial aspect, also supporting the results of these studies.

A widely held assumption has been that FA therapy temporarily accelerates the growth of the mandible, and when the treatment is terminated, the growth rate of the mandible slows down. However, the results from the present study suggest that rather than an increase in mandibular growth rates, mandibular *shape* changes are the primary effect of FA therapy. Consequently, while these shape changes with FA therapy seem to *imply an increase in the **linear length** of the mandible, in actuality the actual size of the mandibles did **not** change* (Figure 15.7).

That is, the posterior tipping of condylar neck and an opening of gonial angle has the *apparent* effect of increasing a linear distance, like Condyle-Po, without actually lengthening the overall size of the mandible. Thus, when such a linear measurement is used for calculating the growth rate, the presence of shape changes as described above could be misinterpreted as a positive growth rate change. Furthermore, if the shape changes found in this investigation are not permanent, but subject to relapse after FA therapy, this may explain the relapse found in mandibular length (DeVincenzo, 1991). This raises the question of how this relapse contributes to the post-FA response, and how can one prevent it? Clearly,

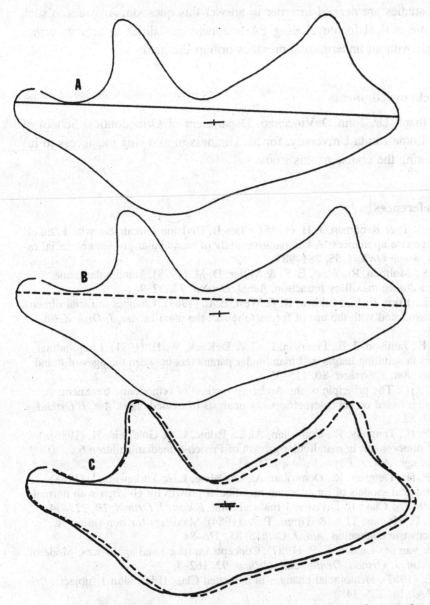

Fig. 15.7 Increase in mandibular length, based only on shape changes. (a) SMEANS2N (before treatment). (b) SMEANS4N (after treatment). (c) Superimposition of the two, SMEANS2N (solid) and SMEANS4N (dotted). Superimposition here is on the condyle head oriented on a line drawn from the condyle head to gnathion.

follow-up studies are needed in order to answer this question. Answers to such questions are critical for developing FA treatment modalities to achieve orthodontic goals without undermining previous orthopedic gains.

Acknowledgments

I wish to thank Dr. John DeVincenzo, Department of Orthodontics, School of Dentistry, Loma Linda University, for his kindness in allowing me access to his FA data during the course of this study.

References

Adenwella, S. T. & Kronman, J. H. (1985). Class II, Division 1 treatment with Frankel and edgewise appliances: A comparative study of mandibular growth and facial esthetics. *Angle Orthod.* **55,** 281–98.

Baumrind, S., Molthen, R., West, E. E. & Miller, D. M. (1978). Mandibular plane changes during maxillary retraction. *Am. J. Orthod.* **74,** 32–9.

Baumrind, S., Korn, E. L., Molthen, R. & West, E. E. (1981). Changes in facial dimensions associated with the use of forces to retract the maxilla. *Am. J. Orthod.* **80,** 17–30.

Bishara, S. E., Jamison, J. E., Peterson, L. C. & DeKock, W. H. (1981). Longitudinal changes in standing height and mandibular parameters between the ages of 8 and 17 years. *Am. J. Orthod.* **80,** 115–35.

Bjork, A. (1951). The principle of the Andresen method of orthodontic treatment: A discussion based on cephalometric x-ray analysis of treated cases. *Am. J. Orthod.* **37,** 437–58.

Buschang, P. H., Tanguay, R., Demirjian, A., La Palme, L. & Goldstein, H. (1986). Sexual dimorphism in mandibular growth of French-Canadian children 6 to 10 years of age. *Am. J. Phys. Anthrop.* **71,** 33–7.

Buschang, P. H., Tanguay, R., Demirjian, A., La Palme, L. & Turkewicz, J. (1988). Mathematical models of longitudinal mandibular growth for children with normal and untreated Class II, Division 1 malocclusion. *Europ. J. Orthod.* **10,** 227–34.

Caldwell, S. F., Hymas, T. A. & Timm, T. A. (1984). Maxillary traction splint: A cephalometric evaluation. *Am. J. Orthod.* **85,** 376–84.

Carels, C. & van der Linden, F. P. (1987). Concepts on functional appliances: Mode of action. *Am. J. Orthod. Dentofacial. Orthop.* **92,** 162–8.

Carter, N. E. (1987). Dentofacial changes in untreated Class II Division 1 subjects. *Brit. J. Orthod.* 14, 225–34.

Cohen, A. M. (1983). Skeletal changes during treatment of Class II/I malocclusions. *Brit. J. Orthod.* **10,** 147–53.

DeVincenzo, J. P. (1991). Changes in mandibular length before, during, and after successful orthopedic correction of Class II malocclusions, using a functional appliance. *Am. J. Orthod. Dentofac. Orthop.* **99,** 241–57.

DeVincenzo, J. P., Huffer, R. A. & Winn, M. W. (1987). A study in human subjects using a new device designed to mimic the protrusive functional appliances used previously in monkeys. *Am. J. Orthod. Dentofac. Orthop.* **91,** 213–24.

DeVincenzo, J. P. & Winn, M. W. (1987). Maxillary incisor intrusion and facial growth. *Angle Orthod.* **57,** 279–89.

DeVincenzo, J. P. & Winn, M. W. (1989). Orthopedic and orthodontic effects resulting from the use of a functional appliance with sufficient amount of protrusive activation. *Am. J. Orthod. Dentofac. Orthop.* **96**, 181–90.

Eirew, H. L. (1976). The functional regulator of Frankel. *Brit. J. Orthod.* **3**, 67–74.

Eirew, H. L. (1981). The bionator. *Brit. J. Orthod.* **8**, 33–6.

Enlow, D. H. (1975). *Handbook of Facial Growth.* Philadelphia: W. B. Saunders.

Forsberg, C. M. & Odenrick, L. (1981). Skeletal and soft tissue response to activator treatment. *Europ. J. Orthod.* **3**, 247–53.

Frankel, R. (1984). Concerning recent articles on Frankel appliance therapy. *Europ. J. Orthod.* **85**, 441–4.

Gianelly, A. G., Arena, S. A. & Bernstein, L. (1984). A comparison of Class II treatment changes noted with light wire, edgewise, and Frankel appliances. *Am. J. Orthod.* **86**, 269–76.

Gianelly, A. G., Brosnan, P., Martignoni, M. & Bernstein, L. (1983). Mandibular growth, condyle, position, & Frankel appliance therapy. *Angle Orthod.* **53**, 131–42.

Graber, T. M. (1966). *Orthodontics: Principles and Practice.* Philadelphia: W. B. Saunders.

Graber, T. M. & Neuman, B. (1975). *Removable Orthodontic Appliances.* Philadelphia: W. B. Saunders.

Hamilton, S. D., Sinclair, P. M. & Hamilton, R. H. (1987). A cephalometric, tomographic, and dental cast evaluation of Frankel therapy. *Am. J. Orthod. Dentofac. Orthop.* **92**, 427–36.

Harris, J. E. (1962). A cephalometric analysis of mandibular growth rate. *Am. J. Orthod.* **48**, 161–74.

Harvold, E. P. & Vargervik, K. S. (1971). Morphogenetic response to activator treatment. *Am. J. Orthod.* **60**, 478–90.

Haynes, S. (1986). A cephalometric study of mandibular changes in modified function regulator (Frankel) treatment. *Am. J. Orthod. Dentofac. Orthop.* **90**, 308–20.

Hiniker, J. J. & Ramfjord, S. P. (1966). Anterior displacement of the mandible in adult rhesus monkeys. *J. Prosthet. Dent.* **16**, 503–12.

Ishioka, S. (1991). Effects of functional appliances on craniofacial structures: Photoelastic analysis. Master thesis, UCLA.

Jakobsson, S. O. (1967). Cephalometric evaluation of treatment effect on Class II, Division 1 malocclusions. *Am. J. Orthod.* **53**, 446–55.

Joho, J. P. (1968). Change in form and size of the mandible in the orthopedically treated *Macaca irus. Europ. Orthod. Soc. Rep. Congr.* **44**, 161–73.

Knight, H. (1988). The effects of three methods of orthodontic appliance therapy on some commonly used cephalometric angular variables. *Am. J. Orthod. Dentofac. Orthop.* **93**, 237–44.

Kuhl, F. P. & Giardina, C. R. (1982). Elliptic Fourier features of a closed contour. *Comp Graph. Imag. Proc.* **18**, 236–58.

Lestrel, P. E. (1989a). Method for analyzing complex two-dimensional forms: Elliptical Fourier functions. *Am. J. Hum. Biol.* **1**, 149–64.

Lestrel, P. E. (1989b). Some approaches toward the mathematical modeling of the craniofacial complex. *J. Craniofac. Genet. Dev. Biol.* **9**, 77–91.

Lestrel, P. E. & Kerr, W. J. S. (1992). Shape changes due to functional appliances. *Calif. Dent. J.* **20**, 30–6.

McNamara, J. A. Jr. (1985). The role of functional appliances in contemporary orthodontics. In *New Vistas in Orthodontics*, ed. L. E. Johnston. Philadelphia: Lee & Febiger.

McNamara, J. A. Jr. (1986). On the possibility of stimulating mandibular growth. In

358   *Won Moon*

Orthodontics: State of Art, Essence of the Science, ed. L. W. Graber. St. Louis: The CV Mosby Company.

McNamara, J. A. Jr., Bookstein, F. L. & Shaughnessy, T. G. (1985). Skeletal and dental changes following functional regulator therapy on Class II patients. *Am. J. Orthod.* **88,** 91–110.

Meach, C. L. (1966). A cephalometric comparison of bony profile changes in Class II, Division 1 patients treated with extraoral force and functional jaw orthopedics. *Am. J. Orthod.* **52,** 353–70.

Nielsen, I. L. (1984). Facial growth during treatment with the function regulator appliance. *Am. J. Orthod.* **85,** 401–10.

Pancherz, H. (1984). A cephalometric analysis of skeletal and dental changes contributing to Class II correction in activator treatment. *Am. J. Orthod.* **85,** 125–34.

Pancherz, H. (1985). The Herbst appliance: Its biological effects and clinical use. *Am. J. Orthod.* **87,** 1–29.

Remmer, K. R., Mamandras, A. H., Hunter, W. S. & Way, D. C. (1985). Cephalometric changes associated with treatment using the activator, the Frankel appliance, and the fixed appliance. *Am. J. Orthod.* **88,** 363–72.

Ricci, A. J. (1983). Twenty years of functional appliances. *J. Pedodont.* **7,** 161–83.

Righellis, E. G. (1983). Treatment effects of Frankel, activator and extraoral traction appliances. *Angle Orthod.* **53,** 107–21.

Robertson, N. R. (1983). An examination of treatment changes in children treated with the function regulator of Frankel. *Am. J. Orthod.* **83,** 299–310.

Vargervic, K. & Harvold, E. P. (1985). Response to activator treatment in Class II malocclusions. *Am. J. Orthod.* **88,** 242–51.

Wieslander, L. (1984). Intensive treatment of severe Class II malocclusions with a headgear-Herbst appliance in the early mixed dentition. *Am. J. Orthod.* **86,** 1–13.

Woodside, D. G., Metaxas, A. & Altuna, G. (1987). The influence of functional appliance therapy on glenoid fossa remodeling. *Am. J. Orthod Dentofac. Orthop.* **92,** 181–98.

16

Size and Shape of the Rabbit Orbit: 3-D Fourier Descriptors

PETE E. LESTREL, DWIGHT W. READ AND CHARLES WOLFE

UCLA School of Dentistry
Department of Anthropology, UCLA.
Wolfe Consulting Software Engineers, Sylmar, California

The latest authors, like the most ancient, strove to subordinate the phenomena of nature to the laws of mathematics.
Sir Isaac Newton (1642–1727)

16.1 Introduction

The form of the vertebrate skull is dependent, in part, on the size and shape of the orbit and its contents. The orbit has been an object of interest to surgeons dealing with the growth of the human eye and its surrounding structures (Sarnat, 1981). Experiments have demonstrated that removal of the orbital contents slows orbital growth as well as affecting other regions of the skull (Sarnat and Shanedling, 1970; 1972). The orbit is bounded superiorly by the supraorbital process attached to the frontal bone and inferiorly by the zygomatic arch. It represents a morphological structure that is not planar, extending significantly into all three planes, and requiring that the orbital margin be modeled as a curve in 3-space (Figure 16.1).

Irregular morphological forms such as the orbital margin are particularly difficult to characterize with conventional numerical methods, hence the lack of information on orbital size and shape changes with either growth or treatment.

Elliptical Fourier functions (EFFs) were chosen to numerically describe the shape changes in the growing orbital margin. This is a curve-fitting procedure based on a converging trigonometric series. The specific descriptor utilized was developed by Kuhl and Giardina (1982), as a *parametric two-dimensional* solution; that is, a pair of equations as functions of a third variable. The coordinates (x,y) of the predicted points are given in the form $[x(t), y(t)]$, $0 \leq t \leq T$ (T usually taken as 2π) and the parametric function $x(t)$, $y(t)$ approximated by a finite Fourier series (see Chapter 2). This approach has been extended here to allow for the third dimension, or z-axis. Any outline, whether in 2- or 3-D, can be approximated with this function. However, it must be reiterated that this 3-D extension only allows for the numerical description of a *curve in 3-space* and not a solid (i.e., containing volume).

359

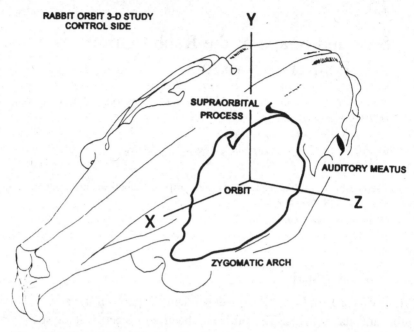

Fig. 16.1 The rabbit orbit visualized as a three-quarter view showing the orbital margin and its orientation in the Cartesian coordinate system.

The Kuhl and Giardina formulations can be extended without difficulty allowing a form to be modeled as a curve in 3-space (Lestrel et al., 1993; Chapter 2). The finite elliptical Fourier series in $x(t)$ is written as

$$x(t) = A_0 + \sum_{n=1}^{N} a_n \cos nt + \sum_{n=1}^{N} b_n \sin nt, \tag{16.1}$$

in $y(t)$ as

$$y(t) = C_0 + \sum_{n=1}^{N} c_n \cos nt + \sum_{n=1}^{N} d_n \sin nt, \tag{16.2}$$

while the series in $z(t)$ is now

$$z(t) = E_0 + \sum_{n=1}^{N} e_n \cos nt + \sum_{n=1}^{N} f_n \sin nt, \tag{16.3}$$

where n equals the harmonic number, N equals the maximum harmonic number and the interval is over 2π.

Any 2- or 3-D outline can be approximated with a polygon which is constructed by connecting the points with lines. Note that the distances between the points defining the vertices of the polygon can now vary in length, which is a decided advantage over conventional Fourier descriptors where equal intervals are generally used to avoid a weighted analysis. The extension to 3-space can be viewed as an orthogonal system in three axes. The polygon that serves as the form will be represented in a spatial domain which is sinusoidal in *x*, *y*, and *z*.

This study was undertaken with three aims in mind: (1) To numerically describe the form of the rabbit orbital margin in 3-space; (2) to document the morphological changes that are present during normal growth; and (3) by controlling for size differences, allow for an independent assessment of changes in shape.

16.2 Materials

A sample of rabbit skulls used in previous studies (Sarnat, 1981) was available through the kindness of Dr. Bernard Sarnat, UCLA School of Dentistry. This cross-sectional study focused on growth of the orbit from infancy to adulthood. These specimens were subjected to various treatment modalities with the right orbit receiving experimental treatment and the left orbit serving as its control[1] (Osman, 1993; Okamoto, 1994). This study is an investigation of size and shape changes of the orbital margin utilizing the untreated control side.[2] Specimens were subdivided into infants (*n* = 14) of mean age in months of 26.43 ± 11.75; juveniles (*n* = 9) mean age of 94.56 ± 33.57 and adults (*n* = 21) of mean age of 210.86 ± 68.85.

16.3 Methodology

Although the extension to 3-space is seemingly straightforward, a number of challenging technical issues arose. These issues revolve around the computation of area and the centroid of a form. The area was needed for scaling purposes to control for size, and the centroid location was used for the superimposition of the boundary outlines. These two aspects cannot be precisely described in numerical terms with the commonly used methods available. This problem will be made clearer in Section 16.3.10.

16.3.1 Preparation of specimens

A bounded curve in 3-space is described with triplets; that is, *x*, *y*, and *z* coordinates for each sampled point. This is attained with the creation of two views, one in the *x*-*y* plane and one in the *x*-*z* plane, from which a curve in 3-space can be constructed. Additionally, the third plane, the *y*-*z*, is computer-generated.

---

[1] The right orbit was subjected to a number of experimental procedures: (1) evisceration, the removal of intraocular contents but leaving the scleral shell intact; (2)] enucleation, the severance of the eye muscles and optic nerve and removing the eyeball; and (3) exenteration, the removal of the entire orbital contents. Comparison of these different experimental treatments will be the subject of a future publication.

[2] A question that arises is whether the treated side influenced the orbital form on the untreated side. Such differences were not visually apparent and if they existed, must be small in magnitude.

Each specimen was carefully placed in a special jig,[3] which held the rabbit skull in a standardized position that allowed rotation along the long axis of the skull. The skull was oriented with the left orbit (control side) toward the camera (snout facing to the left), and adjusted so that a horizontal plane passed through *prosthion* (defined as the alveolar margin at the central incisor) and the inferior aspect of the *auditory meatus*. The choice of the inferior border of the *auditory meatus* was dictated by the fact that the normally used superior aspect was often obstructed. This orientation became the lateral view in the *x-y* plane. The specimen was then rotated 90 degrees, so that the top of the skull was facing the camera.[4] This orientation became the superior view in the *x-z* plane.

Precise orientation was maintained with a fixed thin horizontal black wire placed between the specimen and the camera. This wire was used to orient the skull by superimposing it so that it coincided with *prosthion* anteriorly, and the inferior aspect of the *auditory meatus*, posteriorly. Finally, a 10-cm plastic ruler with 1-mm divisions was attached to the holding jig at the midline (along the long axis of the skull) for scaling purposes.

16.3.2 Photography

The orbital margin of each specimen was carefully marked with a fine black felt pen so that the outline was more clearly seen on the photograph. A Pentax SF1N 35 mm camera with a 200 mm telephoto lens was used. The fixed film-subject distance was 116 mm which resulted in an image magnification of 0.25. Each specimen was photographed twice, generating a lateral view and a superior view. Exposure times were 1/15 sec at f 9.5 using Ilford film (ASA 50) with two 500 W flood lamps. The two negatives were enlarged 2.8x onto 8x10 resin-coated paper[5] to maintain dimensional stability of the image. The size of the image was carefully controlled during the enlarging process (using the 10 cm plastic ruler which was visible in each negative) to insure that the lateral and superior views were scaled the same.

16.3.3 Tracing procedures

Acetate tracings[6] of the orbital outlines were then prepared of each view. Each photograph was carefully traced using a 0.3 mm lead mechanical pencil. Briefly, the tracing procedure involved the following steps: two registration crosses were

---

[3] A jig for holding small specimens with a micrometer adjustment, was made available through the generosity of Dr. Tom James, UCLA Department of Zoology.

[4] Photographs of the rabbit specimens and the specimen holding jig are illustrated in Okamoto (1994).

[5] Kodak Polycontrast III RC, medium weight paper.

[6] Keuffel and Esser (K & E), Herculene matte drafting film, 2 mil thickness, product number 19-1152 C3, Rockaway, NJ 07866.

placed directly on the photograph to insure repositioning of the tracings. Tracings were then taped to each photograph and the crosses traced onto the acetate sheets. A line was traced through the center of the black plastic wire to define the *x*-axis. The zygomatic arch was then traced from its most superior surface starting at the anterior aspect (orbital process of the maxilla), to the posterior aspect terminating at the zygomatic process of the squamosal. The outline was continued along the anterosuperior aspect of the supraorbital process of the frontal bone and finally followed the orbit outline posteriorly along the temporal fossa until it rejoined the zygomatic arch (Figures 16.2 and 16.3). This procedure was carried out for both views yielding a lateral and superior view of the orbit for the control side.

Starting the lateral view (*x*-*y* plane), two vertical lines were then constructed as tangents at 90 degrees to the *x*-axis. One tangent was constructed at the most anterior aspect and the other at the most posterior margin of the orbit. These lines were extended upward. The completed lateral tracing (*x*-*y* plane) was placed below the completed corresponding superior view (*x*-*z* plane) and carefully oriented until the vertical lines of the two tracings each were made coincident.

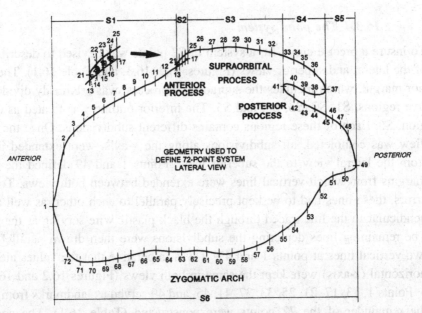

Fig. 16.2 The 72 points used to describe the lateral or *x*-*y* view of the orbit. This view is oriented parallel to a horizontal plane drawn through *prosthion* and the inferior margin of the *auditory meatus*. Reference should be made to Table 16.2 and the text for a description of the point locations.

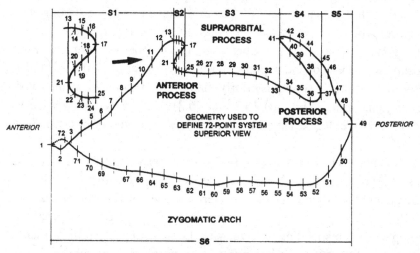

Fig. 16.3 The 72 points used to describe the superior or *x-z* view of the orbit. This view is oriented parallel to a horizontal plane drawn through *prosthion* and the inferior margin of the *auditory meatus*. Reference should be made to Table 16.2 and the text for a description of the point locations.

16.3.4 The point system

To insure a precise curve-fit, a system of 72 points was devised to describe each of the lateral and superior views (Figures 16.2, 16.3, and Table 16.1). The superior margin, which includes the supraorbital process, was arbitrarily divided into five regions, S1, S2, S3, S4, and S5. The inferior outline was treated as one region, S6. Each of these regions contains different subdivisions. Once the lateral view was completed, all subdivisions along the *x*-axis, were extended upward from the lateral view to the superior view. Points 1 and 49 defined the orbital margins from which vertical lines were extended between both views. To avoid errors, these lines had to be kept precisely parallel to each other, as well as perpendicular to the line traced through the black plastic wire serving as the *x*-axis. The remaining lines depicting the subdivisions were then drawn parallel to the two vertical lines at points 1 and 49. It was imperative that the values along the horizontal (*x*-axis) were kept the same in both views (Figures 16.2 and 16.3).

Points 1, 13, 17, 21, 25, 33, 37, 41, 45, and 49 served as landmarks from which the remainder of the 72 points were constructed (Table 16.1). The complete 72-point system was largely derived from bisections. The three columns on the left of Table 16.1 defined the line segments that were bisected, and the three columns on the right represent the bisected points. The asterisks refer to line seg-

Table 16.1 Development of the 72 point system used to describe the orbital margin. The designations S1 through S6 refer to the divisions illustrated in Figures 16.2 and 16.3 (see text).

Rabbit Orbit Point System					
Points defining line segments			Generated points		
S2	S4	S5	S2	S4	S5
13–17	33–37	45–49	15	35	47
13–15	33–35	45–47	14	34	46
15–17	35–37	47–49	16	36	48
17–21	37–41	—	19	39	—
17–19	37–39	—	18	30	—
19–21	39–41	—	20	40	—
21–25	41–45	—	23	43	—
21–23	41–43	—	22	42	—
23–25	43–45	—	24	44	—
S1	S3	S6	S1	S3	S6
1–13	25–33	49–1	7	29	61
1–7	25–29	49–61	4	27	55
1–4*	25–27	49–55	2,3*	26	52
4–7*	27–29	59–52*	5,6*	28	50,51*
7–13	29–33	52–55*	10	31	53,54*
7–10*	29–31	55–61	8,9*	30	58
10–13*	31–33	55–58*	11,12*	32	56,57*
—	—	58–61*	—	—	59,60*
—	—	61–1	—	—	67
—	—	61–64*	—	—	62,63*
—	—	64–67*	—	—	65,66*
—	—	67–1	—	—	70
—	—	67–70*	—	—	68,69*
—	—	70–1*	—	—	71,72*

ments that were trisected, rather than bisected, yielding two points shown on the columns on the right.

16.3.5 Digitization procedures

Once the pair of tracings had been completed, each one was separately digitized using a HIPAD digitizer.[7,8] The center of the digitizer was used as the origin and identified with two fine lines running horizontally and vertically the whole length

[7] See Wolfe et al., (1995) for software description and user's manual. Available from the authors.
[8] Houston Instruments, 8500 Cameron Road, Austin TX 78753.

of the digitizer surface. The tracing of the lateral view was then oriented so that the vertical line constructed earlier through point 1 was aligned with the vertical line inscribed on the digitizer surface. The 72 points of the lateral view were then digitized in order. The superior view was then slid down along the constructed vertical line through point 1, and its 72 points digitized in turn.

Software was written to insure that none of the 72 x-coordinates on the lateral view exceeded their counterparts in the superior view by more than 0.5 mm. Finally, as a check for systematic digitizing errors, a plot was created using a HP 7470A 2-pen plotter[9] and compared visually with the original tracing.

16.3.6 Computation of EFFs

Coordinates from both views, lateral and superior, were then used to generate the y-z plane or posterior-anterior (P-A) view. At this stage the digitized x, y and x, z coordinates generated the x, y, z triplets for the 72-point system. Custom software, written using Power Basic Version 3.0c[10] then computed EFFs of the orbital margin as a curve in 3-space. From the 72 triplets, EFFs were computed with 36 harmonics.

16.3.7 Goodness-of-fit considerations

The goodness-of-fit of the 3-D Fourier descriptor was determined from the mean residual or difference between the actual data and the predicted fit from the function. These 36 harmonics were deemed sufficient as measured by the residual between the observed data points and the predicted values derived from the function. Values averaged over the whole orbital margin, and scaled to reflect actual size, were found to be 0.07 ± 0.02 mm ($n = 44$). These values were deemed satisfactory as they are well below the tracing and digitizing errors.

16.3.8 Maintenance of homology

Bookstein (1978; 1991; and elsewhere) has stressed point homology and the requirement of a one-to-one mapping across specimens. Considerable controversy has ensued over definitions of homology, for example, between structural versus point homology, and what constitutes "true" homologous points. These issues are beyond the scope of this chapter (see Chapters 1 and 4). It is readily admitted that the majority of the 72 points used are not true homologous points; nevertheless,

---

[9] Hewlett-Packard, 16399 W. Bernardo Drive, San Diego, CA 92127.
[10] Power Basic, Inc. 316 Mid Valley Center, Carmel, CA 93923.

operationally, these points can be reliably located on each tracing within reasonable error limits. Such points can be thought of as pseudo-homologous.

A procedure was developed to insure that the homology of the 72-point system was maintained across specimens. Since the EFF is a close analog of the observed form, as judged by the small differences between the observed points and the actual (predicted) points, that is, fractions of millimeters, it is feasible to use the predicted values of these points as if they were homologous. This is equivalent to "replacing," so to speak, the observed values with their computed (predicted) counterparts. This can also be viewed as "moving" the observed values onto the EFF curve. Although some loss in fit is inevitable in regions of sharp curvature, this was generally minor; moreover, the loss was quite localized, affecting only small regions of the form. With point homology now largely maintained, an additional set of points (bisections) between the original 72 points was computed, yielding a total of 144 points. This was done to insure a smooth final image for visual purposes. These 144 points were then used to produce "size and shape" as well as "shape only" data as described in the next two sections.

16.3.9 Size and shape data

EFFs were computed to compare *size and shape*, as well as *shape only* changes, as a function of time. It will be recalled that the specimens were originally enlarged 2.8x onto the Kodalith resin-coated sheets and are not actual size. To reduce the size of the rabbit orbital margin to their original dimensions, reduction factors were calculated based on the ratio between the actual length of the skull and the 10 cm scale which was visible in each photograph. These multiplicative "size" factors ranged from 0.3486 to 0.3584. The size of each specimen was then numerically corrected to actual size by multiplying the x, y, and z coordinates with the appropriate reduction factor.

16.3.10 Area and centroid considerations

As the area is required for size-standardization and the centroid for superimposition and comparison of mean forms, these two properties would seem to be straightforward extensions of the 2-D case (see Chapter 2). Unfortunately, as alluded to earlier, this assumption proved to be incorrect. Since the orbital margin is modeled here as a "curve in 3-space," it has no inherent physical area or surface. Theoretically, this situation also applies to irregular figures in 2-space, but such forms are, by definition, embedded into a plane and the computation of area and centroid is relatively straightforward using line integrals. Unfortunately, this approach cannot be extended to curves in 3-space, since an "area" does not, technically, exist. However, in 3-space, the area within the orbital margin can be modeled as a "minimum surface."

The concept of a minimum surface has held considerable fascination for mathematicians and laymen alike. Soap films are one example of minimum surfaces (Boys, 1959; Almgren, 1982; Almgren and Taylor, 1976). The modern interest can be traced to the work of Joseph Plateau (1801–1883), a Belgian physicist who demonstrated that minimization problems could be represented with analog solutions using soap films (Isenberg, 1992). This has led to what has been termed the Plateau problem; namely, whether every closed surface in space can be spanned by at least one minimum surface (Hildebrant and Tromba, 1985). Computationally, this translates into determining the minimum area contained within any closed boundary. Mathematicians dealing with the Plateau problem have now proven that for simple closed curves, that is, those that do not contain intersections, there exists a surface of least area (Hildebrant and Tromba, 1985). The discovery of new minimum surfaces has become an exciting area in mathematics and has led to the development of new computational techniques based on differential geometry and topology.

Consider a thin flexible ring used to blow soap bubbles. If this ring is first distorted by twisting it so that it becomes three-dimensional and then dipped into a soap solution, a smooth surface will always appear. The force that is responsible for the shape of the soap film is surface tension. A soap film consists of a sandwich with two surfaces within which is contained a thin layer of water (or some other liquid). The two surfaces are composed of ions carrying both negative and positive charges, separated by water molecules. Because soap contains both hydrophilic (attracted to water) and hydrophobic (repelled by water) elements, electrostatic forces are set up that create attractive forces which give the film stability. The soap film can vary in thickness from 50 to 200,000 Ångstroms (Isenberg, 1992). The film will be thickest at the moment it is formed and then thinning until it reaches a state of equilibrium. At that point in time, the surface tension will be constant at all points on the surface. It turns out that the soap film is not only seeking to minimize surface tension, but also minimizing the bounded area. Until quite recently, the computation of such minimal surface areas represented major challenges. One solution has been to view this minimization of surface tension in terms of energy. It can be shown that the energy (called free energy) is proportional to the area of the bounded surface. Thus, if, and only if, the soap film has reached equilibrium (surface tension constant at all points) will the free energy be minimized which in turn insures that the area will also be minimized. This approach has led to software called the Surface Evolver[11] which minimizes the sur-

---

[11] We are greatly indebted to Dr. David Hoffman, Dept. of Mathematics, University of Massachusetts, for providing the Evolver program and to its author, Dr. Ken Brakke, of the Geometry Center, University of Minnesota, for kindly adding the code needed to calculate the coordinates for the centroid in 3-space.

face area by using a sophisticated recursive process based on energy considerations (Brakke, 1992).

Thus, the rabbit orbit, treated as a closed curve in 3-space, has no physical or real area, only boundary, and this bounded orbital margin can be visualized and modeled as a minimum surface using the Surface Evolver.

16.3.11 Standardized for size data

To standardize the orbital margin for size, the set of 144 predicted points on the outline, as derived from the EFF, were submitted to the Surface Evolver which returned the bounded area and coordinates for the centroid. These values allowed precise superimposition of the mean infant, juvenile, and adult traces on a common neutral center, the centroid, as well as correction for size by standardization on area. This approach provided for a numerical as well as visual assessment of the "shape only" orbital changes in growth from infant through juvenile to adult.

16.3.12 Computation of selected distances

Once the centroid coordinates were available, 12 distances (D1–D12) from the centroid to the boundary, were computed to characterize selected aspects of the 3-D orbital margin. These distances, in 3-space, were distributed roughly evenly over the orbit, although some emphasis was placed on specific areas of interest such as the supraorbital process (Figure 16.4).

In addition to the 12 distances mentioned above, their components in the x, y, and z directions were also computed. This was predicated on the grounds that form differences could be significant in a component direction in contrast to the actual distance itself (Figure 16.4). This yielded a total set of 48 variables. These 48 variables were then subjected to statistical analysis[12] to remove variables with high correlations[13] (Pearson's $r \sim 0.8$). The reduced data set was then submitted to a MANOVA program to test for size and shape, as well as shape only, differences between the infant, juvenile, and adult samples.

16.4 Results

16.4.1 Size and shape differences

The orbital margin data are illustrated as a curve 3-space in Figure 16.5. Also shown are the lateral, superior, and P-A views, embedded into their respective orthogonal planes. Displayed are the mean curve plots of the infant, juvenile, and

---

[12] StatSoft, 2325 East 13th Street, Tulsa, OK 74104.
[13] The presence of high correlations led to an ill-conditioned matrix (singularity) when the full dataset was submitted to the MANOVA routine.

RABBIT ORBIT DISTANCES

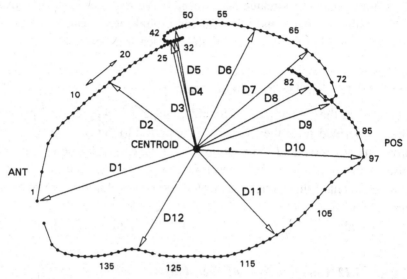

Fig. 16.4 The twelve distances (D1–D12) from the centroid to the 3-D orbital margin. Although for clarity this is the lateral or 2-D view, the distances were computed in 3-space.

adults. All three groups (solid symbols) are shown superimposed on the 3-D centroid.[14] The MANOVA of the reduced set of 11 distances and components (reduced from 48) yielded statistically significant differences between the groups (Wilk's Lambda = 0.056, Rao's R = 9.13, p < 0.001) substantiating the visual impression. To assess where these orbit differences are localized, univariate F-tests were computed (refer to Figure 16.4). One distance and six components were found to be statistically significant. These were: **D1** (p < 0.001), **Y2** (p < 0.001), **Y10** (p < 0.001), **Z2** (p < 0.001), **Z4** (p < 0.001), **Z6** (p < 0.005), and **Z10** (p < 0.02).

It must be emphasized that the statistics using the initial 12 distances (Figure 16.4) are based on the 3-D data; visually, however, the data became more accessible when broken down into the more familiar 2-D orthogonal views. Consequently, these were separately reproduced for each orthogonal view (Figures 16.6, 16.7, and 16.8). For the lateral or *x-y* view (Figure 16.6) pronounced, statistically significant differences in size were present (D1, Y10, etc.), but differences in shape were not clearly apparent due to the overwhelming effect of size. Only in the *x-z* or superior view (Figure 16.7) is it possible to see a rotational

---

[14] Note: it may not be readily apparent from Figure 16.5 that the centroid of the orbital margin in 3-space (solid symbols) cannot be used to compute the 2-D centroids of the three orthogonal views (open symbols). Those centroids must be computed separately.

RABBIT ORBIT - SIZE AND SHAPE

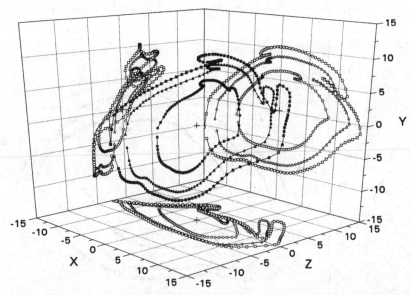

Fig. 16.5 Computer-generated 3-D view of the orbital margin. The three group means: (1) infants (filled circles); (2) juveniles (filled triangles); and (3) adults (filled squares), are superimposed on the 3-D centroid. Also shown are the orthogonal projections in the lateral, superior, and P-A planes. These are identified with open symbols. Both size and shape differences are involved here.

component with growth around the long axis of the rabbit skull. The *y-z* or P-A view (Figure 16.8) yielded a more complex picture with the presence of crossings, or technically, self-intersections. These will be examined in more detail in Section 16.4.3. Since shape differences are of primary interest here, the EFF data were standardized for size differences.

16.4.2 Shape only differences

When the orbital margin was standardized for size by making the bounded area a constant, shape differences could be more clearly displayed. Statistically significant differences between the age groups remained after the effect of differences in size was controlled for by standardizing for area (Wilk's Lambda = 0.053, Rao's R = 6.69, p < 0.001). The univariate F-tests of three distances and six *x*- and *y*-components were found to be statistically significant. These are: **D1** (p < 0.001), **D2** (p < 0.001), **D12** (p < 0.005), **X2** (p < 0.05), **X10** (p < 0.001), **Y10** (p < 0.05), **Z1** (p < 0.005), **Z3** (p < 0.01), and **Z6** (p < 0.001).

LATERAL VIEW - SIZE AND SHAPE

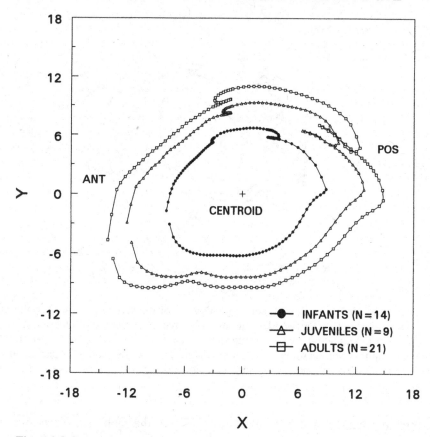

Fig. 16.6 Computer-generated 2-D plot of the lateral or *x-y* view of the three group means. Size and shape data.

The differences in shape are evident in all three orthogonal projections. Focusing on the lateral or *x-y* view (Figure 16.9), the size-standardized data show a significant anterior elongation of the orbit (D1) with a concomitant elongation at the posterior margin (D10, X10, and Y10) as a function of growth. Also notable is the posterior migration of the posterior aspect of the supraorbital process (D7) as well as superior movement of the anterior aspect of the zygomatic margin (D12) as a function of age. A similar anteroposterior pattern is present in the *x-z* or superior view (Figure 16.10). Especially striking is the rotational aspect (also evident in the size and shape data discussed earlier) toward the long axis of the skull from infant to adult. This movement is especially pronounced at the posterior aspect of the supraorbital process (Z3 and Z6). This rotational component

SUPERIOR VIEW - SIZE AND SHAPE

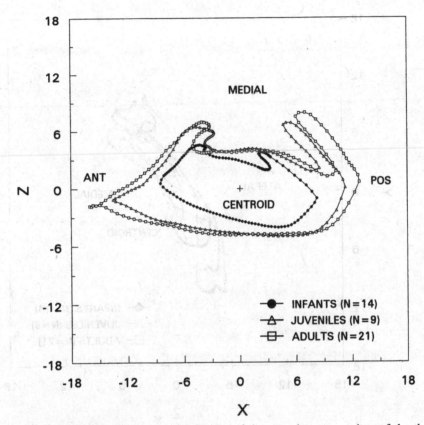

Fig. 16.7 Computer-generated 2-D plot of the superior or *x-z* view of the three group means. Size and shape data.

represents the distal rotation of the anterior aspect of the orbital margin around the central axis of the skull to accommodate the growing eyeball. Finally, the P-A view (Figure 16.11) with its complexity, merits a separate discussion.

16.4.3 *The P-A view*

The P-A projection in the *y-z* plane presented an unexpected picture with the appearance of the "crossings" or intersections.[15] Subsequently, it was discovered that crossings were also present in the *x-y* view (primarily at the supraorbital

---

[15] The presence of crossings as seen in the P-A view of the orbital margin (and in retrospect also in the *x-y* view) vanish if the orbit is viewed as a curve in 3-space. This can be considered as a justification for a 3-D analysis of such forms.

P-A VIEW - SIZE AND SHAPE

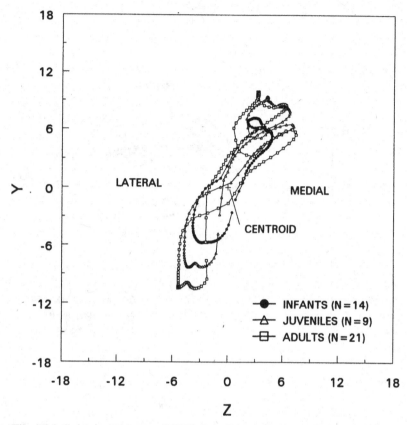

Fig. 16.8 Computer-generated 2-D plot of the A-P or *y-z* view of the three group means. Size and shape data.

process) requiring re-computation of all EFFs. These crossings required special mathematical handling prior to the area and centroid calculations. The procedure consisted of writing a special algorithm which "uncrossed" the two-dimensional data prior to the area and centroid computations (see Appendix). A computer-generated trace (Figure 16.12) depicts the visual changes that the orbital margin underwent as a function of a systematic rotation, starting with the lateral view and ending with the P-A view. The apparent differences in form between infants and adults was primarily due to the more developed supraorbital process in adults. In the infants this difference was expressed with the anterior aspect of the margin "crossing" the posterior aspect earlier (at 75 degrees) compared to the adults (at 83 degrees). In contrast, the adult margin did not display an intersection until the rotation was almost at the full P-A view. This difference between infants and

LATERAL VIEW - SIZE STANDARDIZED

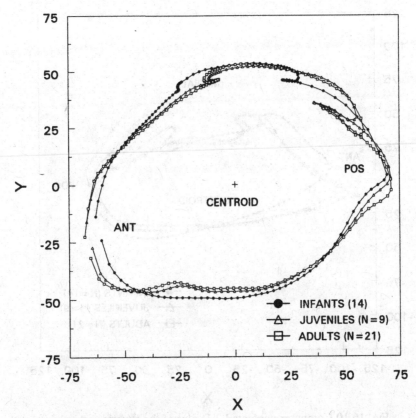

Fig. 16.9 Computer-generated 2-D plot of the lateral or *x-y* view of the three group means. Size-standardized data.

adults can be discerned by following the movement of point 1 as the skull is rotated from 60 to 90 degrees.

16.5 Discussion and conclusions

In conclusion, if we ask which of the 2-D orthogonal projections provides the most information, the answer is not at all apparent, as all three views generate essential data, and all three are necessary for an adequate description of the shape changes that the rabbit orbital margin undergoes during growth. Moreover, the usefulness of the view of the orbital margin in 3-space is also difficult to assess at the moment, because of the customary expectation and familiarity with the readily recognized and accepted orthogonal projections. Work is currently underway

SUPERIOR VIEW - SIZE STANDARDIZED

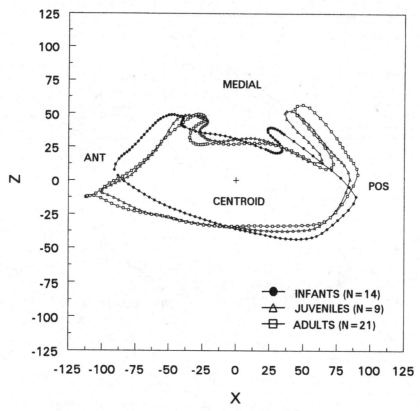

Fig. 16.10 Computer-generated 2-D plot of the superior or *x-z* view of the three group means. Size-standardized data.

to establish the rotational angle(s) which may provide the most meaningful views of the orbit in 3-space. Ultimately, a single view containing all the information present in the three separate orthogonal views is a more efficient descriptor and hence, preferable.

Any irregular outline that can be represented as a curve in 3-space can now be more fully investigated with this morphometric approach. Future studies are contemplated to explore the applicability of EFFs to numerically characterize morphologies in a craniofacial context, including the curve of Spee, masticatory chewing cycles, and so on. Needless to say, few biological structures are actually planar in nature. This work represents the initial attempt to extend Fourier descriptors to visually and numerically characterize morphological shape changes, in the way they *really* present themselves; that is, in three dimensions.

P-A VIEW - SIZE STANDARDIZED

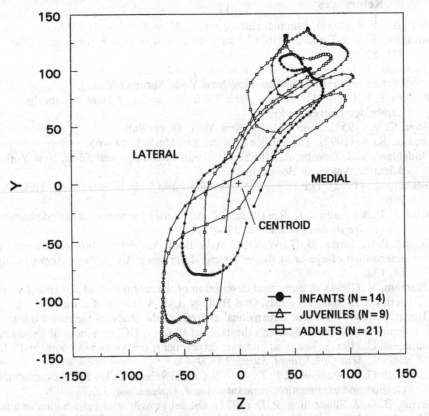

Fig. 16.11 Computer-generated 2-D plot of the A-P or *y-z* view of the three group means. Size-standardized data.

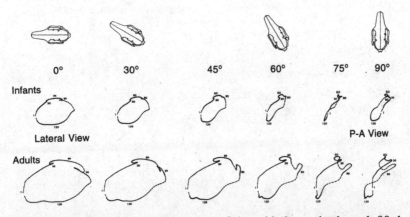

Fig. 16.12 A depiction of the rotation of the orbital margin through 90 degrees from the lateral view to the P-A view (top row) illustrating how the pattern of crossings (intersections) arose in the infant data (middle row) and the adult data (lower row).

377

References

Almgren, F. J. (1982). Minimal surface forms. *Math. Intell.* **4**, 164–72.

Almgren, F. J. & Taylor, J. (1976). The geometry of soap and soap bubbles. *Sci. Amer.* **235**, 82–93.

Bookstein, F. L. (1978). *The Measurement of Biological Shape and Shape Change.* Lecture notes in biomathematics. New York: Springer-Verlag.

Bookstein, F. L. (1991). *Morphometric Tools for Landmark Data.* Cambridge: Cambridge University Press.

Boys, C. V. (1959). *Soap Bubbles.* New York: Dover Pub.

Brakke, K. A. (1992). The surface evolver. *Exp Math.* **1**, 141–65.

Hildebrant, S. & Tromba, A. (1985). *Mathematics and Optimal Form.* New York: Scientific American Books, Inc.

Isenberg, C. (1992). *The Science of Soap Films and Soap Bubbles.* New York: Dover Pub.

Kuhl, F. P. & Giardina, C. R. (1982). Elliptic Fourier features of a closed contour. *Comp. Graph. Imag. Proc.* **18**, 236–58.

Lestrel, P. E., Sarnat, B. G., Wolfe, C. A. & Bodt, A. (1993). Three dimensional characterization of eye orbit shape: Fourier descriptors. *Am. J. Phys. Anthrop. Suppl.* **16**, 134.

Okamoto, V. (1994). A numerical description of the rabbit eye orbit: Control versus experimental. Master's thesis, Oral Biology, UCLA School of Dentistry.

Osman, A. (1993). A three dimensional analysis of the shape of the rabbit orbit: Fourier descriptors. Master's thesis, Oral Biology, UCLA School of Dentistry.

Sarnat, B. G. (1981). The orbit and eye: Experiments on volume in young and adult rabbits. *Acta Ophthalmol. Suppl.* **147**, 9–44.

Sarnat, B. G. & Shanedling, P. D. (1970). Orbital volume following evisceration, enucleation and exenteration in rabbits. *Am. J. Ophthalmol.* **73**, 787–99.

Sarnat, B. G. & Shanedling, P. D. (1972). Orbital growth after evisceration or enucleation without and with implants. *Acta Anat.* (Basel) **82**, 497–511.

Wolfe, C. A., Lestrel, P. E. & Read, D. W. (1995). *EFF23 2-D and 3-D Elliptical Fourier Functions.* Software description and User's manual. PC/MS-DOS Version 2.5.1.

17

From Optical to Computational Fourier Transforms: The Natural History of an Investigation of the Cancellous Structure of Bone

CHARLES E. OXNARD
The University of Western Australia

17.1 Introduction

The remarkable patterns exhibited by cancellous bone, and the likelihood that it is related to mechanical stress and strain, have been known for a long time (e.g., Todd and Bowman, 1845; Wyman, 1867; Wolff, 1892). During most of that period, however, the nature of the cancellous patterns themselves remained based upon visual impressions of bone sections and later, bone radiographs (e.g., Thompson, 1917; Murray, 1936; Evans, 1957). It was not until after the middle of the present century that the idea of characterizing these complex bony patterns with Fourier analyses came about.

In 1968, J. C. Davis showed me how useful optical Fourier transforms could be in the analysis of thin sections of rocks (Davis, 1970). In so doing he put into my mind the idea that the method might be excellent for the study of sectional and radiographic information in bones. Using his specialized equipment, we produced the first transforms of cancellous architecture in sections of human vertebrae (Oxnard, 1970a; 1970b; 1972a; 1972b; 1973). The 1973 volume may have been the first time that a Fourier transform was figured on the cover of a book.

Later H. C. Pincus (e.g., Power and Pincus, 1974) demonstrated to me the Rank Image Analyser 2,000 used for similar purposes. I then applied the technology to the study of radiographs of both human and ape vertebrae (eventually published in Oxnard and Yang, 1981). The findings made it obvious that the approach had promise for elucidating many problems of bone architecture.

However, it was also clear that this promise had to lie alongside the use of biomechanical techniques in the elucidation of the biological underpinnings, the functional implications, of the anatomical images being studied (e.g., Oxnard, 1973).

17.2 Optical Fourier transforms of sections of vertebrae

The earliest of these Fourier transform investigations demonstrate just how useful and sensitive the technique is for studying bone architecture.

379

First, the cancellous structure of the second human lumbar vertebra could be clearly distinguished from that of the fourth. Yet the visual patterns of these two vertebrae are so similar that most anatomists would find it hard to distinguish them, except on the basis of their external form (Figure 17.1).

Second, the particular form of the transform of the human second lumbar vertebrae differs from that of the fourth because it indicated the presence of a preferential orientation at about 60 degrees to the horizontal. While not at all obvious in

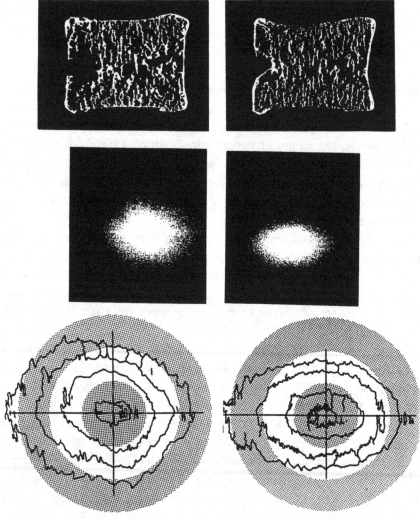

Fig. 17.1 *Above*: sagittal sections of second and fourth human lumbar vertebrae. *Middle*: Fourier transforms of each. *Below*: densitometry of Fourier transforms (expansion of densitometry plots to the left is an artifact of the early television system used in the display). The orientational differences between transforms and the densitometric displays are much more obvious than in the original sections.

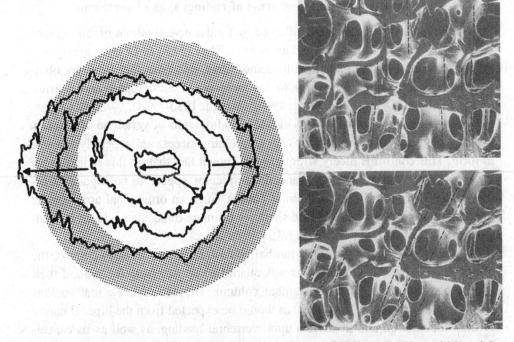

Fig. 17.2 *Left*: densitometry for second lumbar vertebral transform with special 30 degree to the horizontal left components indicated by an arrow. *Right above*: electron microscopy of second lumbar vertebrae section. *Left*: vertical and horizontal elements. *Right below*: approximately 30 degree to the vertical right elements.

the section, it is evident in the transform (Figure 17.1) and even more readily visible in the densitometric analysis (Figure 17.2 — left frame). Yet it can be distinguished in the original through the use of scanning electron micrographs (taken in the same plane). Thus, the 60 degree to the horizontal information in the transform seems to relate to the 30 degree to the vertical information in the radiograph (Oxnard and Yang, 1981; Oxnard, 1983/1984; Figure 17.2 — right frames).

Functional implications of this finding were posited (e.g., Oxnard, 1973). The particular angle of the additional elements was similar to the angle of flying buttresses in cathedrals. It seemed likely that their presence might relate to lateral support for longer vertical trabeculae. This could be because of the eccentric loading schedule of the second lumbar vertebrae. This vertebra, in comparison with the fourth, lies more posteriorly offset from the line of gravity in humans. The more centric loading schedules that presumably act, on average, in the fourth lumbar vertebra because of its closer alignment to the line of gravity in bipedal creatures such as humans, may not require such extra support. But this biomechanical explanation was merely inferred from the results. No morphological or mechanical studies were carried out at that time to test the idea.

17.3 Optical Fourier transforms of radiographs of vertebrae

The above sectional studies were followed by similar investigations of radiographs of lumbar vertebrae of humans and great apes. They show, to an even greater degree, the abilities of the technique to demonstrate structures that were not obvious either to inspection by eye or to the more usual radiographic and densitometric methods (Oxnard and Yang, 1981; Oxnard, 1983/1984).

For example, Fourier transforms of lateral radiographs of vertebrae of humans, gorillas, chimpanzees, and bonobos (pygmy chimpanzees) are generally cruciate in form. This conforms nicely with the well-known fact that, in this radiographic view, the great majority of the cancellous spicules appear to be organized orthogonally in a vertical and horizontal pattern. Such an orthogonal arrangement seems obvious in the superimposed shadows of many trabeculae evident in a lateral vertebral radiograph (Figure 17.3).

In an attempt to study the entire lumbar column in an evolutionary context, the orthogonal nature of the transforms was quantified by assessing the ratio of their length to width along the entire lumbar column. The data indicate that humans are completely different from apes, as would be expected from the bipedal nature of their locomotion and its effects upon vertebral loading, as well as its cancellous adaptive response that it presumably elicits (Figure 17.4).

The data also show that all three African apes, gorillas, chimpanzees, and bonobos (pygmy chimpanzees), have length-to-width ratios that, though they differ somewhat from one species to another, are basically similar among all three, as would be expected from three creatures having common locomotor activities such as knuckle-walking, even though they differ somewhat in their degrees of arboreality.

Finally, the data show that the patterns in the orangutan are completely different from those of either African apes or humans as, presumably, befits a creature that is different from the others in demonstrating a completely different form of acrobatic activity in the trees.

However, in addition to confirming the prior findings of a general orthogonality of structure from sections of humans, the studies of ape and human radiographs went further. Thus, notwithstanding that on visual inspection there is very little difference between the lateral radiographs of orangutans, and those of any of the other large hominoids, the Fourier transforms of the orangutans indicated that a totally different structure must be present. These transforms had a "star-burst" pattern and this shows that the radiograph must have the shadows arranged, in some way, at many different angles in addition to the orthogonal (Figure 17.5).

This could have resulted from many different orthogonal networks being superimposed upon one another within the three dimensions of the whole bone. It could also have been due to a single orthogonal network that had a different pri-

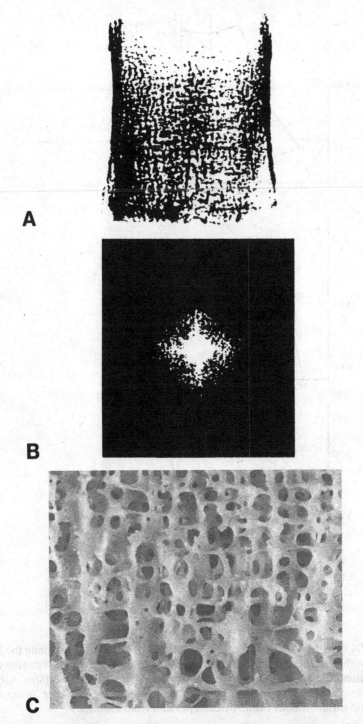

Fig. 17.3 *Above*: vertebral radiograph. *Middle*: corresponding optical Fourier transforms. *Below*: sample section of vertebra that is typical of radiographs of humans, chimpanzees, bonobos, and gorillas.

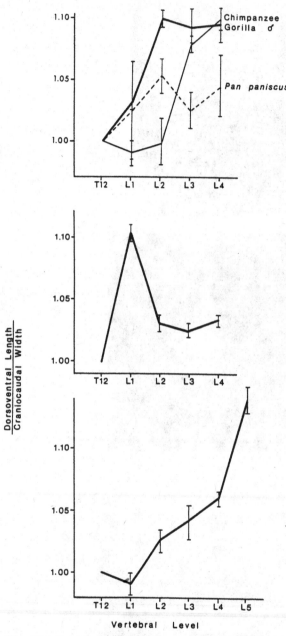

Fig. 17.4 Comparisons of differences in optical Fourier transforms along the lumbar vertebral column in humans, African apes, and orangutans as visualized by the vertical/horizontal ratio of the transforms. *Upper frame*: the three African apes show overall similarities. *Middle frame*: humans. *Lower frame*: orangutans are quite different both from African apes and from humans.

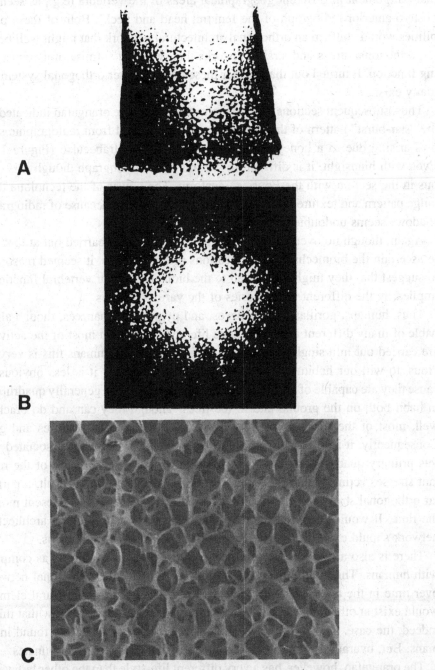

Fig. 17.5 *Above*: vertebral radiograph. *Middle*: corresponding optical transform. *Below*: section of orangutan vertebra. Though the radiograph does not appear visually much different from that in Figure 17.3, the transform and the section are totally different.

mary alignment in different geographical areas of the vertebra (e.g., as seen in a postero-anterior radiograph of the femoral head and neck). Both of these possibilities would confirm an orthogonal architectural network that might well be like the orthogonal stress and strain networks that, we may surmise, must exist during function. It turned out that neither these nor any other orthogonal systems actually exist.

Thus, subsequent sections of specimen vertebrae of the orangutan indicated that the "star-burst" pattern of the Fourier transform resulted from radiographic shadows arising due to a honey-comb arrangement of the trabeculae (Figure 17.5). Even with hindsight, it is difficult to see this in the radiograph though it is obvious in the section with the Fourier transform. The power of the technique to divulge pattern and texture information that appears hidden because of radiographic shadows seems undoubted.

Again, though no overt mechanical investigations were carried out at that time to ascertain the biomechanical implications of the findings, it seemed reasonable to suggest that they might be related to the biomechanics of vertebral function as implied by the different life histories of the various species.

Thus, humans, gorillas, chimpanzees, and pygmy chimpanzees, though all capable of many different activities, have life-styles in which most of the activities are carried out in a single major postural framework. In humans this is very obvious, to wit, our habitual bipedality. In the African apes it is less obvious because they are capable of climbing. However, these apes are generally quadrupedal in habit both on the ground and in the trees. Though they can and do brachiate well, most of their locomotor time is spent in quadrupedal postures and gaits. Consequently, it is not unreasonable to assume that the loadings associated with this primary quadrupedality will predominate in the determination of the resultant stresses acting within individual vertebrae over time. As a result, a particular orthogonal stress and strain network might be expected to be present most of the time. It would, therefore, not be surprising that an orthogonal architectural network should exist even if the overall posture differs from humans.

There is also a wider range of other loading possibilities in apes as compared with humans. This would result in a wider variance of the orthogonal networks over time in the apes. Thus, it would be likely that some architectural elements would exist at other angles. The Fourier transforms for the apes show that this is, indeed, the case. There are more elements at other angles than are found in humans. But, overall, a primary vertical and horizontal pattern predominates.

The orangutan, however, has a very different life-style than the other large apes. This has been known for many centuries as is evident by its being described, not as quadrupedal, but as quadrumanal, in habit. Although it does sometimes walk on all fours like African apes, with very peculiar fist-walking and crutch-running gaits, and though it does spend some time on two legs, but not much like humans, it spends most of its time in acrobatic activities in the trees. Here, it frequently adopts a wide variety of postural and locomotor patterns. Suspension is by one or both upper limbs, by one or both lower limbs, by one upper and one lower, by any or all combinations of three limbs, and by all four limbs together.

Although no biomechanical studies have so far been carried out, it seems reasonable to assume that there would not be a single orthogonal network of stresses that would predominate over a day's activity. It seems more likely that there will be many such transform patterns at all different angles to one another, approximately equally represented due to the changing loads on the animals over time. Under such circumstances an architectural pattern more suited to protecting grandmother's best china in the mails, that is, a honey-comb arrangement, might be most useful. Though this is not at all evident from the radiograph, it is what the transform suggests; and is exactly what is found in the sample vertebral sections from this animal. The association between the complex architecture and the complex biomechanics seems especially evident, even though it is still only inferential.

Since those days we have studied cancellous networks in a number of different animal species. For example, in whale vertebrae in sagittal section, there is a remarkable checkerboard combination of orthogonal and honeycomb zones (Oxnard, 1991). Together with other structural information about the bony and ligamentous architecture of the spine as a whole, it seems likely that this can be associated with the torques that result from the undulating and turning movements that, far more than simple flexion and extension, characterize cetacean swimming. These animals rarely swim in an absolutely straight line; maneuvering is their forte.

Another remarkable architectural arrangement can be found in the long bones of giant ground sloths. The internal cylindrical marrow space typical of most mammals living in air is filled here entirely by a complex spiral cancellous network. Moreover, in the central part of the shaft, the marrow space is so infilled as to appear pseudo-cortical (Oxnard, 1990). Consideration of the second moment of area in these bones suggests that this is related to the different balance of muscular loads, as contrasted with gravitational loads, that may have existed in these presumably extremely slow-moving animals. Similar infilling of the cylindrical marrow cavity with cancellous bone is present in other slow-moving creatures (e.g., other edentates, sloths, and some of the Australian mega-fauna; Oxnard, 1991).

17.4 Biomechanical confirmation of optical Fourier transform findings

In spite of the biomechanical inferences that have been drawn above, there are only two investigations so far in which the Fourier transform findings have been followed up by biomechanical studies.

One of these relates to the 30-degree elements mentioned earlier. Thus, both the transforms of sections, and radiographs of human vertebrae, show the orientational difference, as already outlined above, between the second and fourth lumbar vertebrae. That is, in the second lumbar vertebra but not in the fourth, a preferential set of elements are orientated at about 30 degrees to the vertical on the right of the lateral radiograph (i.e., on the anterior aspect of the whole vertebra). These were not evident in either the sections nor the radiographs. But they are clear in both sets of transforms, and identifiable in scanning electron micrographs (Figures 17.1 and 17.2).

Since those findings, stress analyses suggest that a functional implication, though not the specific original idea — flying buttresses — that was earlier proffered, is likely correct. This became clear when both photoelastic stress analysis (work carried out with C. Hoyland Wilkes) and finite element analysis (work carried out with C. K. Runnion using the special program FLAC: Fast Lagrangian Analysis of Continua — Runnion et al., 1991) were applied to simple models of cancellous bone. It was found that both the overall mechanical efficiency in stress bearing of the basic orthogonal network, and the special mechanical efficiency in eccentric loading of the 30-degree elements were closely related (Oxnard, 1993). This is most easily seen by a sequence of analyses in which cancellous structures are modeled with increasing verisimilitude.

For example, if we treat the vertebra as though it were (in lateral view) a homogeneous block of bone and load it axially, then the distribution of the stress contours is relatively evenly spread across the material; that is, the material has no mechanically inefficient areas bearing very high stresses (Figure 17.6, first frame).

If we then model trabeculae within the vertebra as though the spaces between them were circular (a first simplistic approximation), it becomes immediately obvious that enormous stress concentrations (inefficiencies) are introduced. Improving the level of simulation so that the spaces between the trabeculae are elliptical (somewhat more like reality) reduces the stress concentrations as compared with the "circular spaces" model (Figure 17.6, 2nd and 3rd frames). Refining the simulation even further, so that the spaces are approximately rectangular with rounded corners (even more realistic) also reduces the stress concentrations (Figure 17.6, 4th frame).

However, the addition of "flats" oriented at thirty degrees to the horizontal in the rectangular holes with rounded corners (though, of course, changing the pattern of stress contours), does not further reduce their number and concentration. This alteration, under the loading patterns of the previous models, produces neither a less nor a more efficient set of stress contours than the simulation of rectangles with rounded corners alone.

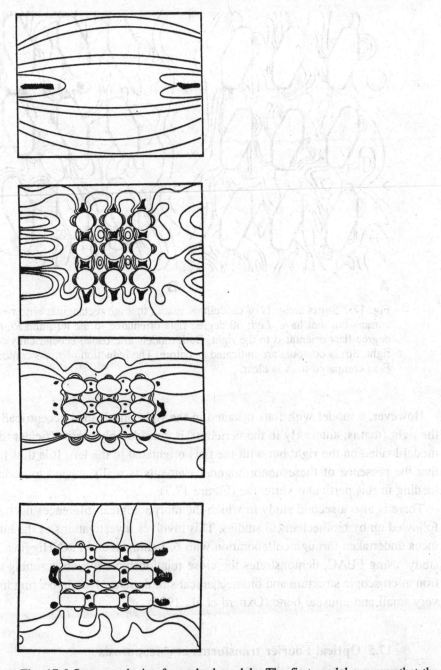

Fig. 17.6 Stress analysis of vertebral models. The first model assumes that the vertebral body is solid, the second model assumes that cancellous spaces are circular, the third model assumes that spaces are elliptical, and the fourth model assumes that spaces are rectangular with rounded corners. All models are loaded axially. Stress contours are indicated (through pattern of isochromatic contours).

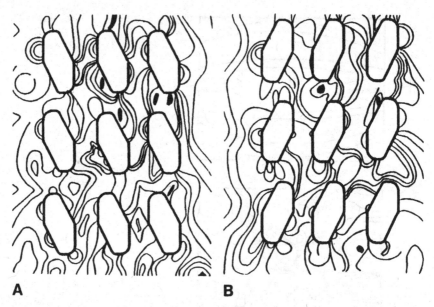

A   **B**

Fig. 17.7 Stress analysis of cancellous spaces that are rectangular with rounded corners but that have, *Left*: 60 degree flats orientated to the left; and *Right*: 60 degree flats orientated to the right. Both models are loaded eccentrically on the right. Stress contours are indicated as before. The reduction of stress contours in B as compared to A is clear.

However, a model with flats oriented to the right and loaded eccentrically on the right (that is, anteriorly in the vertebra), is considerably more efficient than a model loaded on the right but with the flats orientated to the left. It is thus likely that the presence of these nonorthogonal elements is really related to eccentric loading in this particular vertebrae (Figure 17.7).

There is also a second study in which the morphological inferences have been followed up by biomechanical studies. This involves investigations of the human incus undertaken through collaboration with F. Lannigan and P. O'Higgins. This study, using FLAC, demonstrates the close relationship between scanning electron microscopic structure and biomechanical simulations of strain bearing in this very small and unusual bone (Oxnard et al., 1995).

17.5 Optical Fourier transforms in osteoporosis

Understanding the sectional and radiographic textures present in functional and evolutionary investigations of bones obviously implies usefulness in studying medical imaging pictures of bone. For example, Pfeiler (1969) used filtering of the optical Fourier transforms for removing soft tissue shadows to reconstruct pictures with sharpened edges in chest and skull radiographs. Becker et al., (1969) likewise used the filtering technique to reveal a small ivory osteoma 1 mm in size in radiographs of a tibia specimen. But it was also clear that the Fourier transform itself might be useful for the analysis of radiographs in medical situations.

We, therefore, applied optical Fourier transforms to radiographs of thick sagittal sections of lumbar vertebrae from a range of human subjects with and without overt osteoporosis (work with A. Buck; Oxnard, 1989; 1990). The transforms indicated the generally orthogonal arrangement that prior studies have shown to exist in lateral radiographs. They also indicated differences in individuals with reference to age and osteoporosis. For example, the reduced sizes of the trabecular shadows in osteoporosis are evident in increases in the sizes of the transforms. And the degradation of the orthogonal pattern of trabeculae in osteoporosis is evident from the loss of the cruciate form of the transforms.

However, we also found patterns similar to those in overt osteoporosis in younger subjects who do not have the clinical condition but who are sufficiently postmenopausal that the incipient disease could be present. And a few even younger females that clearly do not have the condition at all, nevertheless show a mild stigmata of the advanced condition (Oxnard, 1989; 1990; Figure 17.8).

17.6 Computational Fourier transforms of human vertebrae

With such results, one might pose the question: Why have we not applied optical Fourier transforms more widely to animal forms and to every range of human subject, every normal, every disease process? The answer is that until rather recently, these transforms were difficult to produce in large numbers, needed much preliminary testing and calibration, and required additional secondary analysis in order to render quantitative the visual impressions that they provided. Recent developments with personal computers and fast Fourier transform software now allow the whole process to be carried out computationally and we have used this method in the last four years (work with A. Buck and M. Hull).

First we must ask the question: Do the computational transforms provide the same result as the optical method? This question is easily answered in the affirmative. Thus, Figure 17.9 shows that the stigmata provided by computational Fourier transforms, of sections from aged individuals, are similar to the optical transforms of Figure 17.8. It shows, further, that these features are lacking in young individuals. It also shows that, though in middle-aged males the stigmata are absent, in middle-aged females they are already present.

Because it is now much easier to analyze large samples in a reasonable period of time, the method has been applied to the full suite of data resulting from radiographs of thick sections of human lumbar vertebrae. These comprise materials

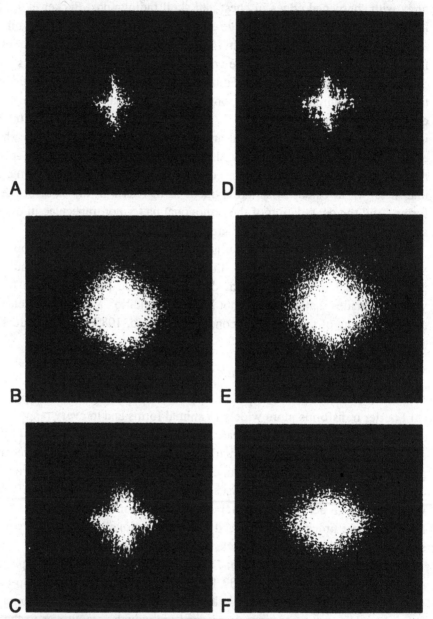

Fig. 17.8 Optical Fourier transforms of radiographs of vertebral sections in humans. Left column: females; right column: males. *Above*: young individuals of each sex. *Center*: middle-aged individuals of each sex. *Below*: aged individuals of each sex. The difference between middle-aged males and females is obvious through these transforms; it is not evident in visual examination of the corresponding radiographs.

Females   Males

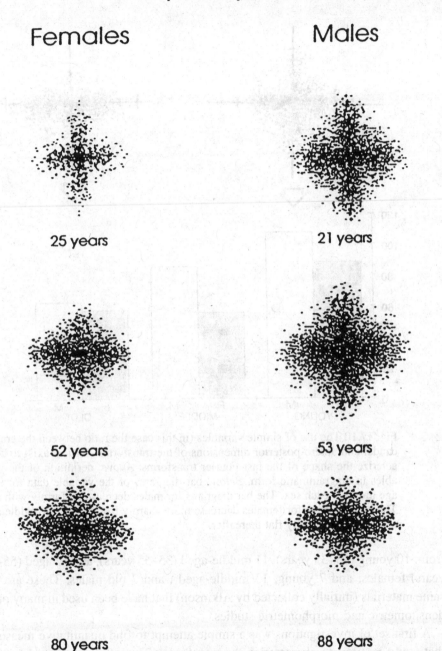

25 years   21 years

52 years   53 years

80 years   88 years

Fig. 17.9 Computed fast Fourier transforms of radiographs of vertebral sections in humans. Left column: females; right column: males. *Above*: young individuals of each sex. *Center*: middle-aged individuals of each sex. *Below*: aged individuals of each sex. The similarity with Figure 17.8 is clear though the analyses do not show the identical specimens. That is, the middle-aged female resembles the old female, the middle-aged male resembles the young male.

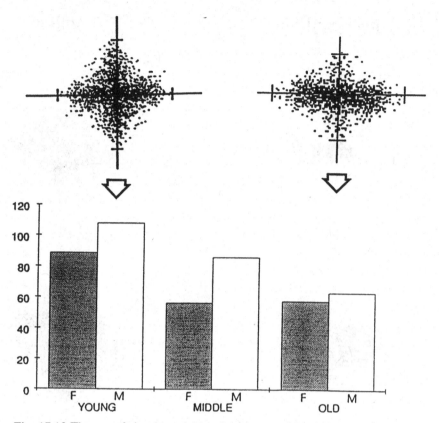

Fig. 17.10 The use of simple variables (in this case the ratio between the cranio-caudal and antero-posterior dimensions of the transforms–vertical axis) to characterize the shape of the fast Fourier transforms. *Above*: definition of the variables in diagrammatic form. *Below*: bar diagrams of the variable data for each age group in each sex. The bar diagrams for males decrease uniformly with age. The bar diagrams for females decrease more sharply from young to middle age, and remain generally flat thereafter.

from: 10 young (15–35 years), 11 middle-aged (35–55 years), and 12 aged (55–80 years) females; and 7 young, 13 middle-aged , and 7 old males. These are the same materials (initially collected by Atkinson) that have been used in many prior densitometric and morphometric studies.

A first set of investigations was a simple attempt to find quantitative measures that could be used to characterize the transforms. Thus, the degree and type of orthogonality were assessed, as with the optical transforms, by simply measuring the ratio of their vertical and horizontal extents (Figure 17.10). Other measurements are possible, for example, we defined the degree and extent of nonorthogonal elements by measuring a second ratio: the chord depth ratio of the transforms (see, e.g., Buck et al., 1992, not shown here). Both these ratios show (see Figure

17.10 for the vertical-horizontal ratio) what would be expected. There is a relatively linear change with age in males, whereas in females, the differences are non-linear, declining most rapidly near the time of the menopause, but being in the oldest women, similar to the oldest men.

However, because there is such a large amount of information contained within the Fourier transforms that is not assessed by such simple measures (however useful they might be in a clinical situation), more detailed analyses were carried out.

17.7 The FFT analyzed using multivariate statistics

Instead of the crude ratios, the transforms were divided into sixteen regions as indicated in Figure 17.11. The numbers of pixels in each region generated sixteen variables containing considerably more information about the transform than the simple ratios. These new variables were then examined using multivariate statistical analyses.

Principal components analyses were used in a group-finding mode to answer the question: Do these data indicate the existence of groups? Canonical variates analyses were used in a group-separating mode (because groups that are not de-

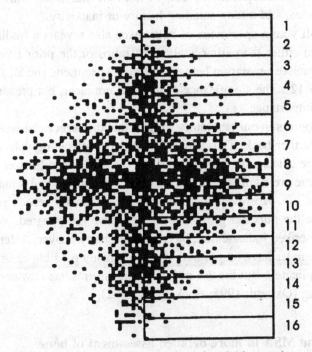

Fig. 17.11 The use of a complex of variables to characterize the Fourier transform somewhat more completely. The sixteen new variables comprise the pixel counts in each of the rectangular areas defined.

pendent upon the data do truly exist, for instance, sex and age groups) to answer the question: What are the separations of the groups?

The analyses demonstrate what is obvious, that indeed, far more information is present within the transforms than is revealed by the simple ratios. Principal components analysis of all specimens at the oldest ages confirms the ratio analysis (and much other work on osteoporosis) indicating that the oldest males and females are relatively similar (Figure 17.12). Almost all specimens showed greater or lesser degrees of osteoporosis.

Likewise, principal components analysis of all female specimens at all ages confirms the results of the ratio analysis. This indicates that, indeed, young, middle-aged, and old individuals can be separated from one another (Figure 17.12).

Additional information is, however, also provided. For example, the specimens from middle-aged females are clearly divided into two groups: younger individuals (premenopausal) being completely separated from older ones (postmenopausal). Again, among the oldest females, there are several outlying specimens that, in each case, coincide with the possession of clinical conditions of varying kinds that have in common the production of extreme physical inactivity over long periods. Also, there is one single young specimen that lies close to the middle-aged postmenopausal specimens. This is from an individual with mitral valve stenosis, dyspnoea, and a long-standing history of inactivity.

Finally, study of all young specimens of both sexes also reveals a finding totally unexpected from either the earlier analysis of ratios or the prior literature. That is, there is an absolute separation between all male specimens and all female specimens (Figure 17.12). The significance of this was not clear; but presumably it is of fundamental importance.

Other studies in progress in our laboratory suggest that differences between age groups extend even further back. Thus, Fourier transforms of radiographs of the cancellous bone of the calcaneus (work carried out by L. Ellison; Ellison et al., 1993) suggest that there are significant differences between males and females in the first two years of life. However, these differences gradually recede so that sometime well before the pubertal growth spurt they have disappeared. Yet the investigations just described indicate that, by the time puberty is over, differences between males and females have reappeared. There are many possible causal factors that might be implicated. But the use of Fourier transforms has now certainly established the finding (Oxnard, 1995; Buck et al., 1992).

17.8 FFT and MSA in more detailed assessment of bone

The above studies have thus shown both expected and unexpected differences between radiographic shadows of cancellous bone in specimens from differently aged individuals. These differences could be due to a systemic condition that grad-

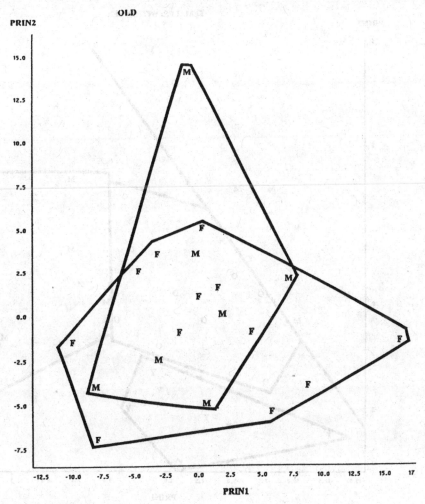

PERCENTAGE OF VARIATION

PRIN1 = 60%

PRIN2 = 34%

M = MALE

F = FEMALE

Fig. 17.12 Principal components analyses of transforms for each subject when defined through the sixteen variables of Figure 17.11. *First plot*: the first two principal components for all old individuals; there is major overlap between the sexes. *Second plot*: the first two principal components for all females of each age group; the age groups are clearly defined as indicated (Y = young, M = middle-aged, O = old). The middle-aged group is divided, slanted line, into those before the menopause on the right, and those after the menopause on the left close to the old group. *Third plot*: the first two principal components for all young subjects; there is an unexpected separation between all males and all females.

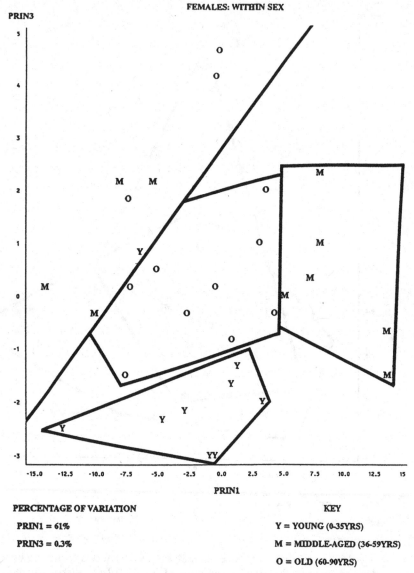

PERCENTAGE OF VARIATION

PRIN1 = 61%

PRIN3 = 0.3%

KEY

Y = YOUNG (0-35YRS)

M = MIDDLE-AGED (36-59YRS)

O = OLD (60-90YRS)

Fig. 17.12 *Continued*

ually affected the entire vertebra. Although still the result of a systemic condition, they could also result from changes that started in one anatomical location and then gradually extended to others. Most recently of all, therefore, we have attempted to see if regional differences in pattern exist within vertebral radiographs. The question being examined here is: Are the changes with aging (and, possibly, the onset of osteoporosis) an overall phenomenon, or do they commence in one particular region of the vertebral body and then spread?

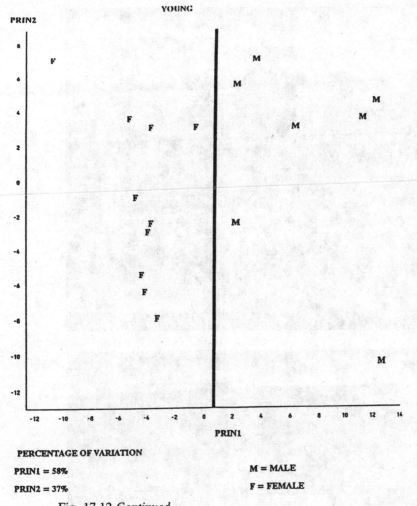

Fig. 17.12 *Continued*

Fourier transforms were, therefore, calculated for each individual quadrant of each radiograph. Visually, the results are what one might expect. Thus, Figure 17.13 demonstrates an example for a young individual. The transform for each quadrant is basically similar to the transform for the entire radiograph: a cruciate structure. This implies that a similar orthogonal network of shadows exist throughout the vertebra, even though there are some interesting small regional differences in orthogonality that can be distinguished. For example, the basically orthogonal pattern of the main body of the vertebra changes somewhat in the region of the attachment of the pedicle. This could have an obvious biomechanical association.

Similarly, Figure 17.13 demonstrates the result for an old individual. The transform for each quadrant is also basically similar to the transform for the entire ra-

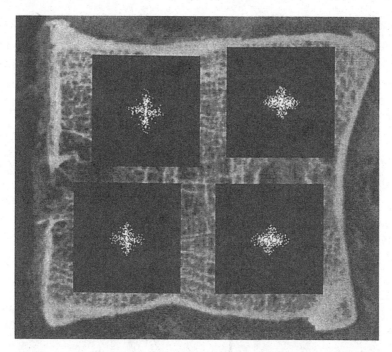

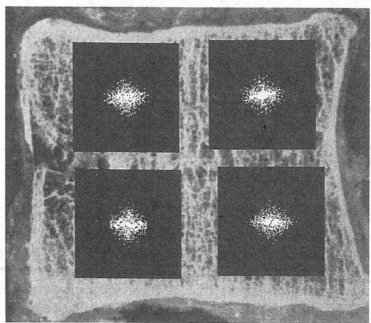

Fig. 17.13 Fourier transforms of quadrants of radiographs. The upper frame is from a typical young individual, the lower from a typical old individual. The cruciate form of the upper and the American football form of the lower is characteristic of all radiographs of all specimens in each age group.

diograph; but in this case the Fourier transform is shaped like an American football rather than a Celtic cross. This implies for each quadrant, as it did for the whole vertebra, that a degradation of the orthogonal network of shadows, characteristic of both general aging and osteoporosis, is found in all quadrants. Again, there are some interesting smaller regional differences that can be distinguished.

However, multivariate statistical analysis of the full suite of data, in this case using canonical variate analysis, supplies information about age and sex, additional to that noted above.

First, although principal components analysis of the entire vertebral transform suggested little difference between old males and old females (Figure 17.12) these new canonical variates studies demonstrate that there are regional differences between the oldest age groups. Thus, quadrant for quadrant, the means for males are always separated in a coherent manner from the means for females. Study of the contributions of the variables to these mean quadrant positions implies that, though not obviously significant at the level of the whole vertebra, there are significant differences between the quadrants. On average, each female quadrant demonstrates a somewhat greater degree of the aging/osteoporosis stigmata than does each male quadrant (Figure 17.14). In the examination of the whole vertebra, this level of significant finding is not so obvious.

Second, the finding that there is a large difference between middle-aged males and females (Figure 17.12) is further explicated at the level of individual quadrants. Thus, in addition to a separation of the sexes, there is also a separation of the quadrants (Figure 17.14). Again, consideration of the contributions of the transform variables to the multivariate analysis makes it clear that this has to do with the two anterior quadrants. The stigmata of aging/osteoporosis are more advanced in females in the two anterior quadrants than they are in males. It is presumably this difference that is picked up in the separation of males and females when the whole vertebra is examined.

Finally, the finding that there is a surprising difference between young males and young females (Figure 17.12) is also further explicated in the examination of individual quadrants. In this case the canonical plot (Figure 17.14) shows that the difference is basically confined to a single quadrant, the anterior superior. Again, this is quite clear in the visual examination of an individual transform (Figure 17.15).

The overall picture that is thus provided is surprising. For this particular vertebra (second lumbar) the features (the American football form of the transform) presumed to appear in later life diagnostic of osteoporosis, seem to commence in the anterior superior quadrant in the quite young individuals. They are later present in both anterior quadrants in the somewhat older individuals. Later still, they encompass all four quadrants in the oldest individuals, and especially those with

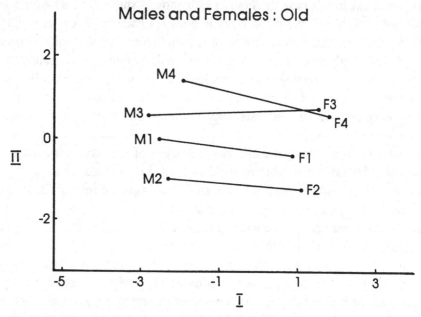

Fig. 17.14 Plots of the first two canonical variates of the analysis of sixteen variables for each quadrant in each subject. M and F = males and females; 1, 2, 3, and 4 = postero-superior, postero-inferior, antero-superior, and antero-inferior quadrants for each sex. *First plot*: analysis of data from old subjects. There is a coherent difference between all quadrants in males as compared with females. This information is not obvious in the analysis of the whole vertebral transforms (Figure 17.11). *Second plot*: analysis of data from middle-aged subjects. There is a coherent difference between all quadrants in males as compared with females; but this difference also incorporates a coherent difference between both posterior quadrants as compared with both anterior quadrants. Study of the individual variables indicates that it is the anterior quadrants that are more affected. This is, again, information additional to that in the analysis of the whole vertebral transforms (Figure 17.11). *Third plot*: analysis of data from young subjects. There are only small and randomly orientated differences between three of the quadrants (1, 2, and 4) in males as compared with females. However, for quadrant 3, there is a large and significant difference between the sexes. This quadrant is the anterior-superior quadrant. Study of the individual variables indicates that it is the anterior quadrant in females that is different. This is yet further additional information that is not obvious in the analysis of the whole vertebral transforms (Figure 17.11).

the worst degrees of osteoporosis. Does this finding mean that at least one component of osteoporosis is simply due to phenomena associated with the ways in which many people age? There seems little doubt that the process begins well before the time of the menopause, that it is accelerated by the menopause, and that a similar but less acute, more linear, change also occurs in males. Does this find-

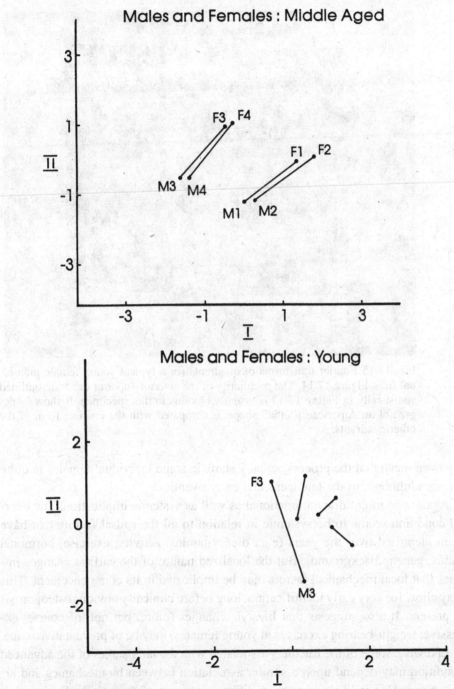

Fig. 17.14 *Continued*

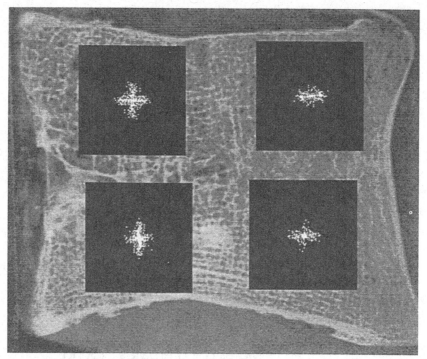

Fig. 17.15 Fourier transforms of quadrants for a typical young female individual from Figure 17.14. The peculiarity of the anterior-superior quadrant outlined statistically in Figure 17.13 is visually obvious in this specimen. It shows a degree of an American football shape as compared with the cruciate form of the other quadrants.

ing even mean that the process actually starts in some individual females in quite young adulthood, in the late teens and early twenties?

Again, one might draw a functional as well as systemic implication. The overall condition seems to be systemic in relation to all the causal factors that have been identified over the years (e.g., diet, vitamins, activity, exercise, hormonal status, genetic background). But the localized nature of the earliest changes implies that local mechanical factors may be implicated in its commencement. This may allow for very early identification, long before clinically obvious osteoporosis is present. It also suggests that lifestyle changes (more, but not, of course, excessive, weight-bearing exercises in young females) may be of preventative value.

Likewise, some of the fractures associated with the later stages of the advanced condition may depend upon a similar association between biomechanics and architecture. This is likely to vary in different regions of the vertebral column. It could well result in the special patterns of fracture well known in different vertebrae. The information revealed by such studies may, likewise, provide useful prognostic information about fracture complications.

17.9 Conclusions

It is apparent then that these more complex multivariate statistical analyses of computational Fourier transforms (easily carried out today with personal computers) may be extremely useful. In elucidating evolutionary trends, they may provide structural data that allow functional insights into the nature of biological adaptation. In clinical situations, they may provide normative data, identification of single putatively abnormal individuals, and screening techniques for both incipient disease and early onset of complications. In contrast to many current methods, these are noninvasive, easy to carry out, and cheap to apply.

Acknowledgments

Consultations and collaborations with many individuals are acknowledged in the text. Of especial importance was the early help of J. C. Davis of the University of Kansas at Lawrence, and H. C. Pincus of the University of Wisconsin at Milwaukee. The work of many of my graduate students and post-doctoral colleagues has been most important, especially H. C. L. Yang, A. T. Hsu, A. Buck, C. Hoyland-Wilkes, C. Runnion, and L. Ellison. Discussions with P. O'Higgins have been especially useful.

The work has been supported over the years by funds from the NIH and NSF (USA), the ARC and NH&MRC (Australia), the CSIRO/University of Western Australia, the Raine Medical Research Foundation, the Sylvia and Charles Viertel Charitable Trust, and the Department of Anatomy and Human Biology and the Centre for Human Biology, University of Western Australia.

References

Becker, H. C., Meyers, P. H. & Nice, C. M., Jr. (1969). Laser light diffraction, spatial filtering, and reconstruction of medical radiographic images. *Ann. New York Acad. Sci.* **157**, 465–86.

Buck, A., Oxnard, C. E. & Hull, M. J. (1992). Variations in the trabecular patterns of the second lumbar vertebra. *Pro. Austral. Soc. Hum. Biol.* **5**, 285–89.

Davis, J. (1970). Optical processing of microporous fabrics. In *Data Processing in Biology and Medicine*, ed. J. L. Cutbill. London: Academic Press.

Ellison, L., O'Higgins, P. & Oxnard, C. E. (1993). A radiographic study of developing cancellous bone in the calcaneus of children. *Proc. Austral. Soc. Hum. Biol.* **6**, 22–3.

Evans, F. G. (1957). *Stress and Strain in Bones*. Springfield: Thomas.

Murray, P. D. F. (1936). *Bones. A Study of the Development and Structure of the Vertebrate Skeleton*. London: Cambridge University Press.

Oxnard, C. E. (1970a). Functional morphology of primates: Some mathematical and physical methods. *Burg Wartenstein Symposium*, **48**, 1–42. Also published (1972a) in *The Functional and Evolutionary Biology of the Primates*, ed. R. H. Tuttle. Chicago: Aldine Atherton.

Oxnard, C. E. (1970b). The use of optical data analysis in functional morphology: Investigation of vertebral trabecular patterns. *Burg Wartenstein Symposium,* **48,** 1–30, additional paper. Also published (1972b) in *The Functional and Evolutionary Biology of the Primates,* ed. R. H. Tuttle. Chicago: Aldine Atherton.

Oxnard, C. E. (1973). *Form and Pattern in Human Evolution: Some Mathematical, Physical and Engineering Approaches.* Chicago: University of Chicago.

Oxnard, C. E. (1983/1984). *The Order of Man: A Biomathematical Anatomy of the Primates.* Hong Kong: University of Hong Kong Press (1983); New Haven: Yale University Press (1984).

Oxnard, C. E. (1989). Biomechanics and architecture of cancellous bone. *Proc. Austral. Soc. Hum. Biol.* **2,** 189–212.

Oxnard, C. E. (1990). From fossil giant ground sloths to human osteoporosis. *Proc. Austral. Soc. Hum. Biol.* **3,** 75–96.

Oxnard, C. E. (1991). Mechanical theories of stress and strain at a point: Implications for biomorphometric and biomechanical studies of bone form and architecture. *Proc. Austral. Soc. Hum. Biol.* **4,** 57–110.

Oxnard, C. E. (1993). Bone and bones, architecture and stress, fossils and osteoporosis. *J. Biomech.* **26,** 63–79.

Oxnard, C. E. (1995). Bone, a good age and osteoporosis. *Perspectives Hum. Biol.* (in press).

Oxnard, C. E., Lannigan, F. & O'Higgins, P. (1995). The mechanism of bone adaptation: Tension and resorption in the human incus. In *Bone Structure and Remodelling.* ed. A. Odegaard & H. Weinans. *Volume 2 of Recent Advances in Human Biology,* Series Editor, C. E. Oxnard. London: World Scientific.

Oxnard, C. E. & Yang, H. C. L. (1981). Beyond biometrics: Studies of complex biological patterns. *Symp. Zool. Soc. (Lond.)* **46,** 127–67.

Pfeiler, M. (1969). Image transmission and image processing. In *Automatic Interpretation and Classification of Images.* ed. A. Graselli. New York: Academic Press.

Power, P. C. & Pincus, H. J. (1974). Optical diffraction analysis of petrographic thin sections. *Science* **186,** 234–9.

Runnion, C. K., Oxnard, C. E., Robertson, W. V. & Windsor, C. R. (1991). Biomechanical modelling of vertebrae using experimental stress analysis. *Proc. Austral. Soc. Hum. Biol.* **4,** 125–33.

Thompson, D. W. (1917). *On Growth and Form.* Cambridge: Cambridge University Press.

Todd, R. B. & Bowman, W. (1845). *The Physiological Anatomy and Physiology of Man.* London: Parker.

Wolff, J. (1892). *Das Gesetz der Transformation der Knochen.* Berlin: Hirschwald.

Wyman, J. (1867). On the cancellated structure of the bones of the human body. *Boston J. Nat. Hist.* **6,** 112–34.

Appendix to Chapter 17

17.A.1 Optical Fourier transforms

The earlier analyses employed optical Fourier transforms carried out first with a custom-built optical bench and later with the commercial Rank Image Analyzer 2,000. The power spectrum resulting from optical transformation is useful in the interpretation of biological patterns as it contains size and directional information. This spectrum can be thought of as a bell-shaped volume (or flux) of light which is made up of bands (size information) and sectors (directional information). The size information is transformed inversely, so that the information (spatial frequencies) relating to large periodicities in the original texture appears small in the transform and vice versa. The directional information is also transformed, in this case by rotation through 90 degrees; that is, information regarding elements in the vertical direction of the original texture appears horizontal in the power spectrum (e.g., Figures 17.1 through 17.5).

The equation for the continuous (optical) power spectrum of the Fourier transform is as follows:

$$f(x',y') = \left| \iint f(x,y) exp^{-j[wx+wy]} \, dx \cdot dy \right|^2, \tag{17.1}$$

where:

$f(x,y)$ = continuous input signal, $[wx + wy]$ = cosine vector, and
j = complex constant (Oxnard, 1973).

17.A.2 Computational Fourier transforms

The computational generation of the power spectrum provides an essentially similar analysis to that presented by the optical diffraction method. The computational method is used in the later parts of the present study because of its speed, efficiency, and reliability (e.g., Figures 17.9–17.11, 17.13, and 17.15). The computational method, however, calculates the Fourier transform from a discrete set of points, a_{ij}, rather than from a continuous input signal.

Because the gray scale matrix, upon which the computational calculations are based, is a discrete set of points, the discrete Fourier transform was employed. The fast Fourier transform (FFT) version of this was used because it reduces the calculation time. The formula for the discrete FFT is as follows:

$$F = \sum \begin{matrix} 1 & 1 & 1 & 1 \\ 1 & W & W^2 & W^{n-1} \\ 1 & W^2 & W^4 & W^{2(n-1)} \\ 1 & W^3 & W^6 & W^{3(n-1)} \\ 1 & W^4 & W^8 & W^{4(n-1)} \\ \vdots & \vdots & \vdots & \vdots \\ 1 & W^{(n-1)} & W^{2(n-1)} & W^{(n-1)(n-1)} \end{matrix}, \tag{17.2}$$

where: $W = e^{-i2\pi/n}$.

Once the transform was obtained, a threshold filter was used to remove gray scale elements related to the grain of the radiographs. Thresholding involved identification and removal of gray scale levels in which the highest-frequency information was stored. When the highest-frequency components were removed, the remaining frequency components, which contained information relating to the larger osteological shadows, especially superimposed trabecular shadows, were examined and compared between specimens. Further studies are required to refine the thresholding technique.

The image of the entire vertebral body was first sampled over a wide area (255×255 pixels) which excluded the cortical shell but represented most of the internal cancellous structure. Quadrants of vertebrae were examined in smaller areas (127×127 pixels) to evaluate the degree of homogeneity of the cancellous shadows throughout the vertebral bodies.

Replicability of the results was tested by an ANOVA of six specimens tested six times. The results showed that little or no intraobserver error existed when capturing and analyzing the images.

All fast FFT calculations were carried out using the software package Image-Pro Plus, Media Cybernetics, 1991.

17.A.3 *Multivariate statistical analyses*

The statistical examination of the variables defining a large portion of the form of the power spectrum was performed using multivariate statistical analyses (e.g., Figures 17.12 and 17.14). First, the data were examined to see what groups (if any) might exist in the data. This involved the application of principal components analysis which is the solution of the following matrix algebra equation with e as the canonical roots and a as the canonical vectors:

$$(\mathbf{S} - e\mathbf{I})\mathbf{a} = 0, \tag{17.3}$$

where $\mathbf{S}$ is the dispersion matrix and $\mathbf{I}$ is the identity matrix.

Once the groups had been determined, the data were reexamined by canonical variates analysis. This requires the solution of the matrix algebra equation with f as the canonical roots and c as the canonical vectors:

$$(\mathbf{B} - f\mathbf{W})\mathbf{c} = 0, \tag{17.4}$$

where $\mathbf{B}$ is the among-populations dispersion matrix and $\mathbf{W}$ is the within-populations matrix.

Both group-finding and group-arranging methods require display techniques. When the results are relatively simple, then two-dimensional plots and three-dimensional models are adequate, as in this study. But when the dimensionality is considerably greater than three, then techniques to examine the large number of principal components, canonical variates, or the full matrix of generalized distances, such as dendrograms, minimum spanning trees, and high-dimensional plots become important.

18

Epilogue: Fourier Methods and Shape Analysis

NEAL R. GARRETT

West Los Angeles VA Medical Center and UCLA School of Dentistry

18.1 Introduction

The previous chapters have dealt with some of the theoretical aspects of the development of Fourier descriptors (FDs) as well as numerous applications reflecting very diverse research areas. They give the reader a sense of the complexities attendant to the numerical characterization of form as well as indicating the relative infancy of these research methods. Although the basic mathematical underpinnings of FDs have been available for over 150 years, significant practical developments and applications in biology have only begun to appear in the last thirty-five years. One feels that there is increasing interest as well as tremendous potential in the use of these methods. However, major breakthroughs in Fourier techniques and applications have been made by individual investigators working in the frontiers of their respective fields. Previously little effort has been made to communicate their findings, techniques, and problems across these specialization boundaries. This book represents the first attempt to do so.

It is interesting to note that many of the authors in this volume exploring Fourier analysis for evaluation of aspects of form have had to develop their own methods and computer programs in the process of trying to answer specific research questions. In fact, the time line for the increase in interest and use of these methods closely parallels the broader capabilities now available to researchers in all fields with respect to computer programming and processing power, which are critical for the practical use of these computationally intensive techniques. These algorithms and computer programs for application of FDs as seen in the preceding chapters, are now available to investigators with the interest in, but without the resources for, development of Fourier analytic techniques. This should lead to an expansion of the application of these methods as researchers become aware of FDs as easily obtainable analytical tools which can provide boundary information previously not amenable to analysis.

The reader who is not already familiar with or employing FDs in their own work may be asking several questions at this point. First, they may want to know how FDs work; second, how the use of FDs will help them in answering their research questions; and third, how do they go about getting the software.

As far as acquiring the software and knowledge to apply these techniques to the reader's question of interest, the Appendix and references will provide a good beginning, and most if not all of the authors in this book are receptive to inquiries and communication.

18.2 Applications

Applications of FDs presented in this book are unusually broad in scope and vary from conventional FDs to assess boundary considerations to the use of Fourier transforms to characterize internal structure. Research topics include analysis of bone structure (Oxnard, Chapter 17) and anatomy (Lestrel et al., Chapter 16; Ohtsuki et al., Chapter 9; Jacobshagen, Chapter 10), ostracode form (Kaesler, Chapter 5), cellular and subcellular (nuclei) structures (Pesce Delfino et al., Chapter 11; Diaz et al., Chapter 13), and even human gait (Uetake, Chapter 12). There are evaluations of processes such as ecological change (Healy-Williams et al., Chapter 6), treatment effects (Lestrel and Huggare; Chapter 14; Moon, Chapter 15), growth (Read, Chapter 3), genetics (Johnson, Chapter 7; Richards et al., Chapter 8) and discussion of evolutionary changes (O'Higgins, Chapter 4; Read, Chapter 3). Although the bulk of the chapters emphasize the application of FDs to morphological forms, the use of FDs to analyze human gait represents a quite different application. This illustrates shape quantification of physiological functions over time, as in the case of movement, auditory speech signals, or electrophysiological responses. Time series analysis of auditory and electrical signals have long been associated with Fourier analysis, but the use of FDs in quantifying the shapes and envelopes of movements have only begun to be recently explored. One needs only to ask if the data to be analyzed can be "drawn" to determine if FDs can be used to describe it. However, Read (Chapter 3), taking a cue from D'Arcy Thompson, points out the need for the study of form to be analyzed in light of growth and other forces acting on structure. As he states, the suitability of these methods for a particular question can only be determined from the biological insight they provide. This theme is also independently echoed by O'Higgins (Chapter 4).

Although space precludes a detailed evaluation of all the chapters, several interesting and innovative points should be noted. Kaesler (Chapter 5) has shown that phase information, which is commonly ignored, can be very informative and discriminatory. He has provided an equation for this purpose, and gives an example which makes it apparent that this information should not be discarded. Richards et al. in Chapter 8 have discussed an interesting approach of correlating the radial vectors to the amplitudes. This may permit better exploration of the relationships of the underlying elements of the form with the FDs. Pesce Delfino et al. in Chapter 11 contribute two novel methods. First, by smoothing the fine irregularities of the form, noise is reduced and the FDs are simplified. In turn, more attention can be focused on the basic shape. However, care must be taken since the degree of smoothing may lead to loss of relevant information in the form. Second, they have developed a new approach which permits the separate evaluation of the overall symmetry (or asymmetry) of a form independent of the boundary or contour perturbations. In considering the treatment effects of functional appliance therapy, Moon (Chapter 16) shows why it is necessary to consider a morphological form in its entirety, rather than each part in isolation. Analysis of the FDs revealed that the apparent lengthening of the mandible (presumed by some dental researchers to be associated with accelerated growth) was due to physical bending of the condyle, and not growth. Ohtsuki et al. (Chapter 9) demonstrate the ability to trace past historical relationships among forms with FDs, and Oxnard (Chapter 17) uses them to predict osteoporosis in women long before symptoms are clinically manifest.

Johnson (Chapter 7) and Richards et al. (Chapter 8) point out the utility of FDs in relating environmental and genetic factors to morphology. Current research has begun to show the significant effects of genotype on development of phenotype. For example, in many organisms (e.g., Drosophila) it has been determined that an overwhelming percentage of the structural components of the organism is genetically determined. The current emphasis and advances in gene research has obvious implications for human studies, in that it suggests that the morphology, as viewed in this volume, including human anatomical structures, may well have a major genetic component. This is not a new concept, but it leads to the question of how to relate aspects of the genome directly to the morphology. Subsequently, differences in the genome should lead to measurable changes in the phenotype. However, these subtle phenotypic changes in shape would require a sensitive approach, potentially supplied by FDs, to discriminate these differences. Introduction of environmental factors (i.e., Johnson, Chapter 7) and growth and development (i.e., Read, Chapter 3) on the phenotype could then be evaluated independent of the genetic factors.

The reader who is actively involved in the development and use of FDs in their research may be interested in how the methods can be improved and extended to provide them with more comprehensive tools to work with. Future development of FDs largely depends on the ability to provide the researcher with answers to questions that they cannot get with other methods such as the conventional metric approach (CMA; see Chapter 1 for an extended discussion). It appears that when we begin to formulate plans to answer our hypothesis, we are operating with at least a subjective intuitive model, based upon background research and experience, which we use to "select" subsets of measures to describe the facets of the form we think are important to our question. If a formal model for the question has already been posited with extensive supporting data, a reduced subset of variables may be adequate. However, the inherent subjectivity in variable selection could lead one to possibly miss discovering new information that might alter existing models of process or discrimination. The less we know about the structure or process under investigation, the more likely we are to miss relevant and possibly critical information by under-sampling with a reduced dataset.

18.3 Need for a formal model of form

Lestrel has presented a model in Chapter 1 which describes a variety of factors, including shape, size, texture, color, and others, which may be seen as composing form. At one point in the continuum, where we have little background information, one might even recommend that all possible factors of form be analyzed at the highest resolution possible. This extreme circumstance would not be very likely, since rarely are we devoid of some relevant evidence to reduce the uncertainty as to what factors of form are critical. For example, our question might involve structures that are identical in color or have certain parts of their boundary outline which have been found to be the same in previous research. However, it points out the wide range of levels of analysis, from application of CMA to a complete modeling of form, and the difficulty in determining what factors to measure and what sampling rate is sufficient. Even if we feel that we need the boundary information of a form between landmarks to answer a particular question of shape change, how do we determine sampling rate? Currently we rely on accuracy of fit of the function, but we still deal with sufficiency of fit in a relatively subjective fashion. Further development of the model proposed by Lestrel and others is necessary to develop a framework for integration and standardization of techniques, definitions, and results from multidisciplinary applications of FDs. The development of such models may also help us to verbalize criteria of sampling sufficiency for a wide variety of shapes and circumstances.

18.4 Conclusions

In summary, this volume makes a strong case for the necessity of gathering more extensive information on form, at multiple levels, in order for the relationships between form and underlying factors of process such as function, growth, evolution, and so on, to be more fully understood. The further development of models of form, as outlined by Lestrel in Chapter 1, can provide theoretical focus and integration of parallel research efforts from many scientific domains.

This first attempt at covering the basic application of Fourier methods to the description of form and its use in a variety of problem areas has indicated an overlap of methodologies, but points out that there are still a variety of techniques, terminology, and interpretations of results. Increased communication within and among disciplines on the development and description of Fourier procedures can lead to greater standardization, acceptance, and cross-utilization of these methods.

Appendix

Software for Elliptical Fourier Function Analysis

CHARLES A. WOLFE

Wolfe Consulting Software Engineers

A.1 Introduction[1]

This appendix describes a system of computer programs called EFF23, for the IBM Personal Computer and compatibles, that will calculate the elliptical Fourier function (EFF), a Fourier descriptor (FD), for both 2- and 3-D boundary outlines. The 3-D computation, viewed as a curve in space (not a solid/volume), is an extension of the Kuhl and Giardina (1982) algorithm (described below). EFF23 will also compute a number of descriptive statistics, several types of distances, as well as amplitudes, power spectra, phase angles, selected ellipse parameters, and can produce a number of different types of plots. Further, most data files may be exported for use with third-party software such as statistics, spreadsheet, and plotting.

Applications of EFF23 may be found in Chapters 14, 15, and 16. Section A.15, A.15 Resources, contains information about availability, program description, and the user manual (Wolfe et al., 1996).

Although the preferred usage of the term "form" is discussed in Chapters 1 and 4, for convenience here, form will mean a computer-based, numerical representation of a boundary outline either (a) as observed (digitized) points, or (b) by an EFF (harmonics or coefficients), or (c) by predicted points in 2- or 3-D format computed from the EFF. As used here, a form is (stored as) a file, but not all EFF23-created files are forms, as will become clearer.

This appendix has two purposes: first, a fairly high-level, broad view of EFF23's capabilities and, second, a discussion of the problems and decisions that arise in implementing the various algorithms as computer software. Therefore, focus will be on the following:

- Digitization requirements and procedures;
- Computation of EFFs for spatially unoriented and oriented forms;

---

[1] Copyrighted and trademarked names used in this appendix are the property of their respective owners. They are used here only for purposes of identification.

- Size-standardization of forms including computation of areas and centroids for 2- and 3-D forms;
- Problems with forms that contain self-intersections (crossings);
- Computation of other FDs using EFF23;
- Exporting of EFF23 files for utilization by other (third-party) software;
- A short note on program implementation of the 3-D algorithms.

A.2 History of EFF23

We thank F. J. Rohlf for providing the original (2-D) FORTRAN implementation of the Kuhl and Giardina algorithm (Rohlf, 1985). D. Read then translated it to (Power) BASIC for the IBM PC. P. Lestrel wrote the original user interface, digitizing, plotting, statistics, amplitude, and other related routines around Read's translation. The programmatic extensions to 3-D, and so on, were devised and implemented by C. Wolfe, D. Read, and P. Lestrel during 1992–1996. The methodology and programming dealing with crossings (Section A.10) and the formulation and inclusion of homologous points were developed by C. Wolfe with assistance from P. Lestrel and D. Read. It has taken several years to work out the algorithms for 2- and 3-D orientation.

A.3 Overview of EFF23

The multiple functions of EFF23 and detailed activities are shown in Figure A.1. The investigator begins specimen collection (Forms) and devising a point system protocol that is uniform across specimens. Each specimen is digitized to create the initial digitized or observed form containing (x,y) or (x,y,z) data. Next, the digitized form is used to compute the EFF and predicted points (either homologous or evenly spaced; see Section A.6); in addition, an oriented (Section A.7.2) and/or size-standardized (Section A.8) EFF fit may also be computed at this time. Once the appropriate EFFs have been computed for the sample of digitized forms, one can then compute amplitudes, phase angles, power spectra, centroid-to-outline distances, statistical mean forms, and mean EFFs; all of which are discussed later. One can also use various plotting and utility functions (such as exporting data and other miscellaneous data manipulations).

A.4 File naming conventions and related issues

Before discussing EFF23 in detail, brief mention is made of the procedures used to keep track of the data generated for each specimen.

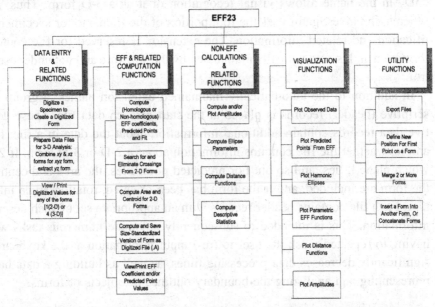

Fig. A.1 EFF23 functional details and data flow. Each box of the second row indicates a major functional category. Each lower-level box indicates one of the actual capabilities of the parent category and is discussed in the text. Many details are omitted from these diagrams. There are over 90 major modules containing several hundred individual subroutines. The total system is comprised of some 32,500 lines of programming code.

A.4.1 The file name schema

A file-naming schema has been implemented which works for MS-DOS-based systems. The full schema is contained in the EFF23 user's guide (Wolfe et al., 1996). Consider the filenames D1001CXY.ΔΔΔ and D1001CXY.PAO, where Δ represents a space (or blank) character. The eight character name D1001CXY is a specimen identifier; in this case, Rabbit skull #D1001, Control side eye orbit, XY view. The extension ΔΔΔ indicates that the first file, above, contains observed (digitized) data points. In the second case, the extension "PAO" means the file contains Predicted, Area-standardized, and Oriented data points computed from the EFF.

Now consider the names D1001T3D.ΔΔΔ and D1001T3D.HΔΔ. Here the eight character name D1001T3D refers to: Rabbit skull #D1001, Treatment side eye orbit, 3D view. The extension ΔΔΔ (3 spaces) indicates, again, that the first file contains digitized (or observed) data points. The extension "HΔΔ" indicates that the second file contains EFF coefficients (harmonics), in this particular case, a 3-D form, that has not been size- (area) standardized or oriented. The use of the

"3D" in the name allows visual recognition of it as a 3-D form. Thus, in this schema, the investigator uses the root portion of the filename for specimen identification and related information. The extension is reserved for file content information and is automatically maintained by the software and not under user control.

In addition to the automatically maintained extension names, a group of descriptive (header) records is placed at the start of each data file. These descriptive data records contain additional information about the data in a file, for example, whether the file contains information about a 2-D or 3-D form. If 2-D, in which plane it lies, and so on. For predicted point files, the area and centroid of the form are included, and if the form has been oriented, rotation-angle information. Each file contains header records with an appropriate set of such descriptive information. This is intended to subsequently automate numerous tasks without having to repeatedly ask the user to (re-) input information at the keyboard, and significantly decreases data processing times, as well as building a data base for representing and analyzing the boundary outlines of objects or forms.

A.4.2 Schema justification

It is very unlikely that one would make use of all the capabilities in EFF23 in any given investigation. The potential number and diversity of kinds of data to be stored requires an ordered method to file naming; hence, the need to use the above schema. Utilization of all EFF23 features can produce some 135 disk files per digitized specimen for 2-D analysis. This number increases to over 450 disk files for 3-D work. Theoretically, it is possible to generate 450 *n files, where n is the number of specimens.

In actuality, a typical investigation would generate the following files: (a) a file of digitized or observed points for each specimen (total of n files where n = number of specimens); (b) harmonic and predicted point files for each specimen (2*n files); and (c) a file representing a mean specimen (either a mean harmonics or predicted points file). At this point, one would have (3*n) + 1 files. The computation of distances, amplitudes, and so on, will add to the total number of files generated, but these are only computed when appropriate to the particular experimental design.

A.5 Digitizing boundary outlines

A.5.1 Equipment

EFF23 is designed to process x,y or x,y,z coordinate point sets input from a digitizer. It supports the Houston Instruments HIPAD digitizer and the Summagraphics SUMMASKETCH III and those fully compatible with either of them. Other digitizer support is being added on an as-needed basis.

Frame grabbers, scanners, digital cameras, and other such equipment may be used to capture an initial digital image, which can be handled if the pixel, or pel, values are properly converted and their number reduced in order for EFF23 to process the derived image.

A.5.2 Digitizing for 2-D

EFF23 assumes all digitizing is done by moving the cursor in a clockwise manner. Failure to digitize in a clockwise manner will result in erroneous results. See Section A.10 for discussion of the problems related to digitizing forms which contain self-intersecting "curves" and how EFF23 deals with them.

A.5.3 Digitizing for 3-D

Digitizing 3-D structures adds complexity in that two 2-D forms, the XY and the XZ planar views, must be combined to create the 3-D form and the third planar, YZ, form which is derived from the 3-D data. This requires increased diligence in applying the point system to specimens and digitizing as the program must be able to align the two forms based on their common x-coordinates. EFF23 is designed to handle digitizing errors up to 0.5 mm in x. If the difference in the x-coordinate exceeds 0.5 mm for any of the points, then the investigator will have to redigitize one or both of the views. Minimization of x-direction differences between the two views can be accomplished by careful alignment, one above the other, when locating the points prior to digitizing.

A.6 EFF23 and homology

Lestrel has treated homology in some detail in Chapter 1. Other authors in this volume have also dealt with it (see Chapters 3 and 4).

EFF23 provides the user with the choice of calculating predicted points from the EFF, which are evenly spaced about a boundary, the traditional method, or "homologous" points instead. The investigator must decide which approach is best for the particular set of specimens and the particular questions being asked. Evenly spaced points are perhaps most appropriate to boundaries with minimal points, and especially those without any readily identifiable points; many unicellular forms are of this category.

EFF23 implements what can be termed a relaxed definition of *biological homology*, not mathematical, or point, homology. The formulation, given below,

places the emphasis on precisely defining the boundary. As an *operational definition of homology* it can be considered as a first step toward the elucidation of
functional and/or biological processes, that is, focus on the extraction of the biological information residing in the boundary across specimens (forms).

An advantage of FDs is their ability to characterize global as well as local elements of form to reduce subjectivity inherent in point selection (see Chapter 1),
provided that boundary points are sampled with sufficient density. With EFF,
points do not need to have equiangular separation, unlike traditional Fourier analysis in polar coordinates (used to avoid a weighted analysis). Thus for digitizing,
a point system is created which incorporates actual homologous points (landmarks)
and points that satisfy operational criteria ("pseudo-landmarks"). In the study of
boundaries, these points will form an equivalence class with, mostly, Bookstein's
Type 1 and Type 2 landmarks with occasional forays into Type 3 (Bookstein,
1991: 64). The points computed from the EFF (the predicted points) are hereafter
called *homologous predicted points*. O'Higgins (Chapter 4) discusses methods by
which one might possibly create such a set of "pseudo-homologous" landmark
points for a variety of boundary shapes.

The first digitized point needs to be fixed at, or mapped to, the same location
on all specimens. The first homologous predicted point is computed to be at the
same location on the Fourier approximating (interpolating) function as the first
digitized point is on the digitized form (observed data file). The second and subsequent homologous predicted points are computed so that they have the same
arc length from the first computed point as their counterpart (pseudo-) landmarks
do from the first digitized point on the original, digitized curve. This in effect
maps the pseudo-landmark points from the digitized curve onto the EFF.

The user can also request that a fixed number (or none) of additional predicted
points be computed between the homologous ones; they are computed as evenly
spaced along the arc between successive homologous predicted points; for example, requesting a single additional point yields a point half way between each
pair of homologous ones. This allows for smoother plotting of a form and may
provide additional useful information about the boundary outline of a specimen.
Figure A.2 shows a digitized form, its predicted points computed as homologous
points and as points evenly distributed along the boundary.

A.7 2-D and 3-D EFF computations

A.7.1 *Untransformed forms*

Once an untransformed (for size and/or orientation) EFF has been computed, the
goodness-of-fit can be tested with the calculation of residuals. The mean residual

error is defined to be the mean Euclidean distance between a set of computed points on the curve and their corresponding digitized or observed points. The computed points have arc-lengths equivalent to the corresponding observed points. Values in tenths of a millimeter are generally obtainable with 20 ~ 40 harmonics. Residuals can only be meaningfully computed using untransformed data (i.e., no positional-orientation or size-standardization), otherwise results will be in error.

A.7.2 Positional-orientation of forms

A.7.2.1 The problem

If one digitizes a boundary outline, in 2- or 3-D, without regard to orientation in space, then in order to compare it with another one, the two must be brought into a common orientation. In addition to methods that phase angles, another approach

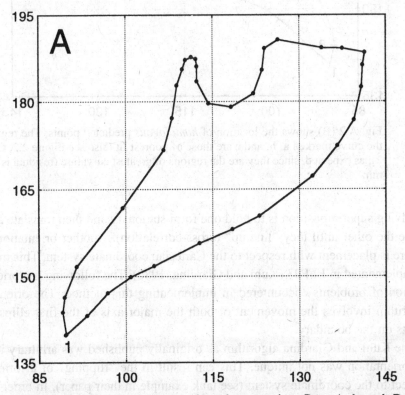

Fig. A.2 Homologous versus non-homologous points. Parts A through D each show the same boundary outline with the first point indicated by a 1 at the lower left portion of the respective panels. (A) is the original digitized file showing the location of the digitizing points being fitted with straight lines.

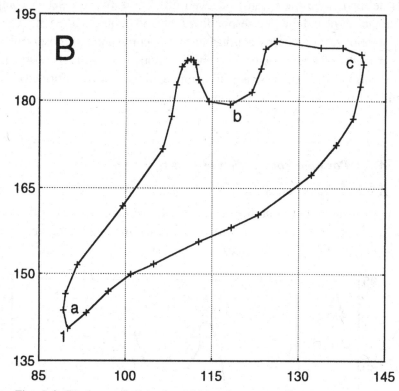

Fig. A.2 (B) shows the location of *homologous* predicted points. The regions of the curve marked **a**, **b**, and **c** are those of poorest fit (also see Figure 2.2, Chapter 2); as expected, since they are the regions of greatest curvature (residual is ≈0.23 mm).

involving superimposition is to hold one form stationary and then translate and/or rotate the other until they "line up" (cross-correlation). Another orientation procedure is placement with respect to the Cartesian coordinate system. This method is implemented in EFF23 (Kuhl and Giardina, 1982). The following is a brief discussion of problems encountered in implementing this method. The orientation algorithm involves the movement of both the major axis of the first ellipse and points on the boundary.

The Kuhl and Giardina algorithm as originally published was arbitrary in that the orientation was not unique. This can result in the "flipping" of a form with respect to the coordinate system (see tank example in their paper). In order to arrive at a corrected orientation algorithm, Kuhl and Giardina's approach has been modified. Use of the unmodified algorithm can, on occasion, appear to yield an image that has been lifted out of the plane, flipped and then placed back into the plane (a mirror image about an axis). On occasion, not only was the imaged flipped

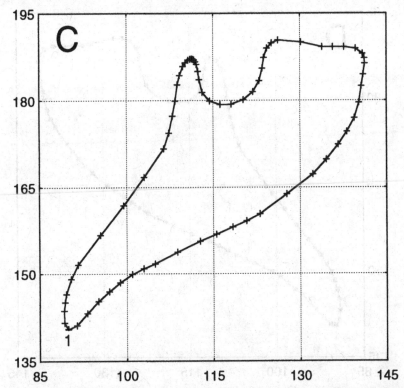

Fig. A.2 (C) is the same as B, but with a nonhomologous point computed between the homologous predicted points; note that the drawing is smoothed when the additional points are incorporated.

about one axis but also about the other axis (in effect, a double flipping). In 3-D, one could have this phenomenon occurring with respect to any one of the three planes. The errors, inherent in the Kuhl and Giardina algorithm, have been corrected in EFF23 by D. Read. Each of the orientations, 2- and 3-D, are taken up separately.

A.7.3 2-D orientation

Orientation in 2-D is done by translating and rotating the first ellipse of the EFF. This results in the major axis of the first ellipse being made parallel to the x-axis and the center of the ellipse translated to the origin (0,0). By orienting the first ellipse in this manner, the higher-order ellipses are also oriented, which results in the actual form being oriented.

After calculating the length of each semi-axis, to identify the major axis, the ellipse is translated and rotated. The program must discern in which quadrant the

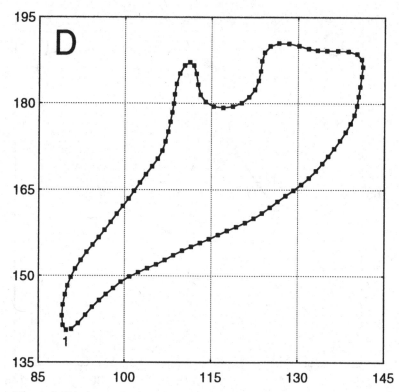

Fig. A.2 (D) shows the location of *evenly-spaced* predicted points. The location of evenly spaced points is dependent on the length of the boundary, whereas the location of homologous predicted points is dependent on the arc-length (distance) of each digitized point from the first digitized point. This cranial base was digitized with 36 points, rather than the usual 54 (see Chapter 14), and was fitted with 18 harmonics.

ellipse axes are directed to always rotate the form in a consistent fashion. The algorithm operates simultaneously on the first ellipse and on the digitized or predicted points comprising the form.

A.7.4 3-D orientation

In E^3 (3-D form), the orientation algorithm must be applied three times to the form. First translation and orientation are done with respect to the XY plane, next with respect to the XZ plane, and finally with respect to the YZ plane. This brings the major axis of the first ellipse parallel to each of the planes (not axes) with the center of the ellipse at the origin of the system (0,0,0). The above methodology will bring all forms into a common orientation and location in space allowing subsequent comparison.

A.8 Size-standardization

In order to size-standardize a form one must first compute the size of a form and then standardize it. Once standardized, one can compute an EFF and predicted points for the standardized form. Lestrel (1989) defined size to be equivalent to area, and EFF23 currently uses area as the measure of size. (The actual computation of area is discussed in Section A.9.)

The current algorithm operates as though the form is being *normalized* to a unit square. After normalization, forms are standardized (resized) to 10,000 square units. Because normalization is not to a true unit square, one does not usually get exactly a 10,000 unit^2 area, although actual experience shows errors to be less than 1%. The reader is referred to van Otterloo (1991) for a detailed discussion of what constitutes a mathematically appropriate size metric.

A.9 Area and centroid computations

A.9.1 General issues

The area of a form is needed to perform size-standardization as noted in Section A.8. EFF23 also uses a form's centroid (a) to compute distances from centroid to boundary; (b) to superimpose the form about a "common" point for statistical and visual comparisons; and (c) for overlay plotting and to center plots on the page.

Biological preparations that can be modeled in 2-D are generally not "simple closed curves" amenable to standard methods for area and centroid computations. Although planar (flat), they need to be viewed as irregularly shaped "wire-frame" models, requiring the use of line integrals derived from Stoke's and Green's Theorems. A wire-frame model has no interior surface, hence one cannot use ordinary integration methods to find the area of the interior and its center of gravity. Line integrals, a topic of advanced calculus, for example, Fulks (1961), integrate along the curve or function itself, not "over" the interior. EFF23 uses an algorithm derived from an algebraic solution of Green's Theorem (Stolk and Ettershank, 1987).

With a 3-D EFF fit of a curve in space, the wire-frame, line integral approach breaks down. As discussed above, the 3-D wire-frame model is only a curve, which has neither interior nor surface points. Therefore it is not a 3-D mathematical solid or surface; it is a (usually) closed 3-D curve in E^3. One can, however, view the 3-D wire-frame model as a minimum (spanning) surface. For example, a soap film covering the wire-frame forms such a minimum surface and it is the area of this surface that is needed for size-standardization (see Chapter 16 for details). Also required is the centroid of the minimum surface. Both the area and centroid are computed with the Surface Evolver developed by Brakke (1992; 1993; see Resources).

A.10 Crossings or self-intersections

A.10.1 Discussion

The line integrals used to compute 2-D area values and centroid coordinates will fail for a body that is made up of several adjacent parts. That is, the form must be a simple closed curve in E^2. Two distinct bodies joined at a point will not appear, mathematically, as a single body, even though visually it appears so. Self-intersections (e.g., a figure-eight) result in such a mathematically composite structure. Therefore, one must either find the area and centroid of each individual "part" and compute (a) a sum of areas and (b) the mean centroid, or alternatively devise a method that will view the object as a single body. The latter is the approach taken in EFF23. It creates a temporary version of the form in which these self-intersections have been removed. For example, the crossing of a figure-eight is "replaced" by a very small space and the curve "reconnected" on each side, resulting in an uncrossed figure. In other words, at each crossing the traverse of the form is treated as though a barrier had been encountered and the traverse is continued along the boundary by always turning to the right.

When searching for crossings, the program examines each of the line segments making up the form, that is, the line segments between each of the closely-spaced points, for an intersection with any other such segment.

The phenomenon of self-intersection is independent of the manner in which the point set is obtained. It is inherent in the form. Any methodology that requires the computation of area, volume, or center of gravity needs to deal with potential self-intersections. Figure 16.12 (Chapter 16) reveals the presence of crossings as a 2-D form is rotated. These crossings disappear when the eye orbit is viewed as a 3-D form because the intersections become portions of the curve in different planes — have different z-values — and thus, apparent intersections vanish. These plots indicate one reason for recasting a problem into a higher dimensional space: self intersections, "knots," and so forth, tend to vanish. Although there may be more variables to deal with, some of the problems often disappear.

Crossings can also occur in 3-D forms. Self-intersections may not be obvious from either of the digitized 2-D forms but appear in either the (generated) YZ form or in the 3-D EFF. EFF23 does not yet handle 3-D crossings.

A.10.2 Digitizing considerations

No portion of the outline should be traversed more than once if at all possible. However, if any portion of the curve is traversed more than once, and if crossings also occur, then as crossings are undone previous crossings can be reintroduced into the intermediate version of the form used for area and centroid computations. One should note that multiple traverses of the curve, whether intersections exist or not, can cause area and centroid values to have indeterminate and undetectable error terms introduced into them.

One may want to consider digitizing separate, identifiable structural components of a specimen and then utilizing the merge or concatenation utilities provided in EFF23 (see Section A.14). Using this approach, one can obtain useful information about each structural component while also dealing with the totality of such components.

A.11 Other EFF23 functions and computations

The range of information that can be computed from an EFF, or a set of EFFs, and derived point files by EFF23, is described here.

A.11.1 Amplitudes, power spectrum, and phase angles

These computations are based on Davis (1986). Among other topics, Davis includes a discussion of Fourier methods and derivations of the equations for amplitudes, and so on. The second edition corrects errors in the equations in the first edition.

A.11.2 Distances

EFF23 can compute several distance measurements from the predicted point files. These distance functions are computable for both 2- and 3-D forms. All distances are Euclidean distances in 3-space (E^3); in E^2, one of the vector components will be zero and "fall out" of the relevant equation. In most instances, the components (x, y, and z directions) of the distances can also be saved to disk files.

A.11.2.1 Centroid to boundary outline

This will give a set of measurements from the centroid of a form to the predicted points on its boundary. With homologous predicted points, one can obtain distance comparisons between identical points on the entirety of a form's boundary, or portions thereof. Let (x_c, y_c, z_c) be the coordinates of a form's centroid, C, and (x_p, y_p, z_p) be the coordinates of a boundary point, P. Then the distance between C and P is

$$d_p = \sqrt{(x_p - x_c)^2 - (y_p - y_c)^2 - (z_p - z_c)^2}. \tag{A.1}$$

A.11.2.2 Baseline to boundary outline

This can be used to compute a set of perpendicular distances from a fixed line, the base, to the boundary outline. This is particularly useful with forms such as the facial profile, where the posterior aspect of the head is bounded with straight lines, since the interest is focused on the profile. With homologous predicted points one can compare distances between forms. Let (x_l, y_l, z_l) be the coordinates of a point, L, on the baseline perpendicular to a boundary point P whose coordinates are (x_p, y_p, z_p), then the distance between L and P is

$$d_p = \sqrt{(x_p - x_l)^2 - (y_p - y_l)^2 - (z_p - z_l)^2}. \tag{A.2}$$

A.11.2.3 Base to target

These are the differences between (a) the distances from the centroid to boundary of one form (the base) and (b) the same measures on a second form (the target). This procedure yields a set of distance differences between the two forms and can be used to display shape differences between them. This should only be used with EFF23 homologous predicted point data; otherwise, there is no guarantee the distance difference function is representing the difference in distance to the same (biological) points for any particular value it might take on. It only makes sense when there is a 1-to-1 correspondence between base form and target form point locations. With evenly spaced points, for example, the tip of the nose, point 10 on one form may be point 13 on another; this function may not yield precise information in that case. If the distance sets, d_{FormA} and d_{FormB} are defined as in Section A.11.2.1, then this function will compute the set

$$BT = \{\delta_i : \delta_i = d_{FormA_i} - d_{FormB_i}\}, \tag{A.3}$$

where $1 \leq i \leq$ number-of-points.

A.11.3 Ellipse parameters

Currently limited to 2-D, this option enables the user to calculate a set of parameters associated with each of the (harmonic) ellipses. They are: (a) the angle the ellipse makes with the x-axis (measured in a counterclockwise direction); (b) the lengths of the major and minor axes; (c) the ratio of minor to major axis; (d) the area of the ellipse; (e) the natural (base e) log of the ellipse area; (f) the common (base 10) log of the ellipse area; and (g) a special estimate of the area of the ellipse (Diaz et al., 1989): If I is the number of the ith harmonic, then the Diaz index, A_{DIAZ}, for the associated ellipse is defined as

$$A_{DIAZ} = (I)^{1.5} * (\text{Area-of-Ellipse}). \tag{A.4}$$

These estimates can be viewed on the screen, printed, and stored in a disk file.

A.11.4 Statistics

A.11.4.1 EFF23 statistical computations

EFF23 can compute means, standard deviations, and standard error of the means for each of the various data types, for example, harmonics (EFF coefficients), observed and predicted points, amplitudes, distances, and so on. One can readily compute mean EFFs and mean observed and predicted point forms.

A.11.4.2 Using other software

By exporting data of interest, in an appropriate format, the investigator can utilize such software as NCSS, SYSTAT, SPSS, STATISTICA, and so on. See Section A.13 for details on exporting data files.

A.12 Plotting with EFF23

A.12.1 Introduction

EFF23 provides several plotting options. At present, all plotting is limited to 2-D. For 3-D plotting the authors recommend using third-party commercial software such as Psi-Plot and EasyPlot (see Section A.15).

There are a number of problems involved in plotting, affecting either hardware or programming or both. A first principle of programming for plotters is: there is no true device-independent plotting. Unless a plotting device's manufacturer provides (a) a subroutine library for controlling the device; (b) the aspect ratio correction factors for each plotting device; and (c) the plotter's x– and y– direction multipliers required to make objects plot at actual size, one needs to experimentally derive these values and write the plotter device driver code. Additionally, one needs to develop programming logic to drive the plotter.

Given the preceding, EFF23 currently supports Hewlett–Packard (HP) 7470 and 7475 plotters and HP LaserJet III printers, and devices 100% compatible with these. Output can also be saved as a set of HPGL (HP Graphics Language) commands in a disk file. Most plots can also be displayed on a (VGA) monitor screen.

A.12.2 Digitized and predicted data plotting

When plotted on paper, the user can request that plots be drawn larger or smaller than actual size to either make the image clearer or fit on the page, and for illustration purposes.

A.12.3 Parametric plots: x(t), y(t), z(t)

Because the 3-D extension of Kuhl and Giardina maintains the parametric formulation (see Chapter 2), the EFF can be viewed as the construct: EFF = x(t) + y(t) + z(t). It is often informative to plot the individual parametric functions. The ability exists to plot one or more of them. They can be plotted separately or as overlays on the same page.

A.12.4 Harmonic ellipses, amplitudes, and distances

Harmonic ellipses can be plotted by themselves and on top of the predicted form. EFF23 can also produce plots of a form's amplitudes and the various distances described above.

A.13 Exporting EFF23 data

A.13.1 Exportable items

EFF23 can export the following data:

- Observed data files
- Predicted point files
- Centroids
- Harmonic Coefficients
- Standard Deviations

- Amplitudes
- Distances
- Standard Error of the Means
- Residuals from the "goodness-of-fit" test

A.13.2 Exported data formats

Most statistics, spreadsheet, graphics, and similar software can read numeric data when recorded in ASCII format along with an indication of how the data are arranged, for example, "comma-delimited" which indicates that values are separated by commas or "space-delimited" which indicates that data values are separated by blanks.

EFF23 can export most data files in any of the following file formats:

- Comma-delimited with or without line numbers with one datum per record — column format. Each variable for a case forms a separate record;
- Comma-delimited with all the data in one record — so called row order format which most statistics packages tend to prefer. All of the variables for a case are in a single record;
- With or without identifying header records in export files;
- Surface Evolver input format.

A.14 Miscellaneous utility functions

The need may arise where data already collected (digitized) require the addition of points not originally contemplated, or the insertion of points of a subsequent structure. The functions described here provide ways to manipulate the data for these purposes. Complex structures may need to be treated as both a whole and as a group of individual structures. In merging (insertion) and concatenation operations, original data files are retained.

A.14.1 Renumbering points

Renumbering of points may be useful when the starting point is inappropriate for further processing and needs to be altered at a later date. Renumbering the points has the effect of "sliding" the point numbers along the digitized image so that there is a "new" first point. The figure, when plotted, will now appear to have been digitized starting at the coordinates of this new first point. All points are at their original locations, only their sequential number has changed. Consider computed forms that need to be plotted with a given first point at a fixed location on the page for overlay purposes. If, for example, digitized point 10 should be the "first point" for analysis, after renumbering, the digitized form's point 10 will be point 1. This may make the visual analysis of the plotted forms more meaningful. Amplitude, power, and phase angle calculations are, however, not affected (refer to Wolfe et al., 1996 for details).

A.14.2 Merging forms

Merging, or inserting, causes "substructure" forms to be combined, creating a single form. For example, starting with separately digitized mandible, maxilla, and cranial base structures, it is possible, using this option, to merge them into a single overall structure (as a single file). The result of this operation is a form, where the data points of one form are inserted between two data points of another form. This is a different operation from concatenation (see below).

A.14.3 Combining forms

Combining (concatenation) will allow, for example, the digitization of mandible, maxilla, and cranial base as separate structures, then, using this option, combining them into an overall structure file. However, concatenating substructures creates a file which contains the input files appended to each other with header records separating them. This is different from merging (above) because here the individual boundaries are recoverable as separate entities.

A.15 Resources

This section contains sources for both EFF23 and third-party software products. The information is correct at the time of its compilation.

- **EFF23** — P. Lestrel, 7327 DeCelis Avenue, Van Nuys, California 91406, United States. Phone: (818) 781-8499 (voice and FAX), Email: **plestrel@ucla.edu**; *or* C. Wolfe, 13376 Dronfield Avenue, Sylmar, California 91342, United States. Phone: (818) 367-6798. Arrangements can usually be made for delivery, installation, and training on use of the software. Contact them for details as to supported hardware, license fees, and so on. The software currently is a DOS version with minimum requirements of MS-DOS 3.1; an 80386 central processing unit (CPU) with a coprocessor (80387); 640 kilobytes of random access memory (RAM) with 550+ kilobytes available for user programs; and a VGA color monitor. *A Windows version* (3.xx and 95) is under development.
- **Standard Fourier Analysis** — A standard Fourier package can also be provided which has been ported from the IBM mainframe environment. It is a 32-bit, MS-DOS environment program which can handle relatively large point samples.
- **VISIO** — [USA/Canada] Shapeware Corporation, 520 Pike Street, Suit 100, Salt Lake City, UT 98101-4001, United States. Phone: (206) 521-4500; FAX (206) 521-4501; or [Outside of USA/Canada] Shapeware International Limited, 20-22 Lower Hatch Street, Dublin, Ireland. International Phone: +353 1 6612036; International FAX: +353 1 66112047. Figure A.1 was prepared using VISIO Professional 3.0 which requires Windows 3.1 or later.
- **Psi Plot** — Poly Software International, P. O. Box 526368, Salt Lake City, UT 84152, United States. Phone: (801) 485-0466; FAX: (801) 485-0480. Excellent 2-D and 3-D plotting capabilities in a large number of formats with the capacity to output to a large number of devices and disk file formats. Some of the plots in Chapters 14, 15, and 16 were created using Psi Plot. A DOS version and a Windows version are available.
- **EasyPlot** — Spiral Software, 15 Auburn Place, Brookline, MA 02146, United States. Phone: (617) 739-1511; FAX: (617) 739-4836. Very good 2-D plotting capabilities; recently has added reasonable 3-D capability. Both DOS and Windows versions are available. Many figures in Chapters 2, 14, 15, 16, and Figure A.2 were prepared using EasyPlot.
- **Surface Evolver** — A DOS environment product. The following is excerpted from Chapters 1 and 2 of the Surface Evolver manual (Brakke, 1993). The order of the material has been altered and layout greatly condensed here.

The Surface Evolver is an interactive program for the study of surfaces shaped by surface tension and other energies ... Evolver was written as part of the Geometry Supercomputing Project (now The Geometry Center), sponsored by the NSF, the DOE, Minnesota Technology, Inc. and the University of Minnesota. . . . The program is in the public domain, and there is no charge for it. . . . Connect to the Internet and use ftp, do an ftp login to **geom.umn.edu** (128.101.25.31) as "anonymous" with any password. . . . For those with PC compatibles, there is a 32-bit DOS version in **pub/evolver.zip**. This PKZIP archive contains **evolver.exe** and the DOS extender program needed to run it. . . . It was made with Free Software Foundation's DJGPP compiler. . . . The manual in TEX DVI format is in **manual.dvi**[2] . . . If you cannot use ftp, you can get Evolver on a floppy disk. Send email to **brakke@geom.umn.edu** or write to

Kenneth A. Brakke, The Geometry Center, 1300 South Second Street, Minneapolis, MN 55454 USA.

- **TEX** — is available in several commercial versions. A German shareware version (with documentation in English and German), including a DVI file printer, is available via **ftp** from the Internet site **hobbes.nmsu.edu**. See also the *HOBBES OS/2 Archived* and *HOBBES OS/2* CD-ROMs from Walnut Creek CD-ROM, Suite 260, 1547 Palos Verdes Mall, Walnut Creek, CA 94596 United States. Phone: 510 674-0783; FAX: 510 674-0821; Email **info@cdrom.com**. Both MS-DOS and OS/2 versions are at **hobbes** and on the CD-ROMs. Walnut Creek also has a CD-ROM devoted to TEX.

Acknowledgments

I want to thank Pete Lestrel and Dwight Read for the hours of excellent and insightful conversation about mathematics, biology, anthropology, and software design. I also thank them for their comments, suggestions, analysis, and coding efforts for program extensions and improvements. And lastly, for assistance in the removal of the design and coding "bugs" that always seem to appear. Programming of this sort is truly a team effort.

References

Bookstein, F. L. (1991). *Morphometric Tools for Landmark Data*. New York, Cambridge University Press.
Brakke, K. A.(1992). The surface evolver. *Exp. Math.* **1**, 141–65.
Brakke, K. A. (1993). *Surface Evolver Manual*, Version 1.91. Minneapolis: The Geometry Center.
Davis, J. C. (1986). *Statistics and Data Analysis in Geology*. (2d ed.) New York: John Wiley and Sons.
Diaz, G., Zuccarelli, A., Pelligra, I. & Ghiani, A. (1989). Elliptic Fourier analysis of cell and nuclear shapes. *Comp. Biomed. Res.* **22**, 405–14.
Fulks, W. (1961). *Advanced Calculus, an Introduction to Analysis*. New York: John Wiley and Sons.

---

[2] To print the manual, you must have access to the TEX system, at least a DVI file printer. If you do not have access, a hardcopy may be requested from Mr. Brakke — *CAW*.

Kuhl, F. P. & Giardina, C. R. (1982). Elliptic Fourier features of a closed contour. *Comp. Graph. Imag. Proc.* **18,** 236–58.

Lestrel, P. E. (1989). Some approaches toward the mathematical modeling of the craniofacial complex. *J. Craniofac. Genet. Dev. Biol.* **9,** 77–91.

Rohlf, F. J. (1985). Private communication to P. Lestrel.

Stolk, R. & Ettershank, G. (1987). Calculating the area of an irregular shape. *BYTE*. Feb. 1987, 135–6.

van Otterloo, P. J. (1991). *A Contour-Oriented Approach to Shape Analysis.* New York: Prentice-Hall.

Wolfe, C. A., Lestrel, P. E. & Read, D. W. (1996). *EFF23 2-D and 3-D Elliptical Fourier Functions Software Description and User's Manual.* PC/MS-DOS Version 2.6. June, 1996. Sylmar, California: Wolfe Consulting Software Engineers.

Glossary

CHARLES A. WOLFE

Part I contains symbols and notations. Part II contains terminology. Terms deemed to be in general use are excluded. **Bold face** words *in a definition* are defined here. Lower case italics are used for coordinate axis and variable names, e.g., *n*, *x*. Upper case italics are coordinate plane names, as in *XY* plane. Matrices and vectors are in bold face, e.g., **M, j**.

PART I

[ ] **interval**, closed.

( ) **interval**, open.

∞ infinity.

≈ is approximately equal to.

∈ is a member of or belongs to.

! factorial, $n! = (n)(n-1)\ldots(1)$.

$\partial x/\partial z$ partial derivative.

Σ summation of terms, often those of a **series**.

Π (upper case pi) a product of terms (multiplication).

$f(x)$ value of the **function** f at the point x.

∫ Integral.

$|expr|$ either absolute value, determinant, or **norm** (size) of *expr*.

$\|expr\|$ the **norm** (size) of *expr*.

E^n Euclidean real *n*-space, where *n* indicates the dimensionality, e.g., E^2 is the (XY) plane; E^3 is 3-space (XYZ space). Also called a *Cartesian space*.

σt (or σ_t) In oceanography, an expression for density of ocean water ($\sigma_{s,\theta,p}$, where s = salinity, θ = temperature, and p = pressure). As σ_t, $\sigma_t = (\rho - 1)*1000$, where ρ = density, t = temperature in C° and σ_t is computed from σ_0 to correct for atmospheric pressure and salinity. If $\rho = 1.0266$ then $\sigma = 26.6$ and is written $26.6\sigma_t$.

435

PART II

2-D See **Two-dimensional**.

26.6σt See σ_t in Part I of Glossary.

3-D See **Three-dimensional**.

3-D digitizer A device that can record a set of **digital** x-, y-, and z- coordinates of a 3-D object. See **digitizer**.

Abscissa Horizontal axis (labeled x) in a 2-D Cartesian (Euclidean) **coordinate system**.

Affine geometry One in which distances are only compared (measurable) on parallel lines with no notion of perpendicularity. Contrast with Euclidean geometry. (Also see **affine transformation**).

Affine transformation A **transformation** in an affine space. It preserves collinearity, hence parallelism and straightness. Examples are *rotation*, *translation*, and *reflection*.

Algebra A symbolic language of generalized operations on a **set**. Arithmetic is *one* algebra for manipulating the set of real numbers.

Allometry Concerned with *differential growth rates* either within or between structures (organs). For example, a change in leg cross-sectional area as a function of weight as one goes from mice to elephants. Compare with **isometry**.

Allophenic (1) Characters (allophenes) in a given cell system, controlled by genes in other cell systems of the organism. (2) Individuals displaying concurrent, alternative cellular phenotypic allophenes.

Alveolar margin Border describing the **lateral** view of a tooth socket.

Amplitude The maximum difference between the value of a **periodic function** and its mean, within a single period or oscillation. In more common terms, the maximum value (or height) above or below the **abscissa**.

Analog (1) A model of, or analogous to, something. (2) Biol. Similar in function, but not necessarily alike in genetic relationship. (3) Math./Engr. Representation of a continuous **function** by a continuously varying measure (e.g., a speedometer *needle*). Compare with **discrete**.

Analytic (1) In mathematics, using algebraic instead of geometric methods. Also see **analytic function**. (2) Breaking a whole into its parts, analyzing an entity; often for identification and/or measurement.

Analytic function/expression (1) Math. Having **derivatives** of all orders and agreeing with its **Taylor series** locally. Also, functions with power series expansions are analytic. Sine and Cosine are power series, hence Fourier series are analytic. Often used in this book in the sense of def. (2) in **analytic**.

Anisotropic Biol. Responding differently to the same external stimulus in different parts of an organism. Compare with **isotropic**.

Anode-to-midsagittal plane distance The distance from the X-ray anode terminal (the tube) to the midsagittal plane. The ratio of this distance to the **midsagittal plane-to-film distance** determines the magnification in the **radiograph**. Multiplication by this ratio is used to correct dimensions/distances; see **cephalometrics.**

Anterior Situated before or looking toward the front; in anatomy the forward part of a specimen.

Anterior Nasal Spine (ANS) The anatomical point that defines the most anterior aspect of the maxillary medial process at the lower margin of the anterior aperture of the nose.

Approximating function A **function** used to approximate another function. The original function is often known only from a table of values. A **Fourier series** fitted to a **boundary outline** is an approximating function for that outline. See also **interpolating function, interpolated** and **interpolant.**

Arc (1) See **Curve.** (2) As a prefix: *arc-*, used to indicate an inverse function, e.g., arc-tangent(t) (written as $\tan^{-1}(t)$, the angle whose **tangent** is t).

Arc length The measure of the length of an **arc** of a **curve** equal to the distance *along* the curve between two points; not a chord.

Arch, dental The curve formed by the dental units (teeth) as distinct from the mandibular border.

Arch depth Dentistry. Any perpendicular distance from the midpoint of the central incisors to a line drawn between right/left dental units.

Arch taper Dentistry. The right/left angulation between anterior and posterior dental units.

Arch width Dentistry. The maximum width between right/left dental units.

Area The space within the interior of a bounded object in a plane. The boundary is distinct from the interior.

Articulare point The intersection, in the lateral view, of the posterior mandibular border with the inferior **occipital** bone.

Aspect ratio The ratio of horizontal to vertical distance.

Auditory meatus Either the internal or external acoustic meatus (passage way) of the ear.

B-point Dentistry. The center of a concavity present at the **anterior** aspect of the **lateral** view of the mandible.

Baculum *Os penis.*

Barycenter Broadly defined, another term for **center of gravity** or **centroid.**

Basion the most inferior point on the **anterior** margin of the foramen magnum in the **lateral** or **midsagittal plane.**

Bending energy From bending moment, which is the *torque* of the *couple* that, together with the tension and shearing force, is equivalent to the total force at a point in a thin elastic "beam." The *couple* is a pair of parallel forces of equal magnitude but opposite direction acting along different lines.

Benthic Region where the ocean environment is in contact with the ocean floor from the high-tide mark to the bottom of the trenches. The *pelagic* includes the entire mass of water.

Bi-euryonic breadth The greatest width of the skull in the lateral view between the *euryon points*.

Biomorph Biological individuals who are members of a class having a similar shape. Although *-morph* is translated as **form**, it is not equivalent to **shape** (see Chapter 1).

Biorthogonal grids The deformation of triangular elements interpolated over the forms' interior to derive a smooth map of shape changes over these elements (Bookstein's (1978) extrapolation of D'Arcy Thompson's transformation grids).

Biostratigraphy The classification, correlation, and interpretation of stratified rocks based on contained fossils.

Bizygomatic breadth The distance between right/left zygomas (cheekbones) in the frontal view (see **zygomatic**).

Bone deposition/resorption Refers to the dual cellular processes by which bone is laid down and removed. Both processes are necessary for growth.

Boundary (outline) (1) Biol. The visual limits of a form. (2) Math. A form is composed of two subsets (boundary and interior); therefore, the boundary is the subset of the set of points comprising a form, which excludes the interior subset.

Boundary Condition Constraints placed on a differential equation to obtain a particular solution. They may be stated as **initial conditions**.

Brachiate Swinging in trees using the upper limbs.

Bregma The junction of the coronal and sagittal sutures.

Buccal/lingual Tooth position or cusps nearest the cheeks/tongue.

Calcaneus Refers to the heel bone.

Calcareous Composed of, or containing, calcium carbonate ($CaCO_3$).

Calvarium (Calvaria) The upper part of the skull.

Cancellous Characterized by a lattice structure such as the spongy tissue of bones.

Canonical axis One of the basis **vectors** in a canonical decomposition of a set of original variables. The set of such vectors forms the basis for a **coordinate system** where each vector's direction defines an **orthogonal** axis.

Carapace The shell-like exoskeleton covering of the head and thorax of crayfish and similar organisms.

Catenary The equation $y = a \cdot \cosh(x/y)$, with a the point of intersection with the y-axis; symmetric about the y-axis. Cosh is the hyperbolic cosine. For example, a uniform chain, freely suspended at both ends.

Center of Gravity The physical point (center of mass) where the object's weight is concentrated no matter how the body is oriented. See **centroid**.

Centriole A body usually found in the centrosome and frequently considered to be the active, self-perpetuating, division center of the cell.

Centroblast A noncleaved follicular center cell (zygomatic cell in a gastric gland).

Centroid The location of a geometrical point within a representation of an object (neutral center), which is invariant to rotation. It is independent of the coordinate system.

Cephalometrics The use of numerical data derived from a cephalometer; an X-ray machine with fixed distances to the head.

Cervical Pertaining to the neck region.

Characteristic root See **eigenvalue**.

Chimeras Tissues, within the same individual, composed of two genetically different cell lines derived from different zygotes (See Chapter 7).

Cladistic A classification of successive evolutionary splittings demonstrating organismal divergence from an ancestor. The **dendrogram** displaying these relationships is called a *cladogram.*

Clivus The slanting **dorsal** surface of the **sphenoid** between *sella turcica* and basilar process of the **occipital.**

Condyle head The rounded eminence of the mandible where it joins the *glenoid fossa* of the **cranial base** (condylar process).

Continuous function Informally, the gradual changes of a dependent variable with independent variable (or variables). Compare with **discrete, piecewise continuous** (see Chapter 2).

Convergence Math. The manner in which a **series** or **integral** is **convergent** to a finite limit.

Convergent (1) With respect to a **series**, the partial sums of the series approach a limit; the limiting value is the sum of the series. (2) Informally, if a **function** is integrable, then the partial sums of an **infinite series,** composed of infinitesimally narrow strips (See **trapezoidal rule**), converge to a finite limit.

Convex function (1) The property that the chord joining any two points in the function's graph lies above the graph. (2) Formally, for arguments x and y in an appropriate abstract space, and t in the **interval** [0,1], we have the inequality $tf(x) + (1 - t)f(x) \geq f(tx + (1 - t)y)$. (See Borowski and Borwein, 1991.)

Convex shape A shape having no interior angle greater than 180 degrees. All lines joining any two points on the boundary of the shape lie entirely inside the shape (See Borowski and Borwein, 1991).

Coordinate-free Any method of representing a **function** or **curve** which is independent of a particular **coordinate system**.

Coordinate system A system for locating points by their coordinates with respect to some set of reference points, lines, directions, and so on. There are numerous systems. A *Cartesian (Euclidean)* system is the familiar rectangular system. A *polar* coordinate system is one in which a pair of coordinates, written (r, θ), locate a point in a plane by means of the length, r, from the origin or pole to the point and the angle, θ, from the single (horizontal) axis. One can convert between Cartesian and polar. Both of these systems can be extended to 3-D.

Coronal suture The junction of the frontal and parietal bones.

Coronoid process A process projecting from the anterior portion of the upper border of the ramus of the mandible.

Cosine (abr. cos) (1) The ratio of the side adjacent the given angle to the hypotenuse of a right-triangle. (2) The complex power **series** $\cos(z) = \sum_{n=0}^{\infty} (-1)^n \cdot z^{2n}/(2n!)$ is used to define the **Fourier Integral Transform** and complex **Fourier series**.

Cranial base The outer and inner margins of the inferior portion of the skull, forming the floor of the cranial cavity, and composed of portions of the maxilla, palatine, temporal, zygomatic, vomer, sphenoid, and occipital.

Cranial vault See Calvaria.

Crenations Biol. A notched or cogwheel-like appearance of shrunken erythrocytes.

Cruciate Having a cross-like shape.

Cubic spline (fitting) The piecewise approximation of a curve by cubic polynomials (degree or order 3). A cubic polynomial is fitted to each **interval** between the successive pairs of points, producing a family of **curves**. Cubic splines guarantee that the fitted curve will pass through each of the original data points. Contrast with **polynomial curve fitting**.

Curvature (κ) Curvature can be variously calculated and described as the rate at which a paddle-wheel would turn when placed in the "flow" of the curve, i.e., if the curve is viewed as a flowing stream. Formally, the rate of change of inclination of the **tangent** to a **curve** relative to the **arc length**.

Curve (1) A synonym for arc: any continuous section of a graph or geometric figure; formally the image of the unit **interval** under a **continuous function**. (2) Synonym for (the graph of a) function.

Cytoskeleton The structural framework of a cell.

D^2 See **Mahalonobis's D^2**.

Dendrogram A tree-like representation of hierarchical relationships. Used in morphometrics to represent phenograms, phylograms, cladograms, and other generational-like relationships.

Densitometric analysis An analysis based on exposed and processed film opacity measurements from a densitometer.

Dental arch See **arch**.

Dentoalveolar See **alveolar margin**.

Dextral Right–handed or the right side. Contrast with **sinistral**.

Digenesis (Diagenesis) The alteration of the generations.

Digital Refers to a discrete representation using a binary system. For example, coordinates (x,y) stored as binary (0s,1s).

Digitized Point See Observed Point.

Digitizer A device consisting of a pad (grid) and a pointer. The pointing device (crosshairs) transmits the coordinates (x,y) of each point to a computer, in **digital** form. See also 3-D digitizer. *Digitizing* refers to the process of creating a model of an observed form from a **specimen** by entering a **discrete** series of points.

Direction angles The angles, or direction cosines, that a line in space, or **vector**, makes with the positive x-, y-, and z- axis of a Cartesian **coordinate system**. They are sufficient to determine the orientation of the line or vector.

Discrete Refers to a noncontinuous representation of data, in opposition to **analog**. A discrete variable has a fixed number of values over an interval or involves counts such as the number of bones in a hand.

Discrete Fourier Transform See Finite Fourier Transform.

Discrete spectrum See Harmonic spectrum.

Dispersion matrix A matrix consisting of the variances and covariances of n variables. The diagonal contains the variances, the off-diagonal elements are the covariances. These matrices are central to all multivariate morphometric methods such as **PCA**, **MANOVA**, and **factor analysis**.

Distal (1) Away from the point of attachment; the hand is the distal part of the arm. Opposite of proximal. (2) Toward the posterior aspect of the dental arch. See **mesial**.

Dizygous Fraternal twins derived from two fertilized ova at a single birth (DZ).

Dolichocephaly The condition in which the length-breadth index of the head is 75.9 or less.

Dolichomorphic Marked by a long or narrow form or build.

Dorsal Pertaining to the back/top; in human anatomy the **posterior** part.

Dorsum sellae A plate of bone forming the **posterior boundary** of the *sella turcica*.

Driving harmonic The harmonic (Fourier coefficient) with the largest magnitude (See Chapter 11).

Dural/Dura Pertaining to the fibrous membrane forming the outermost covering of the brain and spinal cord.

Dysplasia (1) Abnormal development or growth, especially of cells. (2) The extent to which an individual presents abnormally different components (somatotypes) in different bodily regions.

Dyspnoea Difficult or labored breath.

Ecotone boundary The boundary between two ecological communities.

Ecotypes Populations representing local genetic adaptations based on one or more adaphic, microclimate, or biotic factors.

EFF The elliptical Fourier function representation of a form's shape or outline.

Eigenshape analysis A two-stage process composed of Zahn and Roskies' **FD** followed by **principal component analysis (PCA)**. (See Rohlf and Bookstein, 1990.)

Eigenvalue (1) For a **matrix A**, a root λ of the characteristic equation $\det(A - \lambda I) = 0$, where $\det()$ is the determinant. (2) More generally, for a linear operator **A**, a solution λ of the equation $AX = \lambda X$, $(X \neq 0)$. Statistical techniques, for example, **factor analysis** and **principal components**, are based on eigenvalues and associated **eigenvectors**. See Davis (1986) for a simple geometric interpretation.

Eigenvector For a **matrix** or other linear operator, **A**, a nonzero **vector X**, such that $AX = \lambda X$, $(X \neq 0)$, where λ is an **eigenvalue** of A. See Davis (1986).

Elytra The forewings of beetles.

Endocranial Pertaining to the inner bony margin of the skull.

Endothelial cells Thin, flat, plate-like cells forming the lining of the heart, blood, and lymphatic vessels.

Entropy A measure of the degree of randomness of energy or organization in a "system" (See any text on Information Theory). Every spontaneous process in nature is characterized by an increase in the total entropy of the bodies concerned in the process.

Epistasis The suppression of the effect of one gene by another, as in the suppression of genetically determined pigment variation in albinos.

Epithelial Pertaining to *epithelium*, a tissue composed of contiguous cells with a minimum of intercellular substance. It forms the epidermis and lines hollow organs and various passages.

Euclidean Distance Matrix Analysis (EDMA) The comparison of forms based on matrices of Euclidean distances between equivalent landmarks (see Lele and Richtsmeier 1991).

Even function (1) Fourier analysis. A **function** composed of **cosine** terms only; the **sine** terms vanish (equal to zero). Even functions are measures of **symmetry**. (2) Math. A function which fulfills the criteria $f(x) = f(x + t)$; where t is the function's **period**. See **odd function**.

External auditory meata See **auditory meatus**.

FD Abbreviation for Fourier descriptor.

Factor analysis Several techniques, based on the **characteristic equation** of a rotated **dispersion matrix** where the resulting vectors need not be orthogonal. Results in a smaller number of variables that can be used to explain relationships among the original ones.

Fast Fourier transform (FFT) The name given to a class of methods to simplify the evaluation of the **Finite (discrete) Fourier transform** or its inverse. For example, for a 1024-point sample, the number of computations reduces from four million to 40,000.

Finite (Discrete) Fourier transform Application of the series (or its inverse) $X(k) = \sum_{n=0}^{N-1} x(n)W_n^{kn}$, where $W_n = e^{-j2\pi/N}$, j is $\sqrt{-1}$ and N is the number of observations. See also **fast Fourier transform**.

Finite series See **infinite series**.

Finite element analysis Originally devised for computing stresses on engineering materials, has been adapted to characterize shape changes. Shape differences are described in terms of directions and magnitudes of the **principal strains** (usually given as **vectors** or **tensors**) in the **transformation** of one form to another.

Follicular adenoma A benign tumor of a lymph gland.

Foraminifera A mostly marine amoebae that builds a shell.

Foraminifera side keel Demarcation or juncture of the umbilical (shell) and spiral sides.

Form (1) A model composed of aspects, including size, shape, color, spatial orientation, surface patterning, interior structure, and so on, as detailed in Chapter 1. (2) An imprecise synonym for shape.

Fossa A pit, hole, or depression in a bone.

Fourier analysis The separation of a complex waveform into its sinusoidal components. It consists of **harmonic synthesis** and **harmonic analysis**.

Fourier descriptor (FD) (1) The fitting of a Fourier series to represent outlines as wave forms. (2) Any of the values derived from **Fourier analysis**. For example, **harmonic (Fourier) coefficients**, **amplitudes**, and **power spectra**.

Fourier spectrum See **harmonic spectrum**.

Fourier series An **infinite series** of the form $\frac{1}{2}a_0 + \sum_{i=1}^{\infty} [a_i \cdot \cos(ix) + b_i \cdot \sin(ix)] = \frac{1}{2}a_0 + a_1 \cos x + b_1 \sin x + a_2 \cos 2x + b_2 \sin 2x + \ldots$, where the a_i and b_i are the Fourier coefficients. In practice, the limit of summation (largest value of i) is finite, not infinite, yielding a finite Fourier series.

Fourier transform See **finite Fourier transform** and **integral transform**.

Frame grabber A device that can "grab" an image of a single video frame and store a **digital** copy. Stored as a series of **pixel** coordinates (x,y) and associated color or **gray scale** values.

Frankfort horizontal or plane An anthropometric plane originally for skull orientation, modified for **cephalometrics**. Defined as the line joining the points *porion* and *orbitale*, averaged for right/left.

Frequency The number of times a **periodic function** repeats itself over an interval, usually, $[0, 2\pi]$. Also, the reciprocal of the **period** of the function. See also **fundamental frequency** and **harmonic frequency**.

Frequency domain It is the result of the mapping from either the time or spatial domains. The primary frequency components are **amplitudes** and **phase angles**. It represents frequency aspects of a waveform. Compare with **time domain** and **spatial domain**.

Function Given two **sets**, S and T, a function f is a mapping (rule) from S to T such that for any member of the *domain* set S, one can find a corresponding value in the *codomain* set T. The functions in this text are such that S and T are either the real numbers or occasionally the complex numbers as in the definition of the **integral Fourier transform**.

Fundamental frequency The frequency associated with the first, lowest, harmonic.

Generalized Distance See **Mahalonobis's D^2**.

Genome A complete set of hereditary factors, such as is contained in a haploid chromosome.

Glabella A point in the **sagittal** plane of the bony prominence (forehead) midline to the **supraorbital** ridges, defining the most anterior projection of this region with respect to a plane.

Glial Pertaining to neuroglial; the fibrous and cellular, nonnervous, supporting elements of the nervous system, chiefly derived from ectoderm.

Glenoid fossa A pit or depression on the temporal bone of the **cranial base**.

Gnathion The most anteroinferior point on the chin in the **sagittal plane**. See *menton*.

Gonial angle The mandibular angle between the *gnathion-gonion* line and a line from *gonion* to the most posterior aspect of the **condyle head**. Other definitions are also used.

Gonion The most inferoposterior aspect of the junction of the ascending ramus with the mandibular body.

Goodness-of-fit (1) Stat. The extent to which observed sample values of a variable approximate the values derived from a theoretical distribution. (2) EFF23 software: The extent to which observed sample values of a variable (digitized boundary outline) are approximated by values derived from a theoretical distribution (the elliptical Fourier function fit). See computation of residuals in the Appendix.

Group theory A branch of mathematics dealing with **sets** known as groups. A group is a set that is closed under an *associative binary* operation, which also means that there is a unique identity element within the set. Further, every element in the set has an inverse with respect to this identity element. For example, the integers form a group under addition but not under division; the identity element is 0, $a + 0 = 0 + a = a$; the inverse is $-a$: $a + (-a) = 0$. There is no integer that functions in division as 0 does under addition: If a and b are integers, a/b is most likely not an integer, hence the set would not be closed under this operator. An *associative binary* operator takes two arguments and the bracketing of arguments can be disregarded: $a + (b + c) = (a + b) + c$.

Harmonic Is comprised of a sine and cosine term preceded by their respective coefficients, for example, $a\cos \pm b\sin$.

Harmonic analysis The decomposition of wave forms into separate **orthogonal** Fourier components.

Harmonic distance Composed of both **amplitudes** and **phase angles**. Analogous to the average taxonomic distance coefficient. See Chapter 5.

Harmonic synthesis The summation of terms in a Fourier series.

Heterochrony (1) Variation in time relationships. (2) Departure from the typical sequence in the time of formation of organs or parts.

Heuristic Using or obtained by informal methods or reasoning from experience, often since no precise algorithm is known or is relevant. Contrast with "mechanical" or (strictly) algorithmic.

Hominid The family of modern man and his direct ancestors; adjective form of *Hominidae*.

Hominoid The super-family that contains apes and the genus Homo as well as their ancestors; adjective form of *Hominoidea.* .

Homogenous Biol. (1) Composed of parts all of the same kind. (2) Of or derived from an individual of a closely related or similar strain of the same species. (3) Of two or more individuals, genetically or otherwise identical in certain characteristics.

Homogenous Math. A **function** (on a **vector space**), having the property that $f(tx_1, \ldots, tx_n) = tf(x_1, \ldots, x_n)$ for every nonzero scalar t.

Homologous (1) Math. Playing the same role in distinct but related figures, points, or functions. For example, the corresponding points of a figure and a **projection** (mapping) of that figure. (2) Biol. Possessing the property of **homology**.

Homology Biol. Similarity due to evolutionary origin; applied to structures, organs, and character states resulting from descent from a common ancestry. These entities may or may not have the same function. Refer to Chapters 1, 4, and 5.

Homeomorphism (also **Homomorphism**) (1) Biol. Corresponding in form or external appearance but not in type of structure and in origin. (2) Math. A mapping from one algebraic structure to another under which the structural properties of its domain are preserved in its range in the sense that if * is the operation on the domain and $\otimes$ is the operation on the range then $\theta(x * y) = \theta(x) \otimes \theta(y)$. An **isomorphism** is a one-to-one (bijective) homeomorphism. An automorphism is a homeomorphism where the domain and range sets are identical.

Homoplasy Correspondence(s) in form or structure owing to similar environments.

Hormion The point in the **sagittal plane** between the alae of vomer, where the **vomer** is attached to the body and the sphenoid bone.

Hydrocephalus Excessive water in the brain-cranium space causing enlargement of the head and putting pressure on cranial structures.

Hydrographic conditions Surface water conditions; also hydrographic itself is used with regard to the measurement, description, mapping, and so on, of surface waters.

Hypophyseal fossa A depression in the sphenoid bone containing the hypophysis, an endocrine gland.

Hyperspace Math. Refers to a "space" with more than three dimensions. Such spaces are generated by **principal components, factor analysis**, and so on, where each variable constitutes a dimension (one of the basis variables).

Inferior Situated below or looking up towards the bottom. In anatomy the bottom part of a specimen.

Infinite (Finite) series A **series** arrived at from an infinite (finite) number of terms. If the sequence of partial sums of the series tends to a limit, the series converges and its sum exists. If the series is not **convergent**, then the sum is either undefined or infinite and it is divergent.

Inflection point A point of change in the graph of a function from rising to falling, or falling to rising. Determinable from examination of the second **derivative's** value at the point.

Initial Condition Math. A set of starting values for the variables involved in a **function**.

Instar Refers to larval stages.

Integral transform A **transformation** of a **function** $f(x)$ by an **integral** operator which is defined in terms of an *integral kernel*, for example, the Fourier transform whose kernel in the complex plane is $(1/2\pi)e^{ixy}$ and whose range is $[0,\infty]$. The Fourier transform should not be confused with the **finite or discrete Fourier transform** (nor the **FFT**) which although a **transform**, is not an *integral* transform. Some of the authors in this volume tend to use Fourier transform in the sense of the integral transform, whereas others mean the finite Fourier transform or an FFT.

Interpolating function A function that is fitted to a set of data, allowing for the computation of an approximate value to be used between actual data values. See also **approximating function, interpolant,** and **interpolated**.

Interpolate (1) To estimate a value of a function between values already known. (2) To approximate a function using a simpler function with given interpolating values, for example by Langrangian interpolators, Bezier curve, or splines.

Interpolant (1) A value that results from use of an **interpolating function** which permits one to make a reasonable estimate of the unknown value. (2) An **interpolating function**.

Interval A **set** containing all of the real numbers or points between two given real numbers or points a and b, for discussion we assume $a < b$. A *closed interval*, written $[a, b]$ includes the endpoint values: $a \leq x \leq b$, for all x in $[a, b]$. An *open interval*, written (a, b), does not contain the endpoints: $a < x < b$ for all x in (a, b). An interval may be *half-open* or *half-closed*, containing only one of its endpoints. An interval half-open {half-closed} on, say the right {left}, is written $[a, b)$ and $a \leq x < b$ for all x in the interval {$(a, b]$ and $a < x \leq b$ for all x in the interval}.

Isometric (1) Math. A method of projecting (**projection**) a figure in **3-D** having the three axes equally inclined and all lines drawn to scale. (2) Biol. Indicating or having equal measure.

Isometry (1) Math. An automorphism or **homeomorphism** that preserves distances. (2) Biol. (a) Having equal measure in several directions. (b) During growth, maintenance of equal distance, volume, and so on.

Isomorphism/Isomorphic (1) Math. A one-to-one correspondence (mapping) between the elements of two or more **sets** that preserves the structural properties of the domain. See **homeomorphism** and **isometry**. (2) Biol. Having similar or identical structure or appearance.

Isotropic (1) Math. In tensor theory, have components that remain unchanged under an arbitrary change of basis (*very loosely* "coordinate system" representation). (2) Biol. Growth with equal rates in the x and y (and possibly z) directions; compare to **allometry**.

Knuckle-walking Locomotion with the knuckles of the forelimbs in contact with the ground.

Lambdoid suture The junction of the **occipital** with the **parietal**.

Landmark Anatomical points considered to be useful in morphometric studies for phylogenetic and other purposes.

Latent root See **eigenvalue**.

Linearize To place data (often a tabulated **function**) into a linear form via a **transformation**. For example, conversion to logarithms. Analysis is often made easier by linearizing variables (data). Also See **rectify**.

Line integral (also known as a curvilinear integral). Informally, an integral computed by integration along a curve. It can be used to compute the work done moving a particle along a given path, or the area enclosed by a wire frame (See Chapter 16).

Lymphoblasts A blast cell considered to be the precursor to a lymphocyte.

Macrophage A phagocytic cell, not a lymphocyte, belonging to the reticuloendothelial system.

Mahalonobis's D^2 (Generalized Distance) A generalized statistical distance measure of taxonomic distance. It expresses the distance (D^2) between two forms in a Riemannian space where the axes are not necessarily **orthogonal**.

Mandible The lower jaw.

Mandibular Plane A line joining either *gnathion* and *gonion* or *menton* and *gonion*.

Mangabey Any of the several species of the African long-tailed monkeys of the genus *Cercocebus*.

MANOVA Stat. Multivariate Analysis of Variance. An extension of ANOVA for testing the equality of mean vectors. Required for the analysis of two or more samples, each of which contains more than a single variable comparison.

Matrix (1) Math. A rectangular array of elements, usually scalars. The elements are in rows and columns. Matrices are used in the transformation of coordinates, *and*, in statistical techniques classified as linear models: ANOVA, PCA, and so on. Usually indicated by square brackets, [ ] or parentheses (). An $m \times n$ matrix has m rows and n columns. An element is denoted as a_{ij} and is the intersection of row i and column j. A **vector** can be represented as a $1 \times n$ (row) or an $m \times 1$ (column) matrix. See texts on linear algebra, matrix analysis, and statistics. (2) Geol. The finer-grained portion of a rock in which coarser materials are embedded. (3) Biol. The intercellular substance of tissue.

Maxilla The upper jaw.

Medial axis transform A procedure that maps a 2-D outline onto a curve. The locus of points within the interior of a form that lie equidistant from the boundary (See Blum, 1973).

Menton The most inferior part of the chin with respect to the **mandibular plane**. Often confused with *gnathion*.

Mesial Toward the midline of the dental arch. See **distal.**

Metameric series From metamere — one of a series of homologous segments. E.g., the vertebral column. Metamerism refers to a body made up of a succession of metamers.

Microlamines A very small thin plate.

Microvilli (pl.) Any of the nonmotile, free cell-surface evaginations which, depending upon location, may be absorptive or secretory in function.

Midsagittal plane The median or lateral plane (*norma lateralis*).

Midsagittal plane-to-film distance Distance from the X-ray film to the **midsagittal plane.** See **anode-to-midsagittal plane distance.**

Minimum Spanning Surface See **Minimum Surface.**

Minimum Surface A surface of minimum tension. A minimum surface film is one where all surface points are under minimum stress, for example, soap films and bubbles. See Chapter 16.

Miospore Refers to members of the order Microsporidia of the class Sporozoa (parasitic protozoans).

Molt stages See instars.

Moment (1) Stat. The first moment is the mean, the second moment is the sum of the squares of the deviations (variance). (2) Math. Given a function $f(x)$, and its density $\delta(x,y)$ then one can define moments such as $Mass = \iint \delta(x,y)dA$.

Monocyte A large mononuclear leukocyte.

Monotonic increasing (decreasing) A **series**, or **function**, consistently increasing (decreasing) in value so that: for increasing, $f(x_1) > f(x_2)$, for all $x_1 > x_2$, and for decreasing, $f(x_1) < f(x_2)$, for all $x_1 > x_2$.

Monozygous Identical twins developed from a single fertilized egg (MZ).

Morphology The study of the structure and form of organisms.

Morphometrics The measurement of the forms and structures of organisms.

Myelin tubes The tubes about some nerves, which is composed of myelin, a white fatty substance.

N-**Leafed rose** The form of the graph produced by either of the families of curves $a \sin(n\theta)$ or $a \cos(n\theta)$ as n increases. N refers to the number of lobes ("rose petals") in the graph.

N-**tuple** A **set** of numbers and/or names separated by commas and enclosed in bracketing symbols. N is the number of elements in the set. E.g., a point's coordinates, (x,y); a **vector**; the model of form in Chapter 1.

Nasion The point where the midsagittal plane intersects the most **anterior** aspect of the frontonasal suture.

Nasolabial curve The curve between the upper lip and the nose.

Necrolysis Dissolution or disintegration of dead tissue.

Neural canal In embryology, the vertebral canal.

Neutrophile A highly motile and phagocytic leukocyte.

Norma basilaris (also *norma ventralis*)The **inferior** aspect of the skull.

Norma frontalis (also *norma facialis*) The **frontal** aspect of the skull.

Norma lateralis The **lateral** aspect, or profile view, of the skull.

Norma occipitalis The **posterior** aspect of, or view from behind, the skull.

Norma verticalis The **superior** aspect of, or view from above, the skull.

Normal (Normal vector) A **vector** that is **orthogonal/orthonormal** to another vector. A **unit normal vector** is one that has been divided by its original length to have a length of one unit; it has been **normalized**.

Normalize Normalization usually reduces the object's size so that it fits into a unit measure space. Normalization does not affect shape. Irregular objects can be normalized to a unit circle or sphere. There are numerous techniques used to normalize. **Normal**, which implies perpendicularity, should not to be confused with normalize. See also **size-standardization** (See van Otterloo, 1991).

Nosology The science of the classification of disease.

Numerical taxonomy A classification based on measures of affinity composed of a large number of dichotomous or coded characters (See Sneath and Sokal, 1973).

Nyquist Frequency The highest frequency that can be detected to avoid aliasing. It is equal to 1/2 the number of points used to sample the outline.

Observed Point A member of the **set** of **digitized** points that comprise a form as a **tabular function**.

Occipital The posteroinferior aspect of the base of the skull.

Odd function (1) In Fourier analysis, a **function** comprised of **sine** terms, the **cosine** terms being identically zero. Odd functions measure **asymmetry**. (2) In mathematics generally, a **periodic function** that meets the criteria $f(x) = -f(x + t)$, where t is the period. See **even function**.

Ontogeny/Ontogenetic Refers to the developmental history of an individual organism, from initial cellular formation to death. Contrast with **phylogenetic**.

Operational taxonomic unit (OTU) A collection of objects (biological specimens), each member of which is described with a set of measurements which becomes a data set.

Opisthocranion The point in the **midsagittal** plane of the **occipital**, marking the **posterior** extremity of the longest diameter of the skull, measured from the *glabella*.

Optical Fourier Transformation See Chapter 17.

Orbitale The lowest point of the inferior margin of the orbit, used in conjunction with the *porion* to orient the skull in the **Frankfort horizontal plane**.

Ordinate The vertical (y) axis in a 2-D Cartesian coordinate system.

Oriented (form) A form whose representation has been rotated in space to a known, fixed orientation to allow statistical comparison.

Orthodontic Structural alterations primarily affecting teeth. See **orthopedic**.

Orthogonal, orthogonality The mathematical (geometric) notion of being at right angles (90 degrees), or perpendicular. See **orthonormal** for a more formal definition.

Orthonormal Two elements in any **algebra** are orthonormal if their product, however defined, is zero. A possibly infinite set of functions $f_1, f_2, \ldots, f_n$, that satisfy the identity on some interval of integration (a,b), $\int_a^b w(x) f_i f_j = \begin{cases} 0 \text{ if } i \neq j \\ 1 \text{ if } i \neq j, \end{cases}$ where $w(x)$ is a weight function. See Chapter 2.

Orthopantomograms A proprietary X-ray device used in dentistry to make panoramic radiographs.

Orthopedic Structural alterations affecting bone in contrast to teeth. See **orthodontic**.

Osteoma A benign bony tumor.

Osteology The study of bone.

Ostracod A genus of sea-dwelling Arthropods.

OTU See **Operational Taxonomic Unit.**

Overdominance Genetics. The interaction of genes to produce a superior phenotype in a heterozygote. Another term for heterosis or hybrid vigor.

Pan Chimpanzee.

Period (1) Math. The smallest interval after which a **periodic function** takes on identical values, that is, the values repeat from period to period. The constant, k (the period), where $f(s) = f(x + k)$ for all x. For example, $\sin \theta = \sin(\theta + 2\pi) = \sin(\theta + 4\pi), \ldots, k = 2\pi n$ is the period for $\sin \theta$ for all integers, n; its principal period is 2π. (2) Geol. A subdivision of a geologic era.

Periodic Function A **function** with values that are repeated for all **integral** multiples of a constant increment of the independent variable. See **period**.

Perturbation (1) The introduction of a small change in the values of a parameter(s), to obtain a solution in the study of stability in a system. (2) Boundary irregularities (invaginations) have been defined as contour pertubations. See Chapters 11 and 13.

Phase The shifting of one **periodic function** with respect to another. In 2-D either along the horizontal axis (the independent variable) or as rotation in polar coordinates. The difference is referred to as the "lag time" or **phase angle**.

Phase angle The angular value by which **phase** is measured.

Phenetics (1) A classification based on morphological similarity, without consideration of evolutionary relationships. (2) Studies of variation between **OTUs**, the lowest-ranking taxa employed in a given study. The results of such studies are **dendrograms** known as phenograms.

Phylogeny/Phylogenetic Refers to the evolutionary history of a group, not the individual as in **ontogeny**.

Physiognomy Dimensions, color, and the shape of constituent elements (nose, forehead, profile, etc).

Piece–wise continuous function In an interval, a **function** that is continuous and continuously differentiable, except at a point(s) where the derivative has a jump discontinuity.

Pixel (Pel) Acronyms derived from the term "Picture Element." The smallest illuminated area on a video display tube or smallest printed area (dpi) generated by a printer. The size and number of pixels governs the degree of resolution obtainable with a **digital** image.

Planktonic The aggregate of floating organisms in a body of water.

Plantar surface The surface of a limb (foot) in contact with a substratum (ground). *Plantar ground reaction force* is the force of the ground against the plantar surface in walking.

Pleiotropism (Pleiotropy) The occurrence of multiple phenotypic effects produced by a single gene.

Pogonion **(PO)** The most anterior aspect of the chin in the midsagittal plane with respect either to the **mandibular plane** or the Frankfort plane.

Polynomial curve fit Approximating a **function** by means of polynomial equations; usually via a least squares. Compare with **cubic splines**.

Pongo The Orangutan.

Porion The point of the upper margin of the **auditory meatus**. *Porion* and *orbitale* determine the **Frankfort horizontal**.

Position vector A **vector**, the components of which are the coordinates of a given point; the directed line from the origin to that point. A vector definition of **radius**. See also **radius vector**.

Posterior Situated behind or looking toward the back. In anatomy, the rear part of a specimen.

Power spectrum The plot of the variances versus increasing harmonic number (periodogram).The variances are computed as 1/2 the square of the amplitudes. Also called the discrete or raw spectrum.

Predicted Point A member of the set of points computed from a Fourier function (e.g., FD, EFF, FFT). The set of point coordinates represents the equation of the curve fit.

Presacral Lying in front of the sacrum.

Principal Component Analysis (PCA) A **dispersion matrix** decomposition method based on solution of the **characteristic equation**. Each **eigenvector** is assumed to be a canonical representation of some aspect of the specimen sample with no *a priori* assumptions about an underlying model. Contrast with **factor analysis**.

Principal strain Either the **vector** in the direction of the maximum elastic stress or its magnitude.

Procrustes analysis (methods) Techniques applicable for data where there are marked departures in variability across forms in localized regions, for example, forms identical to each other with the exception of an appendage. For such forms, least squares is inappropriate.

Prognathia See prognathism.

Prognathism Having a projecting lower/upper jaw.

Projection Mapping the image of a figure onto a plane via a mapping **function**. There exist several such mappings including orthogonal and perspective mappings known from technical drawing.

Projective geometry Math. Concerned with the study of properties of geometrical figures that are invariant under **projections**.

Prosthion Defined as a point at the alveolar margin of the central incisors.

Pseudo landmarks Landmarks for which homology may not be precisely definable. However, they are operationally useful. See Chapters 1 and 5; also, Sneath and Sokal (1973).

Q-mode correlation A matrix analytic technique for characterizing interrelationships between specimens.

Quadrumanal An animal with four limbs that are all adapted for use as hands (monkeys).

Radiograph Synonym for an X-ray.

Radius Math. A straight line connecting the center of a circle or sphere to any point on its circumference, or the length of this line. See **position vector** and **radius vector**.

Radius vector A *directed* line from the origin of a **coordinate system** to a point in space and sweeping out a **curve**. Related to the notion of **radius** in a polar **coordinate system**.

Ramus A bony projection or process. The upper ramus of the mandible.

Rectify (1) Math. Determining the length of a curve between two points, for example, calculating **arc length**. (2) Geom. Approximating a curve by fitting straight line segments to it. (3) Elect. Eng. Making all values of a periodic function positive. For example, converting of A.C. to D.C. current.

Recurrent Recurring; returning periodically. (1) Anat. turning back in the opposite direction. (2) Math. A multivalued function.

Reflection A planar **transformation** in which the direction of one axis is reversed. For example, $x'=x, y'=-y$ or $x'=-x, y'=y$.

Roentograph Synonym for an X-ray.

Rotation (1) Analytic geometry. A **transformation** resulting from turning the entire plane about a fixed point in the plane. It is a rigid motion that preserves shape. E.g., $x' = x \cos \theta + y \sin \theta$, $y' = -x \sin \theta + y \cos \theta$. (2) Vector analysis. Another term for Curl, the path taken by a "fluid" flowing down a curved surface.

Retrognathia Posterior Deviation of the mandible.

Sagittal plane The median (midsagittal) plane of the human body or any plane parallel to the median plane.

Scaling A change of scale as in **size-standardization** (geometric) or via some other **transformation** (**analytic**).

Sclerotomal Pertaining to the sclerotome, which is either (a) the fibrous tissue separating successive myotomes in some lower vertebrates, or (b) the part of a mesodermal somite which enters into formation of the vertebrae.

Sella turcica The superior portion of the body of the sphenoid bone that surrounds the **hypophyseal fossa**. It includes tuberculum sellae, anterior clinoid processes, and the **dorsum sellae**.

Sequence An ordered set of objects (for example, numbers) capable of being indexed by the natural numbers or an initial segment of them. Compare with **series**.

Series See **infinite series**.

Set Math. A collection of items. Denoted by enclosure with curly brackets. The set may be referred to by a name such as S. Membership in a set is either by enumeration (the members are specifically identified) or by a rule for inclusion. For example, $S = \{x \mid x = $ a user of Fourier descriptors$\}$; read as S is the set of all x where x is a user of Fourier descriptors.

Set, closed Informally, a closed set includes its boundary (points).

Set, open Informally, an open set does not include its boundary (points).

Sexual dimorphism The existence of males and females.

Shape Refers to the outline (boundary) of an object, not to be confused with **form** — See Chapter 1.

Shoaling — isopycnal Collecting, usually fish, from waters having the same density. From the Old English *shoal*, a school of fish.

Sigmoid "S"-shaped.

Simple Closed Curve A continuous curve in a plane that does not cross itself (self intersections), but meets at the ends. E.g., if $y = f(x)$, then $y(t_1) = y(t_2)$ only if $t_1 = t_2$.

Sine (abr. sin) (1) The ratio of the side opposite the given angle to the hypotenuse of a right triangle. (2) The complex power **series** $\sin(z) = \sum\limits_{n=0}^{\infty} (-1)^n z^{2n+1}/$

$(2n + 1)!$ is used to define the **Fourier integral transform**, and the complex **Fourier series**.

Sinistral Refers to left–handed direction. See **dextral**.

Sinusoidal Periodicity A periodic graph that resembles that of a **sine** or **cosine function**.

Size Value of a metric for determining the space occupied by an object.

Size-standardization Procedure to adjust the representations of all members of a set of forms to the same size. Usually accomplished by first **normalizing** the representation to a unit square, circle, or sphere and then uniformly enlarging the normalized results to the desired standard size.

Spatial domain (1) Representing the spatial information inherent in the boundary outline. (2) The mapping is from the spatial domain to the **frequency domain** containing amplitudes and phases. Compare with **frequency** and **time domains**.

Spectral Analysis See **harmonic analysis**.

Spicule A small spike-shaped, or needle-shaped, bone or bone fragment.

Spinous process Any slender, spine-like process (projection).

Stenosis A constricting or narrowing especially of a lumen or an orifice.

Stereology Studies involving either depth (perception), or 3-D.

Stereophotograph A method where two photographs are taken of the same object, one slightly laterally offset from the other. When viewed through a stereo viewer, the object(s) appear to be in 3-D.

Superior Situated above or looking toward the top. In anatomy, the top part of a specimen.

Supraorbital process The superior aspect of the bony ridge that encloses the orbit. See **supraorbital torus**.

Supraorbital torus Extreme development of the superior aspect of the **supraorbital process**.

Surface Evolver A set of programs that can compute minimal surfaces, areas, and centroids. For more information see references to Brakke in the Appendix.

Suture In osteology, a line of junction or closure between bones, as in a cranial suture.

Symmetry/Symmetric (1) Of a figure, identical with its own reflection in an axis of symmetry or having pairs of points identically placed on either side of some line, point, or plane. (2) Of a function, f, with respect to a point c such that for all x, $f(c + x) = f(c - x)$.

Tabular function An array of values, in which one column refers to the dependent variable, the other column(s) represent independent variables. Every dataset can be viewed as a tabular function whose explicit form may be unknown.

Tangent Euclidean geometry: (1) A line that touches a curve at a point and has the same gradient (slope) as that of the curve at the point. (2) Any line, plane, or hyperplane that touches a curve or surface at a point and has the same **normal** at that point (See also *osculating plane* in **torsion**).

Tangent angle See **direction angle**.

Tangent vector The rate of change of the position **vector** when parametrized by **arc length**: $T(x) = dx(s)/ds$. By construction, the tangent vector is a **unit vector**.

Taylor series A power **series** for an infinitely **differentiable function** of the form $\sum_{i=0}^{\infty} \frac{1}{n!}(x-a)^n f^{(n)}(a)$, where $f^{(n)}(a)$ is the nth **derivative** of f at a.

Tensor In one sense, an extension of the concept of an n-dimensional **vector** to a "vector of n–vectors." A tensor can be used as a **transformation** operator on a vector. However, a zero-dimensional tensor is a scalar number; a one-dimensional tensor is a vector. Tensors are **coordinate system** independent. In morphometric analysis they have been used to, for example, convert a vector describing sexual dimorphism of a species into one describing geographic variation.

Test (1) A hardened shell. (2) Calcareous test: accumulation of a series of progressively larger subspherical chambers to form a calcium-based shell.

Thermocline Ecology. In a body of water, the narrow intermediate zone of rapid temperature transition, below which (the *hypolimnion*) there is practically no seasonal variation. The water above is the *epilimnion*.

The curve of Spee The curve of the occlusal path formed by the interdigitation of the upper and lower teeth. It has to be modeled in 3-D.

Thin plate splines The mathematical characterization of a thin (flexible) plate by means of splines. See **Cubic spline**.

Thoracic Pertaining to the thorax; that portion of the trunk that encloses the heart, lungs, and so on.

Three-dimensional (3-D) (1) Possessing three dimensions, for example, length, breadth, and thickness. (2) Three-dimensional Cartesian space created by the joining of two **orthogonal** 2-D planes (the XY and the XZ) yielding an XYZ Cartesian coordinate system. A point is denoted by the (x,y,z) triplet.

Time domain (1) Representing the time-dependent information inherent in a waveform, usually a **Fourier series** or integral. (2) The mapping is from the time domain to the **frequency domain** containing amplitudes and **phases**. Compare with **frequency** and **spatial domains**.

Time series Stat. A **sequence** of data indexed by time, often made up of uniformly spaced observations.

Tooth bud Primordium of a tooth arising from the dental lamina, it is the earliest manifestation of the developing tooth.

Topological transformation A **transform** defined over a **topological space**. See **topology**.

Topology Among many uses: (1) Algebraic. The branch of geometry concerned with describing the properties of a figure that are not affected by continuous distortions such as stretching or knotting. (2) *"Set" topology*. A family of subsets of a given space that constitute a topological space. A *topological space* is a **set** with an associated family of subsets, together with its **open sets** including the universal and empty sets, that is closed under set union and finite intersection. To be *closed* under some operation means that the result is also of the same type as the terms entering into the operation; for example, the integers are closed under addition but not division ($1 + 2 = 3$ an integer, whereas $1/2 = 0.5$ which is not an integer).

Torsion (τ) The rate at which a curve leaves its *osculating plane*. A measure of how much the curve is twisting at any given point. It is computed as $\tau = -\mathbf{N}(s)\mathbf{B}'(s)$ where $\mathbf{N}$ and $\mathbf{B}$ are respectively the *unit principal* and *binormal vectors* to the curve parametrized by **arc length**. The quantity $1/\tau$ is the *radius of torsion*. The *osculating plane* (or circle of curvature) is the plane spanned by the unit **tangent** and unit principal **normal vectors** to a space curve at a given point; the curvature is evaluated in this plane. The unit binormal is the normal vector to the osculating plane.

Trabeculae One of the variously shaped **spicules** of bone in the **cancellous** bone.

Transform Math. Mapping of a **function** from one form to another by the application of a *transforming function* to the original function. For example, logarithms, Laplace transform, finite Fourier transform, Fourier series, biorthogonal grids, linear displacement (translation), and so on.

Transformation grid (Thompson, 1917) in which the *shape changes* in morphology between related specimens, that is, **OTUs**, are described through distortions of a regular grid, which is the transformation grid.

Translate In Euclidean geometry, to move a figure or body laterally without rotation, angular displacement, or dilation (change in size).

Translation In Euclidean geometry, a transformation in which the origin of a coordinate system is moved to another position but the new axes are parallel to the old; a change in variables: $x' = x + a$, $y' = y + b$.

Transverse processes A process projecting outward from the side of a vertebrae, at the junction of the pedicule and the lamina.

Trapezoidal rule A method for accomplishing numerical integration where the areas of a series of narrow trapezoids, constructed between the curve and an axis, are summed. The resulting area is the "area under the curve."

Trigonometric Series A finite or **infinite series** involving trigonometric functions, for example, **sine, cosine, tangent**, and so on.

Two-dimensional (2-D) (1) A figure having two dimensions, for example, length and width. (2) A figure in a two-dimensional Cartesian plane, usually the *XY* plane but may also refer to either the *XZ* or *ZY* plane. A point in this space is usually denoted by the pair of coordinate values enclosed in parentheses, such as *(x,y)*. (3) Lying on a surface, especially a plane; having an area but not a volume; for example a sphere has a **2-D** surface in a **3-D** space.

Unit vector A vector whose length is one unit. It has been **normalized**.

Valve-regulated shunt device A device designed to relieve the intracranial pressures associated with **hydrocephalus**.

Vector (1) In simplest form a quantity that has magnitude and direction, for example, velocity which is speed in a certain direction. (2) An *n*-tuple of real or complex numbers viewed as a member of an *n*-dimensional Euclidean (Cartesian) space. These are used to represent vector quantities, where the length of the vector is the magnitude of the quantity. The direction cosines of the **direction angles** that a line in space, or a vector, makes with the positive *x*-, *y*-, and *z*-axis, respectively, of a Cartesian coordinate system are sufficient to determine the orientation of the line (vector). See also **unit vector** and **normal vector**.

Vector space This is an advanced mathematical concept. A minimalist definition is: A mathematical space consisting of two **sets** and the operations defined over these sets. One set is a **group** with particular properties and whose elements are called vectors, the other set is a field whose elements are called scalars. An additional operation is defined, called scalar multiplication, by which a scalar can multiply a vector yielding another vector; this operation is associative.

Ventral Pertaining to the belly/bottom; in human anatomy the **anterior** part.

Video-digitizer A device that transforms **analog** video images into a **digital** format.

Vomer The thin plate of bone situated vertically between the nasal cavities and which forms the posterior portion of the septum of the nose.

Wavelet analysis Informally, the study of a wave-like, that is, periodic or nearly periodic, function that is well-localized in time and frequency. In wavelet analysis, scaled and displaced copies of the initial wavelet are used to analyze signals and images. It has been called a mathematical microscope.

X-ray A photographic image formed by X-rays on a sensitive medium.

XY plane The standard designation of the **2-D** world. Formed by the intersection of the *X* and *Y* axes in Cartesian 2-space.

XZ **plane** The **2-D** plane formed by the *X* and *Z* axes in Cartesian 2-space.

YZ **plane** See *ZY* **plane**. This is an alternate designation of the same plane. In the EFF23 software, forms are often referred to and plotted with this orientation as it is the one most recognizable by biologists versed in the anatomical tradition rather than the usual mathematical notation. See, for example, Figure 1, Chapter 16.

ZY **plane** The 2-D plane formed by the *Z* and *Y* axes in Cartesian 2-space.

Zygomatic arch The arch formed by the zygomatic process of the temporal, zygoma, and the temporal process of the zygomatic bones.

Zygomatic process of the squamosal The zygomatic process of the squamous (thin anterior and superior portions) of the temporal bone.

References

Blum, H. (1973). Biological shape and visual science. *J. Theoret. Biol.* **38,** 205–87.

Bookstein, F. L. (1978). *The Measurement of Biological Shape and Shape Change.* Lecture Notes in Biomathematics. Berlin: Springer.

Borowski, E. J. & Borwein, J. M. (1991). *The Harper Collins Dictionary of Mathematics.* New York: Harper Collins Publishers.

Davis, J. C. (1986). *Statistics And Data Analysis In Geology.* (2d ed.) New York: John Wiley and Sons.

Lele, S. & Richtsmeier, J.T. (1991). Euclidean Distance Matrix Analysis: A co-ordinate free approach for comparing biological shapes using landmark data. *Am J. Phys. Anthrop.* **86,** 415–428.

Rohlf, F. J. & Bookstein, F. L. (1990). *Proceedings of the Michigan Morphometric Workshop.* Univ of Mich Mus of Zool Special Pub No 2.

Sneath, P. H. A. & Sokal, R. R. (1973). *Numerical Taxonomy.* San Francisco: W. H. Freeman & Co.

van Otterloo, P. J. (1991). *A Contour-Oriented Approach To Shape Analysis.* New York: Prentice-Hall.

Index

2-D, 159, 227, 246, 326, 423
2-D orientation, 421
2-D outlines, 87, 93, 130, 310
3-D, 159, 227, 247–8, 415, 419
3-D digitizers, 74, 227
3-D orientation, 421
3-D outlines, 41, 87
3-way MANOVA, 332
32nd harmonic, importance of,
    297–8, 301–2

A-point, 345
acetate sheets, 363
additive genetic component (V_a),
    194
adults, rabbit, 361
age differences, 332
aging, 401
algorithms, 409
aliasing, 27
allometry, 97, 250
allophenic mice, 168–9
alveolar margin, 352, 362
alveolar prognathism, 200
among-pair mean squares, 194
amplitude, 29–30, 95, 190, 194,
    296–7
    See equations 2.34, 4.15, 8.2
        and 12.7
    spectrum, 115, 139
    values, 135, 262, 267, 278
    versus radial distance,
        191
angles, 6
anthropology, 144
approximation of shape with ad-
        dition of terms, 164
arc length, 59, 67, 420
area, 361, 367, 374
    computation of, 416
    with Stoke's and Green's the-
        orems, 425
    normalization, 256
    of bounded form, 425

allometric, 260–261
isometric, 260
aspect ratio. *See* equation 4.1
aspects, global and localized, 23
asymmetry. *See* odd function
    small sine components, 164
auditory meatus, 214, 362
auricular points, 229
Australian aborigines, 190
Australopithecus aferensis, 53
Australopithecus africanus, 276
average radius, 133
average residual fit, 330

B-point, 345, 347
BASIC, 328, 336, 416
basion, 15, 327, 330
bending energy, 80, 93
between-group variability, 177
bilateral symmetry, 169
binomial classification of Sergi,
    253
biological process, 42
biologically interpretable repre-
        sentations, 65–66
biomechanics, 379, 387–388
biorthogonal grids, 5, 81, 252,
    324
bipedal locomotion, 296
bisections, 328
Blum's medial axis transform,
    89–90
Bookstein, 49, 50, 59, 65, 76,
    78, 90. *See* principal
        strains, deformation models
    exclusion of phase angle in-
        formation, 111
    the line skeleton, 90
    the thin plate spline, 78–9
    truss system of homologous
        points, 60
    use of extremal landmarks, 76
boundary information, 409
boundary morphometrics, 5

boundary-outline representa-
        tions, 324

cadence as cycle per second, 300
cancellous bone, 379, 388
cancellous networks, 387
cancellous spicules, 382
canonical analysis, 168
    of mouse vertebrae, 181–182
canonical axes, 77, 98, 181, 183
canonical plots, 183
canonical variate analysis, 395
carapace, 109
Cartesian coordinates, 30, 159,
    176, 190, 232
Cartesian transformation grids, 83
catenary curve, 64
Caucasians, 195
CB. *See* cranial base
CCD (charge-coupled device
    camera), 227
cell contour, 313
cell shapes, examples of, 307
cellular types, 285
center of area, 159
center of gravity, 133–134
    computation of, 138
    See equations 6.7 and 6.8
centroid, 34, 36, 38, 109, 134,
    138, 159, 217, 326, 330,
    361, 367, 369, 416, 418
centroid coordinates, from
    Surface Evolver, 369, 426
centroid detection
    computational, 258
    See Shape Analytical
        Morphometry
cephalometric radiographs, lat-
    eral, 345
chimeras. *See* allophenic mice
chin, 200, 352
chondrocytes, 314
Cibicidoides pachyderma, 132
circular data. *See* reflected data

cladistic analysis, 75, 85
Class II malocclusions, 340
clival plane, 337
cluster analysis, 4, 117, 221
  Ward's method, 171, 181–2
CMA and its drawbacks, 6–7,
    37, 323, 343, 412,
coherent optical processing, 5
color, 4, 412
comparison of mean forms, 367
complex shapes, 69–70
complexity. *See* fractal dimension
condrocytes, 308
condylar process, 200
condyle head, 353
condylion, 235
confirmatory studies, 78
confocal microscopy, 309
constant or a_0 term, 191
continuous over the interval. *See*
    uniformly convergent
contour perturbations, 254
contour representations, 23, 420
control side
  of the rabbit orbit, 363
controls, 332, 336, 347
Conventional FDs, 37–38
conventional Fourier descriptors,
    36
conventional metrical approach.
    See CMA
convergent series, 24, 138, 325
coordinate data, 77
coordinate-free, 52
coordinate-free representations,
    52, 66–8
coordinate Morphometrics, 5
coordinate-system independent, 51
coordinate system, local, 51, 263
coordinate transformations
    invariance, 8
  of D'Arcy Thompson, 47, 250
coordinates (x, y) or (x, y, z),
    416, 418
coordinates (x, y, z), 68
corners, 69
coronoid process, 352
correlation matrix, 331
correlations between amplitudes
    and radial vectors, 191
cortical remodeling maps, 86
cosine coefficients, 297
cosine components, 164, 169
cranial base, 234, 322–3
  tracing procedures, 327
  point locations, 327–9
craniofacial complex, 338
craniograms
  Circeo 1, Cro-Magnon, mod-
    ern, STS 5, 273–4
  Upper Paleolithic, *Homo erec-
    tus*, Neanderthals,

Petralona, Broken Hill, and
    Steinheim, 277–8
craniometry, 75, 227
cross-sectional study, 345, 361
crossings, 38, 371, 416, 426
CT (computerized tomography),
    227
Cubuscraniophor, 214
curvature. *See* equations 3.10
    and 3.12
curve in 3-D. *See* equation 3.14
curve in 3-space, 359, 425
curve of Spee, 378
cytoskeleton, 308

D'Arcy Thompson, 45–7, 59,
    71, 250, 253, 410
data acquisition. *See* Shape
    Analytical Morphometry
data from pathology, 285
Deformation models, 80
  drawbacks, 63, 66, 80–1
deformations, 5, 14, 63, 78
Delauney triangulation, 81
dendogram, 169, 221, 245–6
densitometry, 381
dental arch form
  in Japanese, Australian
    Aborigines and Caucasians,
    190
  in MZ and DZ twins, 196
dental casts, 190
dental differences, 195
depth ranges of planktonic
    foraminifera, 147
difference graph, 263–7
digital acquisition of boundary
    information, 258
digital cameras, 419
digitization, 415
  of the boundary, 136, 256
digitized points, 420
digitizers
  Houston Instruments HIPAD,
    329, 365, 418
  Summagraphics SUMMAS-
    KETCH III, 419
digitizing, entering data, 190
  in a clockwise direction, 419
dimensionally-stable acetate,
    215
Dioptrograph, 214
Dirichlet conditions, 24
discontinuous wave form, 161
discriminant analysis, 278, 280,
    283
  of mouse vertebrae shape, 166
discriminant functions, 4, 98,
    100, 106, 243–4
distances, 6
  base to target (as differences),
    428

baseline to boundary outline,
    428
centroid to boundary outline,
    427
criticisms of their use, 76
inter-landmark, 76–7
dizygous (DZ) twins, 194
dolichocephaly, 211
domain, as from spatial (or time)
    to frequency domain, 29,
    95
dominance, 171, 173
dominance component (Vd), 194
dorsal clivus, 327
dorsum sellae, 327
double support time. *See* walk-
    ing, measurements
driving harmonic, 268, 270
dual structure model, 222
Duncan's multiple range test,
    212
dural folds, 337

ecological change, 410
EDMA (Euclidean distance ma-
    trix analysis), 75
Edo age period, 211, 213, 220
EFA. *See* EFFs
EFF23
  data exporting capability,
    429–30
  export module, 416
  file naming conventions,
    416–17
  user manual, 415
EFF23, computation of
  combining forms, 431
  merging (inserting) forms,
    431
  renumbering data points, 431
  amplitudes, 415, 427
  descriptive statistics, 415
  distances, 427
  ellipse parameters, 415, 428
  means, 429
  oriented data, 416
  phase angles, 415, 427
  power spectra, 415, 427
  predicted points, 416
  size-standardized data , 416
  standard deviations, 429
  standard errors, 429
EFFs, 39, 42, 51, 349, 359, 367,
    374
  computation of, 415
  Kuhl and Giardina algorithm,
    39–42
  parametric approach, 38, 323,
    359
  See equations 2.42, 2..43,
    14.1, 14.2, 16.1, 16.2, and
    16.3

EFFs, applications, 40–2
EFFs, extension to 3-D, 41
  See equations 2.42, 2.43, and
    2.52
eigenshape analysis, 5, 93, 108,
    325
eigenvalues, 241
eigenvectors, 76, 110
electrostatic forces, 368
ellipse, 311, 312
ellipse parameters, 428
  perimeter multiplied by its
    frequency, 313
Elliptic Fourier analysis (EFA).
    See EFFs
elliptical Fourier functions, 5,
    38, 51, 95, 343, 359
  See also EFFs
elliptical Fourier representations,
    69
endothelial cells, capillary, 314
energy. *See* equation 10.12
  or power spectrum, 235, 243
energy scores, 241
entropy, 142–6
environmental influences (E),
    193
epistatic component (V_i), 194
epithelial tumors of the ovary,
    285
equal angular divisions, 137
equally-spaced points, 32
equally-spaced vectors, 211
error variance. *See* equation 3.31
Escher
  "Circle limit IV" woodcut,
    254, 271
  "Metamorphose" woodcut,
    272–3
ESR (electron spin resonance),
    227
Euclidean 3-space, 67
Euclidean distance, 176
even functions. *See* equation 2.6
evolutionary changes, 410
exploratory data analysis, 140

F-ratio, 173, 186
  also termed coefficient of
    variability (CV), 166
F_1 shape, 170
FA. *See* functional appliances
FA study, mandible, 352–4
FA therapy, 340–1, 345, 347,
    349–51
face height factor, 333
Facial outlines, 276
facial profile, 202–3
facial stereophotographs, 190
factor analysis of Fourier coeffi-
    cients, 241
factor loadings, 242

FAS, 342. *See* functional appli-
    ances
fast Fourier transform. *See* FFT
FDs, 37–9, 42, 409–10, 412
  carapace of ostracodes, 109,
    410
  cellular structures, 410
  elytra of beetles, 113
  human gait, 410
  human skull form, 210
  invertebrate palentology, 113
  osteology, 113
  rigid parts of miospores, 113
feature extraction, 139
feature selection, 129
feedback mechanism, 305
FESA. *See* finite elements
FFT (Fast Fourier Transform),
    395–6
file naming conventions. *See*
    EFF23
filename, 418. *See* header record
finite elements, 5, 80–1, 252,
    324, 388
finite Fourier series, 81. *See*
    equation 2.24
first harmonic amplitude as an
    offset circle, 32, 134, 137
first harmonic ellipse, 313
FLAC (Fast Lagrangian
    Analysis of Continua), 388,
    390
flipping of a form, 422
foot angle. *See* walking cycle
foraminifera, 131, 140, 145
forensic anthropology, 287
form, 4, 415
  biological, 4
  global and local elements, 420
  representation of, 45
  Form = size + shape, 9, 218
  Form = size + shape + struc-
    ture, 9
  attributes, 10–11
  complexity. *See* fractal dimen-
    sion
  factor, 310
  matrix. *See* EDMA
  formal model, 12–14, 412–13,
    415
  numerical representation, 8
  regular, or man-made, 6–7, 269
FORTRAN, 416
Fourier amplitudes, 194–6, 202
Fourier analysis, 15, 23, 75,
    94–5, 130, 132, 158, 175,
    210–11, 263
  of artificial shapes, 269
  dual axis approach, 95, 311
  computation. *See* Shape
    Analytical Morphometry
    (SAM)

closed form, 133, 150
  See equation 6.11
Fourier coefficients, 178, 190,
    267. *See* equations 2.32,
    2.33, 4.11, 12.3, 12.4,
    and 12.5
  biological meaning. *See* re-
    flection
  x-projection, 39
  y-projection, 39
  z-projection, 41
  polar form. *See* equation 2.28,
    4.12, 8.1
Fourier components. *See* equa-
    tion 7.4
Fourier constant (a_0), 191
Fourier descriptors, 5, 23, 29,
    106–7, 189, 323, 325, 409.
    See also FDs
  advantages, 207
  as curve-fitting approaches, 16
  mandibular growth and treat-
    ment, 16
  limitations, 37–8, 158, 325
Fourier harmonic analysis. *See*
    Shape Analytical
    Morphometry
Fourier methods. *See* FDs
Fourier representation, 68
Fourier series, 15, 23, 130, 163,
    177, 235, 243, 278, 360
  constant term, 195
  definition, 24
  See equations 2.1 to 2.4, 9.1,
    and 12.1
  optimum length, 165
  viewed as a deformation, 164
  definite form. *See* equation 2.28
  discrimination, 166
  phase lag representation. *See*
    equations 4.13, 6.3, and 7.3
Fourier transform, 5, 386, 396
  cruciate in form, 382
  complex, real components,
    134
  closed form, 131–3
  internal structure, bones, 410
  computational, 391
  optical, 390–1, 406–8
Fourier's series, 15, 24–5
Fourier, J. B. S. , 22
fourth harmonic as a measure of
    quadrateness, 135
fractal dimension, 93–4, 252,
    313. *See* equations 4.8, 4.9,
    and 4.10
frame grabbers, 74, 176, 419
Frankfort Horizontal, 214,
    229
free energy, 368
frequency, 29
  as inverse of the period, 296

frequency distributions, 142
frontal bone, 363
function, parametric. *See* EFFs
functional appliances (FAs), 340
functional relationships, 247
functions, tabulated, 31
fundamental frequency, 29
fundamental period, 296
fundamental shape, 258–9, 263

gap-coded, 75
generalized distance
 Mahalanobis's D^2 168, 177
genetic and environmental factors, 189
genetic variance (V_g), 194
genetics, 410
genotype (G), 193
genotypes, 169
geometrical homology, 108, 132
Gibbs phenomenon, 27
glabella, 278
GLM procedure
 using SAS, 211
global aspects of form, 420
Globigerenoides ruber, 145
Globorotalia inflata, 140
Globorotalia truncatulinoides, 132, 145
Gnathion, 347
gonial angle, 200
 opening, 353
gonial aspect, 352
Gonion, 347
goodness-of-fit, 27, 218, 330, 348, 366,
 residual, 420
 See equation 2.16
Gorilla gorilla, 230
group theory, 130
growth, 410
growth process of dentition
 as mathematical model, 57–8
 consisting of two-parts, 53

Haar and Walsh coefficients, 132
harmonic amplitudes, 133, 139
harmonic analysis, 28, 110. *See also* Fourier analysis
harmonic components, 164
 of normal and rectified cell outlines, 314
harmonic contributors, 266
harmonic distance
 analysis, 107, 110
 coefficients, 116–17
 matrix of, 117
 See equation 5.1
harmonic ellipse, first, 313
harmonic number, 133, 360
harmonic parameter, 313
harmonic synthesis, 28

harmonics, 28, 108, 296, 325
 interpretation, 108
HC. *See* hydrocephalus
header record, 418
heritability (h^2) estimates, 194, 197, 202
 See equation 8.4
heterochrony, 119
heterozygous, 171
HIPAD digitizer, 329, 348, 365, 418
histology, 144
histomorphometry, 251
holograms, 227
hominid locomotion, 303
Homo sapiens sapiens, 230, 276
homogeneous finite elements
 uniform strain, 81
homologous features (landmarks), 86, 91, 131, 157
homologous point, 139
homologous point representations, 324
 drawbacks, 60, 83–7, 158
homologous points, 51, 132, 416, 420
homology, 14–15, 84, 419–20
 definition of Owen (1848), 324
 mathematical, 14, 108
 operational, 14, 84–5, 420
 map, 87
 biological, 14, 84, 108, 252, 419
 geometrical, 132
 point, 14, 86–7, 252, 419
homozygous, 169
hormion, 235
human gait, 410
human locomotion, 294
human vertebrae, 379
hydrocephalus, 322
hydrophilic and hydrophobic, 368
hypophyseal fossa, 327

image analysis software, 87
image interpretation, 309
Image Signal Processor (or ISP), 258
inbred lines, mice, 174–5, 178, 183
incisal aspect of the mandible, 352
 anterior movement, 353
increase in resolution, 310
indices, 76, 210, 253
individual shape component number, 133
infants, rabbit, 361
infinite Fourier series. *See* equation 2.23
inferior mandibular border, 352

information preserving, 9
information, visual, 252
instars, 119–23
 molt stages, 109
interlandmark distances, 74, 77
interpolant function, 80
interpolation, 137
intersections. *See* crossings
intraclass correlation, 194, 196
invariant properties, 8, 97
invariants, 68
invertebrate paleontology, 113
isometric growth, 55
isomorphic sets of homologous points, 69
isotopes, oxygen, 150

Japanese, 195
Jomon age period, 211, 213, 220
jump discontinuities, 24
juveniles, rabbit, 361

Kamakura age period, 211, 213, 220
Kanto District of Tokyo, 213
Kellner-System, 229
Kendall's shape space, 77
knuckle-walking, 382
Kodalith resin-coated sheets, 367
Kolmogorov-Smirnov test, 278
Kuhl and Giardina algorithm, 39–40, 51, 263, 311, 323, 330, 343, 359, 415
 See equations 2.42 to 2.56

landmark coordinates, 76
landmark configurations, 78
landmark-based approaches, 75
landmark-free method, 158
landmark-independent approaches, 75
landmarks, 227
laser scanners, 229
lateral cephalometric radiographs, 345
lateral view (or x-y plane), 160
least squares fit method, 289
left inferior orbital point, 229
leg dominance, 299
line integrals, 425
line skeleton. *See* medial axis analysis
linear regression, 325
local aspects of form, 420
locomotion, 386
longitudinal study, 345
lower limb length, 304
lumbar vertebrae, 380, 382
lymphocites in peripheral blood, 309–18

macrophages, 308
Mahalanobis's D^2 distance, 48, 98, 168, 177
mandible, 64–5
mandibular plane angle, 345
MANOVA, 160, 331–7, 353, 369
mapping, 13
masseter muscle, 247
masticatory function, 340
maternal effects, 173
matrix of inter-OTU distances, 76
maximum entropy analysis computation, 142
maximum harmonic number, 360
MD versus ML relationship
    See equation 3.1, 57
MD/ML tooth shape measure
    See equation 3.2, 58
mean parental shape, 171
mean residual error, 420
mean shape, 160
mean spectrum, 268
measures of elongation and undulation, 88
    See equations 4.2, 4.3 and 4.4
medial axis analysis, 5, 89–90, 325
menopause, 395, 402
Menton, 347
metameric series, 158
metameric variation, 175–6
mice, inbred strains, 171
mid-sagittal outlines, 233
middle cranial fossa, 327
minimum spanning surface, 425
minimum surface, 368
misclassification rate, 170
mitochondria, 309
MMDs
    Smith's Mean Measure of Divergence, 222
model(s)
    formal, 3, 6, 412
modeling process, 12
Modern age period, 212, 220
Moire contourography, 212, 227
moments. *See* equations 4.4 and 4.5
monozygous (MZ) twins, 194
morphogenesis, 5
morphological discontinuity, 272
morphometrics, 5, 324
multidimensional scaling, 117, 221
multivariate morphometrics, 5–6
multivariate statistical analysis, 395
Muromachi age period, 211, 213, 220

Nasion, 15, 345
natural selection, 59
neural canal, 179
NMR
    nuclear magnetic resonance, 227
nonhomogeneous finite elements, 81
norma basilaris
    or basal view, 214
norma frontalis
    or frontal view, 214
norma lateralis
    or lateral view, 211
norma occipitalis
    or occipital view, 212, 214
norma verticalis
    or vertical view, 212, 214
normalization based on area, 425
normalization based on the first ellipse, 313
normalization, as a scaling procedure
    See Shape Analytical Morphometry
normalization methods, 67
nucleoplasmic ratio, 317
numerical taxonomy, 7, 49
Nyquist frequency, 27, 324
Nyquist sampling theorem, 137, 139

occlusal variation, 194
Ockham's Razor, 8, 12
odd function, 32
one-to-one mapping, 366
ontogenetic processes, 86
open curves, 261
operational taxonomic units
    See OTUs
opisthocranion, 278
orbit, 359
orbital data, 237
    factor analysis, 241
orbital margin, 359, 361
orbitale, 214
orientation, 415
orthogonal network, 386
orthogonal projections, 376
    of the rabbit orbital margin, 372
orthogonal system in three axes, 361
orthogonality, 8, 50, 382, 399
orthogonality, property of
    trigonometric series, 26, 211
osteology, 113
osteoporosis, 390–1, 396, 401, 404, 411
ostracodes, 108
OTUs, 75–80, 84–7, 90, 96, 98
overdominance or epistasis, 171

P-A view, 375
P=G+E, 193
pairwise orthogonal
    See equation 2.9
paleobiology, 144
paleoceanographic parameters, 131
Pan troglodytes, 230
parabola, 260
parametric approach
    See also EFF; EFFs
parent-offspring correlation, 186
partial warps, 80
path analysis method, 194
pathology, 284
pattern recognition, 323
patterning, 4
Pax-1 gene, 174–5
PCA. *See* principal components
pel (pixel), 419
pelvic bone, 280
Penrose's shape distance, 244
percentage explained (variance)
    See equation 2.38
period, 29
periodic function, 296
periodic motion, 29–30
periodic regressions, 36
peripheral blood lymphocytes, 318
pertubations
    of symmetry and contour, 254
pertubations of shape, 164
phase, 29–30, 34, 35, 95, 96, 297
phase angle invariance, 35, 111
phase angles, 133, 135, 139, 297, 235
phase lag components, 96, 164, 296
phase spectrum, 139
phase values, 267, 275, 278
phenetics, 75, 85, 107, 117
phenotype, 145
phenotypic expression (P), 193
photoelastic stress analysis, 354, 388
photogrammetric method, 214
piece-wise continous
    See jump discontinuities
piece-wise monotone, 25
piecewise constant, 39
piecewise-smooth single-valued function, 32
pixel (pel), 395, 419
plasmalemma, 308
planar (2-D) forms, 425
planar curve, parametric representation. *See* equation 3.11
planum sphenoidale, 322
Plateau's problem, 368
pleiotropic genes, 173

pleiotropy, 175
Plesianthropus transvaalensis
(STS 5), 276–7
plotter
Hewlett-Packard HP7470A-2,
348, 366
plotting of form, 415
Plotting software, 429
Pogonion, 342, 347
point homology, 324, 366
point sampling, clockwise, 233
point system protocol, 416
points, equally spaced, 32
points of maximal curvature, 69
polar angle, 133
polar coordinates, 32, 91, 133–4,
159, 164, 172, 232, 237, 263
polar vectors, 134
polygon, 360
polynomials, 263
Pongo pygmaeus, 230
porion, 214
positional-orientation, 35, 312,
326, 421
posterior-anterior view (PA)
or y-z plane, 366
potence ratio, 174
power (or variance), 29, 34
power spectrum, 35, 100
predicted points, EFF23
as evenly-spaced, 416, 419
as homologous, 416, 419
as oriented, 421–3
as size-standardized, 425
predicted points, mandible
150 equally-spaced, 348
presacral vertebra, mice, 176
principal components (PCA), 4,
50, 76–7, 93, 98, 100, 106,
110
principal components analysis,
49, 108, 168, 395, 401
principal ellipse, 332
principal strains, 80, 81
principle of discontinuity, 250
Procrustes analysis, 77, 90
profile, human pelvic bone
used for sexing, 280
prognathism, 210
prosthion, 362
protohistoric Japanese
craniofacial morphology, 210
pseudo homologous landmarks,
86, 90, 420
pseudo homologous points, 325,
367

Q-mode correlation coefficients,
221
quadrumanal or quadrupedal, 386
quantitative genetics, 189
quantitative stereology, 309

radiographic enlargement
correction for, 348
ramus, (mandible) anterior and
posterior aspects, 352
Rao's R, 370, 372
ratios, 6, 210, 396
re-entrants, 159
recessive genes, 170
reciprocal F_1s, 173
rectification
of fundamental curve, 263
rectified cell profile, 314
recursive procedure (polar coordinates)
to generate equal angles, 34
red blood cells (RBC), 316
reference frame
as in coordinate system (e.g.,
Cartesian), 77
reference point system, 229
reflected data, 161, 178
reflection
resulting in a mirror image,
217
repeated measurements
to assess error, 193
repeated waveforms, 296
residual, 330, 348, 366
residual area, 160
resolution, increase in, 310
root mean square error (RMS),
259
Rotacraniophor I
a skull holder with rotational
capability, 214
rotational component
growth of the rabbit orbital
margin, 373
rotational fitting, 160
roughness factor, 267

SAM
See Shape Analytical
Morphometry
sampling of points
along the x-axis, 296
in a clockwise direction, 233,
419
sampling point density, 231
scaling factor, 236, 326, 258
the constant or A_0 term,
164
scaling factors, 36
scanners, 419
scanning electron microscopy,
183–4, 381
sclerotomal cells, 169
SDUs as standard deviation
units, 168, 183
SE. *See* sphenoid registration
point

sella turcica, 322
second harmonic, 32
as a measure of elongation,
134
sedimentology, 112
self-intersections. *See* crossings
separation of size from shape,
211
sexual dimorphism, 243, 332,
337
shape, 4, 9–10
as a carrier of information, 130
Shape Analytical Morphometry
or SAM, 255–62
shape and form
as synonyms, 9
shape descriptors, 130
shape differences
affine and non-affine components, 80
shape distance, 221, 224
shape information, 139
shape only data, 330, 332, 367
shape variation, 241
shape, cranial
numerical description, 253
shape, Sergi's classification of
skull form
See binomial classification
shape, three classification categories, 253–4
shapes as distortions in symmetry and pertubations of contour, 254
shunt function, failure, 322
shunt-treated CB, 323
shunt-treated HCs, 330, 332–6
sigmoid notch of the mandible,
352
sine and cosine terms, 132
sine coefficients, 297
sine components, 164
sine terms equal to zero
See equations 2.29 and 2.30
single support time
See walking, measurements
single-valued function, 23, 133,
263
sinusoidal in x, y, and z, 361
size, 9–10, 97–8, 412
invariance, 97
using the Fourier constant as
scaling factor, 98, 237
size and shape
separation of, 195, 211
size and shape data, 320, 332,
367
size normalization approaches, 36
size-related shape change, 180–1
size-standardization, 35–6, 236,
326, 367, 425
See equation 10.2

size-standardized and oriented, 348
size-standardized vector distances, 200
size-standardized vectors, 212
skeletal genes, major, 173
skeletal remains, 287
skull, classification
    See binomial classification
skull, various anatomical views, 229
skull, rabbit, 359
soap films, 69, 368, 426
Sokal and Sneath
    See numerical taxonomy
spatial domain, 361
species discrimination, 243
specimen identification
    See filename
Spectral analysis, 28, 325
    See Fourier analysis
sphenoid registration point (SE), 327
sphenoidal plane, 337
spinous process, 174, 179
stance phase
    walking, measurements, 294
standardized anatomical positions, 229
starting point
    standardization of, 139, 256
statistics
    multivariate, 4
step length and width
    See walking cycle
stepwise-discriminant analysis
    See discriminant functions
stereocomparator
    for 3-D landmark measurements, 229
stereology, 309
stereophotogrammetry, 229, 231
stereoscopic separation, 229
structure, 10
superimposition, 159, 234, 236, 326, 350–1, 367, 369, 425
    of rabbit orbit margin samples, 370
superior margin
    of the rabbit orbit, 364
superior view, 362
    or x-z plane, 362
supraorbital process, 359, 363
Surface attributes
    texture and color, 11
Surface Evolver, 369
    computation of area and centroid in 3-D, 426
surfaces, convex or concave, 307
surfaces, examples , 11
swing phase
    walking measurements, 294

symmetry, 32
    See equation 9.2
    large cosine components, 164
    manipulation of the fundamental shape, 259
    See Shape Analytical Morphometry
symmetric axis
    See medial axis analysis
Systema Naturae, 250

tangent angle function
    See Zahn and Roskies
Taung, 1276–7
temporal fossa, 363
textural morpometrics, 5
texture, 11, 412
thick sections
    lumbar vertebrae, 391
thin plate splines, 5, 79–80, 324
    bending energy, 80
third harmonic
    as a measure of triangularity, 135
Thom, Rene
    Catastrophism, 5
tibial torsion, asymmetry, 304
tissue differentiation
    responsible for different cell types, 307
tooth growth, two-part model , 58
topological transformation, 217
topology, 368
torsion
    See equation 3.16
trabeculae, 381
transformation grid, 78
    See D'Arcy Thompson
trapezoidal rule, 31
treatment effects, 347, 410
treatment versus controls, 332, 347
trend analysis, 78
trigonometric series, 359
triplets (x, y, and z), 366
tumor nodes in breast cancer, 285
twin and family studies
    See quantitative genetics
Two-part tooth growth model
    See equations 3.4 and 3.6, 58

un/un
    See undulated strain
undulated, 174–5
uniformly convergent, 24
unit area, 164
univariate F-tests, 332–6, 353, 370, 372

van Otterloo, 66, 323, 425
variance analysis
    of arch shape, 196

variance ratio
    See F-ratio
varimax loadings, 148
vector center (polar coordinates)
    as centroid, 34
vector center (VC)
    of the cranial vault, 216
vector lengths
    as deviations from the mean (a_0 term), 237
vertebrae, mouse, 158, 164
vertebral models
    stress analysis, 388–90
video-digitizer, 157
visual patterns
    of Fourier transforms, 380–1, 383, 385, 392, 400
visual stimuli, 4
$V_p = V_g + V_e$, 193

walking cycle
    components of, 294
walking measurements
    theoretical patterns, 297
    experimental, 298–9
walking, musculoskeletal system
    as related to the visual system, 295
wavelength
    as reciprocal of frequency, 29
Wilk's Lambda, 332–6, 353, 370, 372
wireframe model, 425
within-group variability, 177

x- and y-components, 331
x-projection (coefficients)
    See equations 2.44 and 2.45
x-y plane, 361
x-y view, 371–2
x-z plane, 361
x-z view, 371–2
xy and xz plane, 419

y-projection (coefficients)
    See equations 2.46 and 2.47
y-z plane, 361, 371, 374
y-z view, 371
yz plane, 419

z-axis, 359
z-projection (coefficients)
    See equations 2.53 and 2.54
Zahn and Roskies algorithm, 36–7, 68, 93, 95, 311
    See equations 2.40 and 2.41
    See equations 3.17a and 3.17b
    See equations 4.6 and 4.7
zygomatic arch, 359, 363